GUIDE TO
ENVIRONMENTAL ANALYTICAL METHODS
4th Edition

Edited by
Northeast Analytical, Inc.

Chief Editor
Robert E. Wagner

Associate Editors
William A. Kotas
Inga C. Hotaling
T. Christian Hynes
James Daly
Mark F. McTague

Genium Publishing Corporation
117 1 Riverfront Center, Amsterdam, NY 12010 (518) 842-4111

INTRODUCTION TO 4th EDITION

The fourth edition of the *Guide to Environmental Analytical Methods* covers the massive changes that have recently taken place to EPA methods, affecting organic and inorganic determinative methods along with preparative and sampling procedures. Since our last edition of *"The Guide,"* EPA SW846 has been updated with sixty-two new methods; forty existing methods have been revised; and sixteen methods removed. Some of the highlights are inclusion of the new volatile sampling and preparation methods, the separation of Pesticide and PCB analysis into methods 8081 and 8082, and the removal of methods 8020 and 8010, with 8021 covering the analytes of the deleted volatile methods. The changes from SW846 Final Update III are so expansive that a Method Status Table appendix has been included in this edition, listing current promulgated methods.

Along with SW846, the EPA 500 series methods have undergone changes with the removal of methods 502.1, 503.1, and 524.1, leaving methods 502.2 and 524.2 to address those analytes specified in the deleted methods. In this fourth edition we have included newer proposed EPA 600 series methods in the Sampling Procedures and Preparation section for those of you practicing these less frequently requested methods. All 200 series, 500 series, 600 series, CLP, Standard Methods, and other EPA methods contained in the book have been reviewed for changes and revised.

The editors hope that you find this, the fourth edition to *"The Guide,"* a most useful reference tool for your environmental analysis, field, compliance, and auditing efforts.

Robert E. Wagner
Lab Director
Northeast Analytical, Inc.
2190 Technology Drive
Schenectady, NY 12308

HISTORY and ACKNOWLEDGMENTS

The *Guide to Environmental Analytical Methods* was originally developed from a method comparison document called *Variability in Protocols* (VIP) published by the EPA Risk Reduction Engineering Laboratory in Cincinnati, Ohio. VIP provided a useful tool for understanding similarities and differences among commonly used environmental test methods. This concept was greatly expanded by Northeast Analytical's staff with the first publication of the *Guide to Environmental Analytical Methods* in 1992. Now in its fourth edition, *The Guide* has become a popular reference tool for all who work in the environmental field.

Northeast Analytical is an independent environmental testing laboratory offering quality analytical services. If you have questions or comments regarding this publication or need assistance with environmental testing, the editors would like to hear from you. We can be reached by phone [518]-346-4592, fax [518]-381-6055, or E-mail (nelab@aol.com).

Table of Contents

Table of Contents, *continued*

Analytical Methods - Organic Constituents

The numerous analytical methods for organic constituent testing have been organized into five major categories:

- Herbicides (chlorinated) identified by gas chromatography
- Pesticides (organochlorine) and PCBs identified by gas chromatography
- Semivolatile organic compounds identified by gas chromatography/mass spectroscopy
- Volatile organic compounds identified by gas chromatography/mass spectroscopy
- Volatile organic compounds (aromatic and halogenated) identified by gas chromatography

Within each category there is a section called *Method Comparison* that summarizes the key method requirements and components for each testing method. The methods are compared on a side-by-side basis, which allows the user to quickly compare key aspects of the test methods. Where appropriate the user is referred to a table in Appendix A that provides details about an aspect of the test method. The tables in Appendix A are taken directly from the regulations.

For determining if a specific analyte is analyzed by a test method, each category has an *Analyte Listing* of all compounds addressed by the test methods within the category. This is also presented on a side-by-side basis so that the user can quickly see which method(s) cover the specific compound in question.

If you want to find a test method for a specific analyte but don't know what category(ies) it belongs to, refer to Appendix B. Appendix B is an alphabetical listing of all organic analytes specified in the method comparison sections. Use this appendix to identify the category to turn to for additional information.

Determination of Chlorinated Herbicides by GC

Method ⇒ Parameter ⇓	SW-846 Method 8151A	EPA 500 Series Method 515.1	EPA 500 Series Method 515.2	Standard Methods Method 6640
Applicability	Groundwater, soils, sludges, and non-water miscible wastes	Groundwater, and drinking water	Groundwater and drinking water	Groundwater, surface, and drinking water
Number of Analytes (1)	19 total	17 total	14 total	9 total
Method Validation (2)	Extract and analyze 4 replicates of a QC check sample. Compare accuracy and precision results to Table 4 (see page A-118). Note: accuracy and precision limits not listed for PFB derivatives of herbicides.	Extract and analyze a minimum of 4 to 7 replicates of QC check sample. Concentration 10 times the EDL or at a mid-point in the calibration. Analyte recovery must be within ±30% of fortified concentration. RSD of measurements must be ≤30%.	Extract and analyze a minimum of 4 replicates of a QC check sample. Concentration 10-20 times the MDL or at a mid point in the calibration Results must be within 40% of fortified concentration. RSD must be <30%	Extract and analyze 4 replicates of QC check standards. Results must be within ±30% of values specified in Table 6640: VI. (see page A-104) Determine MDL (6)
QC Check Standards/ Samples	Analyze one laboratory control sample per 20 samples or each batch of samples whichever is more frequent. Compare results to laboratory established limits. Laboratory can use 70-130% as interim limits until in house limits are developed.	Analyze one QC check sample (Laboratory Fortified Blank) per 20 samples or one per sample set (all samples extracted within a 24 hr period), whichever is more frequent. Concentration should be 10 times the EDL or a conc. which is near the mid point of the calibration. Compare %R to laboratory established limits, if available. If laboratory limits not available analyte recovery must be ± 30% of fortified value. Laboratory established limits must not exceed ± 30%.	Analyze one QC check sample (Laboratory Fortified Blank) per 20 samples or one per sample set (all samples extracted within a 24 hr period), whichever is more frequent. Concentration should be 10 times the EDL or a conc. which is near the mid point of the calibration. Compare %R to laboratory established limits, if available. If laboratory limits not available analyte recovery must be ± 40% of fortified value. Laboratory established limits must not exceed ± 40%.	Analyze one QC check sample (Laboratory Fortified Blank) per 20 samples or one per sample set (all samples extracted within a 24 hr period), whichever is more frequent. Concentration should be within 2 times the concentration listed in Table 6640: VI (see page A-104). Compare %R to laboratory established limits, if available, or Table 6640: VI and VII (see pages A-104)
Method Detection Limit	MDLs listed in Table 1 (see page A-117) for diazomethane derivatives of herbicides. MDLs for PFB derivatives are not listed in the method.	MDLs and EDLs listed in Table 3 (see page A-32). Extract and analyze a minimum of 7 laboratory fortified blanks. Use Table 3 concentrations as guide. Calculate MDL.	MDLs listed in Table 2 (see page A-33). Extract and analyze a minimum of 7 laboratory fortified blanks. Use Table 2 concentrations as guide. Calculate MDL.	MDLs listed in Table 6640:I (see page A-103). EQLs measured at approximately 5-10 times the MDL.

Notes:

(1) Analyte lists may vary among methods; a smaller list in one method is not necessarily a subset of a larger list in another method.

(2) Initial, one-time, demonstration of ability to generate acceptable accuracy and precision. Procedure may need to be repeated if changes in instrumentation or methodology occur.

Organic

Determination of Chlorinated Herbicides by GC

Method ⇒ Parameter ⇓	SW-846 Method 8151A	EPA 500 Series Method 515.1	EPA 500 Series Method 515.2	Standard Methods Method 6640
Standard Solution Expiration (3)	Stock standards of derivatized acids: 1 year. Stock standards of free acids: 2 months. Store at 4 °C in Teflon sealed vials protect from light. Calibration standards: 6 months.	Stock standards: 2 months. Store at room temperature in Teflon sealed amber vials, protect from light. Calibration standards: not specified.	Stock standards 2 months. Store at 4 °C in Teflon sealed amber vials, protect from light. Calibration standards: not specified.	Individual herbicide stock solutions: 6 months. Stock herbicide mixture: 3 months. Store in Teflon sealed amber vials at -11 °C. Calibration standards: not specified.
Initial Calibration	Minimum of 5 levels, lowest near but above MDL. **Option 1**: If CF or RF %RSD ≤20 or if mean %RSD of all analytes ≤20 then use avg. CF or RF. Alternatively use a cal. curve (see options below). **Option 2**: If CF or RF %RSD>20 or if %RSD of all analytes >20, then use cal. curve: a) linear cal. using least squares regression; b) linear cal. using weighted least squares regression. **Requirements**: r≥ 0.99, not forced through orgin, do not use orgin as a cal. point. **Option 3**: Where instrument response is non-linear over wide working range or above procedures fail acceptance criteria, non-linear cal. may be employed. **Requirements**: COD ≥ 0.99, not forced through orgin, do not use orgin as a cal. point, other requirements exist pending curve fitting model chosen.	Calibration levels required: 20x range: 3 calibration standards. 50x range: 4 calibration standards. 100x range: 5 calibration standards. Lowest level should be 2 to 10 times the MDL. If %RSD <20, linearity assumed and average RF used. Alternatively use a calibration curve. Calibration standards are to be esterified.	Calibration levels required: 20x range: 3 calibration standards. 50x range: 4 calibration standards. 100x range: 5 calibration standards. Lowest level should be near but above MDL. If %RSD ≤30, linearity assumed and average RF used. Alternatively use a calibration curve. Calibration standards are to be extracted the same as samples.	Minimum of 3 levels (5 recommended), use calibration levels listed in Table 6640: IV (see page A-103). If %RSD ≤ 20, linearity assumed and average RF used. Alternatively use a calibration curve or a single point calibration. Single point calibration standard must produce response within 20% of unknowns.

Notes:

(3) Indicates maximum usage time. If comparisons to QC check standards indicate a problem, more frequent preparation may be necessary.

Determination of Chlorinated Herbicides by GC (Continued)

Method ⇒ Parameter ⇓	SW-846 Method 8151A	EPA 500 Series Method 515.1	EPA 500 Series Method 515.2	Standard Methods Method 6640
Continuing Calibration	A mid-level calibration check standard must be analyzed at the beginning of each 12 hour shift, or more frequently if warranted. If average response of all analytes is not within ±15% of predicted response, recalibrate. Samples which contain target compounds above the reportable limit must be bracketed by an acceptable calibration check standard. Standard RT must fall within daily RT window or system is out of control. Samples injected after criteria was exceeded must be reanalyzed.	A calibration check standard must be analyzed at the beginning and end of analysis day (each at a different level). If not within ±20% of predicted response, recalibrate and reanalyze samples. Verify calibration standards at least quarterly using an independent source.	One or more calibration standards analyzed daily. If not within ± 30% of predicted response, recalibrate. Derivatized calibration standards can be used up to 14 days. Verify calibration standards at least quarterly using an independent source.	A calibration check standard must be analyzed at the beginning and end of analysis day. If not within ±20% of predicted response, recalibrate and reanalyze samples.
Surrogate Standards	One/two surrogates added to each sample (avoid use of deuterated analogs) 2,4-Dichlorophenylacetic acid (DCAA) at 0.5 mg/L in extract is recommended. Results must fall within laboratory established limits.	2,4-Dichlorophenylacetic acid (DCAA) at 0.5 μg/ml in extract. %R = 70-130.	2,4-Dichlorophenylacetic acid (DCAA) at 0.1 μg/ml in extract. %R = 60-140.	2,3,5,6-Tetraflourobenzoic acid %R=70-130 compared to the average daily calibration standard.
Internal Standards	Optional. If used 4,4'-Dibromooctafluoro-biphenyl (DBOB) or others if there is DBOB interference. (DBOB) at 0.25 μg/mL in extract.	Optional. If used, 4,4'-Dibromoocta-fluorobiphenyl (DBOB) at 0.25 μg/mL in extract. Sample IS response must be ±30% of daily response	Optional. If used 4,4'-Dibromooctafluoro-biphenyl (DBOB) at .020 μg/ml in extract. Sample IS response must be ±30% of daily standard response.	Optional. If used, 1,2,3-Trichloro-propane. Sample IS response must be ±30% of daily standard response.

GENIUM PUBLISHING CORPORATION

Organic

Determination of Chlorinated Herbicides by GC (Continued)

Method ⇒ Parameter ⇓	SW-846 Method 8151A	EPA 500 Series Method 515.1	EPA 500 Series Method 515.2	Standard Methods Method 6640
Accuracy/ Precision	One MS/MSD per 20 samples or each batch of samples, whichever is more frequent. Compare results to laboratory established limits. (See also: **QC Check Standards/ Samples**.) Laboratory can use 70-130% as interim limits until in-house limits are developed.	One MS (Laboratory Fortified Sample Matrix) per 10 samples or each sample set, whichever is more frequent. Compare %R to laboratory established limits. Analyze a QC sample from an external source at least quarterly. (See also: **QC Check Standards/Samples**).	One MS (Laboratory Fortified Sample Matrix) per 10 samples or each sample set, whichever is more frequent. Compare %R to laboratory established limits. Analyze a QC sample from an external source at least quarterly. (See also: **QC Check Standards/Samples**).	One MS (Matrix with known additions) per 10 samples or at least one per month. Compare %R to laboratory established limits. Analyze a QC sample from an external source at least quarterly. (See also: **QC Check Standards/Samples**).
Blanks	One method blank per extraction batch (up to 20 samples) or when there is a change in reagents, whichever is more frequent.	One method blank with each batch of samples extracted or when new reagents used.	One method blank with each batch of samples extracted or when new reagents are used.	One method blank with each batch of samples extracted or when new reagents are used.
Preservation/ Storage Conditions	If residual chlorine present, add sodium thiosulfate. Store at 4 °C. Protect from light.	Mercuric chloride as biocide. Sodium thiosulfate if residual chlorine present. Protect from light. Store at 4 °C.	Adjust pH to <2 by adding HCL at sampling site. Sodium thiosulfate if residual chlorine present (aqueous). Protect from light. Store at 4 °C.	Store at 4 °C. Protect from light. Sodium thiosulfate if residual chlorine present. Adjust pH to <2 at sampling site.
Holding Time (4)	Extraction: 7 days (aqueous). 14 days (solids). Analysis: 40 days after extraction.	Extraction: 14 days. Analysis: 28 days after extraction.	Extraction: 14 days. Analysis: 14 days after extraction.	Extraction: 14 days. Analysis: 14 days after extraction.
Field Sample Amount (5)	1 liter (aqueous). 4 oz. (solid) glass container. Teflon lined top.	1 liter glass container Teflon lined top.	250 mL amber glass container. Teflon lined top.	40 mL glass container Teflon lined top. Collect samples in quadruplicate.
Amount for Extraction	1 liter (aqueous). 30 grams (low level solid). 2 grams (medium level solid).	1 liter	250 mL	30 mL

Notes:

(4) Unless otherwise indicated, holding times are from the date of sample collection.

(5) Approximate volumes to be gathered for analysis. Additional volumes are required for the generation of QC data.

Organic

Organic

Determination of Chlorinated Herbicides by GC (Continued)

Method ⇒ Parameter ⇓	SW-846 Method 8151A	EPA 500 Series Method 515.1	EPA 500 Series Method 515.2	Standard Methods Method 6640
Other Criteria (Method Specific)	When doubt exists in compound identification, second column or GC/MS confirmation should be used.	Laboratory Performance Check Sample analyzed daily to monitor instrument sensitivity, column performance, and chromatographic performance. Compare results to Table 11 (see page A-32) When doubt exists in compound identification, second column confirmation or additional qualitative technique must be used.	Instrument Performance Check Sample analyzed daily to monitor instrument sensitivity, column performance, and chromatographic performance. Compare results to Table 11 (see page A-33). When doubt exists in compound identification, second column confirmation or additional qualitative technique must be used.	Use at least two columns for identification and quantification. Laboratory Performance Check Sample analyzed daily to monitor instrument sensitivity, column performance, and chromatographic performance. Compare results to Table 6640: VIII (see page A-105).

Notes:

(6) MDL determination for Standard Methods 6000 Series: Analyze a minimum of seven check samples (concentration = 0.2 times MCL or 10 times estimated MDL.). Average percent recovery should be 80% to 120% of true value with %RSD ≤ 35%. Use results to determine MDLs. Broader acceptance ranges exist for some compounds with lower extraction efficiency and are indicated in the specific method.

(7) For SW-846 Method 8151A, either diazomethane or pentaflurobenzyl bromide may be used for esterification.

Determination of Chlorinated Herbicides by GC

Chemical Name	CAS Number	SW-846 Method 8151A	EPA 500 Series Method 515.1	EPA 500 Series Method 515.2	Std. Methods Method 6640
Acifluorfen	50594-66-6	•	•	•	
Bentazon	25057-89-0	•	•	•	•
Chloramben	133-90-4	•	•		
2,4-D	94-75-7	•	•	•	•
Dalapon	75-99-0	•	•		•
2,4'-DB	94-82-6	•	•	•	
Dacthal	1861-32-1			•	
DCPA acid metabolites	N/A	•	•		
Dicamba	1918-00-9	•	•	•	•
3,5-Dichlorobenzoic acid	51-36-5	•	•	•	
Dichloroprop	120-36-5	•	•	•	
Dinoseb	88-85-7	•	•	•	•
5-Hydroxydicamba	7600-50-2	•	•	•	
MCPA	94-74-6	•			
MCPP	93-65-2	•			
4-Nitrophenol	100-02-7	•	•		
Pentachlorophenol	87-86-5	•	•	•	•
Picloram	1918-02-1	•	•	•	•
2,4,5-T	93-76-5	•	•	•	•
2,4,5-TP(Silvex)	93-72-1	•	•	•	•

Organic

Determination of Organochlorine Pesticides and PCBs by GC

Method ⇒ / Parameter ⇓	SW-846 Method 8082 Congener	SW-846 Method 8082 Aroclor	SW-846 Method 8081A	EPA 500 Series Method 508	EPA 600 Series Method 608	Std. Methods Method 6630 (7)	CLP PEST (Organic SOW)
Applicability	Solid and aqueous matrices	Solid and aqueous matrices	Groundwater, soils, sludges, and non-water miscible wastes	Groundwater and finished drinking water	Municipal and industrial discharges	Drinking water, surface and ground-water, municipal and industrial discharges	Water, soil and sediment from hazardous waste sites.
Number of Analytes (1)	19 listed (others possible)	7 total	52 total	38 total (2 qualitative only)	25 total	35 listed (others possible)	28 total
Method Validation (2)	Extract and analyze 4 replicates of a QC reference sample. Method does not specify accuracy and precision limits. Use recoveries of 70-130% in evaluating results. Must demonstrate initial proficiency with each sample preparation and determinative combination.	Extract and analyze 4 replicates of a QC reference sample. Method does not specify accuracy and precision limits. Use recoveries of 70-130% in evaluating results. Must demonstrate initial proficiency with each sample preparation and determinative combination.	Extract and analyze 4 replicates of a QC check sample. Method does not specify accuracy and precision limits. Use recoveries of 70-130% in evaluating results.	Extract and analyze 4 to 7 replicates of QC check sample. Concentration should be 10 times the EDL or at a mid-point in the calibration range. Analyte recovery must be within ± 30% of recovery value specified in Table 2 (see page A-29) RSD of measurements must be ≤ 20%. Determine MDL.	Extract and analyze 4 replicates of QC check standards. Compare accuracy and precision results to Table 3 (see page A-72).	Extract and analyze 4 replicates of QC check standards. Compare accuracy and precision results to Table 6630:V (see page A-102). Determine MDL (6).	Not specified.

(1) Analyte lists may vary among methods; a smaller list in one method is not necessarily a subset of a larger list in another method.

(2) Initial, one-time, demonstration of ability to generate acceptable accuracy and precision. Procedure may need to be repeated if changes in instrumentation or methodology occur.

(6) MDL determination for Standard Methods 6000 Series: Analyze a minimum of seven check samples (concentration = 0.2 times MCL or 10 times estimated MDL.). Average percent recovery should be 80% to 120% of true value with %RSD ≤ 35%. Use results to determine MDLs. Broader acceptance ranges exist for some compounds with lower extraction efficiency and are indicated in the specific method.

(7) Methods 6630B and 6630C are GC methods, Method 6630D is a GC/MS method cross referenced as Method 6410B.

Determination of Organochlorine Pesticides and PCBs by GC (Continued)

Method ⇒ Parameter ⇓	SW-846 Method 8082 Congener	SW-846 Method 8082 Aroclor	SW-846 Method 8081A	EPA 500 Series Method 508	EPA 600 Series Method 608	Std. Methods Method 6630 (7)	CLP PEST (Organic SOW)
QC Check Standards/ Samples	One Laboratory control sample per 20 samples or each batch of samples whichever is more frequent. Compare results to laboratory established limits. Laboratory can use 70-130% as interim limits until in-house limits are developed.	One Laboratory control sample per 20 samples or each batch of samples whichever is more frequent. Compare results to laboratory established limits. Laboratory can use 70-130% as interim limits until in-house limits are developed.	One Laboratory control sample per 20 samples or each batch of samples whichever is more frequent. Compare results to laboratory established limits. Laboratory can use 70-130% as interim limits until in-house limits are developed.	Analyze one QC check sample (Laboratory Fortified Blank) per 20 samples or one per sample set (all samples extracted within a 24-hour period), whichever is more frequent. Concentration should be 10 times the EDL or a conc. that is near the mid-point of the calibration. Compare %R to laboratory established limits. If laboratory limits not available Analyte recovery must be within ±30% of recovery values specified in Table 2 (See Page A-29). Laboratory limits must not exceed ±30% of recovery values specified in Table 2	Analyze one QC check standard with every 10 samples. (Frequency may be reduced if MS recoveries meet QC criteria.) Compare %R to Table 3 (see page A-72).	Spike a minimum of 10% of all samples. If MS results fall outside ranges designated in Table 6630:V (see page A-102), a QC check standard must be analyzed and fall within those ranges.	Not specified.

(7) Methods 6630B and 6630C are GC methods, Method 6630D is a GC/MS method cross referenced as Method 6410B.

Determination of Organochlorine Pesticides and PCBs by GC (Continued)

Method ⇒ Parameter ⇓	SW-846 Method 8082 Congener	SW-846 Method 8082 Aroclor	SW-846 Method 8081A	EPA 500 Series Method 508	EPA 600 Series Method 608	Std. Methods Method 6630 (7)	CLP PEST (Organic SOW)
Method Detection Limit	Laboratory should develop matrix specific MDLs. EQLs range: Water 5-25 ng/L per congener. Soil 160-800 ng/kg per congener.	Laboratory should develop matrix specific MDLs. Suggested MDL range: Water: 0.054-0.90 µg/L Soil: 57-70 µg/kg EQLs listed in Table 1 (see page A-116).	Laboratory should develop matrix specific MDLs. EQLs listed in Table 3 (see page A-115).	Extract and analyze a minimum of 7 laboratory fortified blanks. Use Table 3 concentrations as a guide. Calculate MDL. MDLs and EDLs listed in Table 3 (See Page A-30)	MDLs listed in Table 1 (see page A-71).	Varies with detector sensitivity, extraction/clean-up efficiency and concentrations. MDLs listed in Table 6630:III (see page A-85).	CRQLs for TCL listed in Exhibit C (see page A-1).
Standard Solution Expiration (3)	Stock standards: 1 year. Calibration standards: 6 months. Store at 4 °C in Teflon sealed vials, protect from light.	Stock standards: 1 year. Calibration standards: 6 months. Store at 4 °C in Teflon sealed vials, protect from light.	Stock standards: 1 year. Calibration standards: 6 months. Store at 4 °C in Teflon sealed vials, protect from light.	Stock standards: 2 months. Store at room temperature in Teflon sealed amber vials, protect from light. Calibration standards: not specified.	Stock standards: 6 months. Store at 4 °C in Teflon sealed bottles, protect from light. Calibration standards: not specified.	Stock standards: 6 months. Store at 4 °C in Teflon sealed bottles, protect from light. Calibration standards: not specified.	Stock and calibration standards [except Performance Evaluation Mixture (PEM)]: 6 months. Store all standard solutions at 4 °C in Teflon sealed amber bottles, protect from light. PEM: 1 week.

(3) Indicates maximum usage time. If comparisons of QC check standards indicate a problem, more frequent preparation may be necessary.

(7) Methods 6630B and 6630C are GC methods, Method 6630D is a GC/MS method cross referenced as Method 6410B.

Organic

Determination of Organochlorine Pesticides and PCBs by GC (Continued)

Method ⇒ / Parameter ⇓	SW-846 Method 8082 Congener	SW-846 Method 8082 Aroclor	SW-846 Method 8081A	EPA 500 Series Method 508	EPA 600 Series Method 608	Std. Methods Method 6630 (7)	CLP PEST (Organic SOW)
Initial Calibration	Minimum of 5 levels, Standards should bracket range of detector. **Option 1:** If CF or RF %RSD ≤20 or if mean %RSD of all analytes ≤20 then use avg CF or RF. Alternatively use a cal. curve (see options below). **Option 2:** If CF or RF %RSD >20 or if %RSD of all analytes >20, then use cal curve: a) linear cal. using a least squares regression; b) linear cal. using weighted least squares regression. **Requirements:** r≥ 0.99, not forced through origin, do not use origin as a cal. point. **Option 3:** Where instrument response is non-linear over wide working range or above procedures fail acceptance criteria, non-linear cal. may be employed. **Requirements:** COD ≥0.99, not forced through origin, do not use origin as a cal. point, other requirements exist pending curve fitting model chosen.	Aroclor 1016/1260 mixture Minimum of 5 levels, Standards should bracket range of detector. Other 5 aroclors are analyzed and used to determine a single point calibration factor. (See also Footnote 8). **Option 1:** If CF or RF %RSD ≤20 or if mean %RSD of all analytes ≤20 then use avg. CF or RF. Alternatively use a cal. curve (see options below). **Option 2:** If CF or RF %RSD >20 or if %RSD of all analytes >20, then use cal. curve: a) linear cal. using a least squares regression; b) linear cal. using weighted least squares regression. **Requirements:** r≥ 0.99, not forced through origin, do not use origin as a cal. point. **Option 3:** Where instrument response is non-linear over wide working range or above procedures fail acceptance criteria, non-linear cal. may be employed. **Requirements:** COD ≥ 0.99, not forced through origin, do not use origin as a cal. point, other requirements exist pending curve fitting model chosen.	Minimum of 5 levels, Standards should bracket range of detector. Multi component analytes employ a single point cal. **Option 1:** If CF or RF %RSD ≤20 or if mean %RSD of all analytes ≤20 then use avg. CF or RF. Alternatively use a cal. curve (see options below). **Option 2:** If CF or RF %RSD >20 or if %RSD of all analytes >20, then use cal. curve: a) linear cal. using a least squares regression; b) linear cal. using weighted least squares regression. **Requirements:** r≥ 0.99, not forced through origin, do not use origin as a cal. point. **Option 3:** Where instrument response is non-linear over wide working range or above procedures fail acceptance criteria, non-linear cal may be employed. **Requirements:** COD ≥ 0.99, not forced through origin, do not use origin as a cal point, other requirements exist pending curve fitting model chosen.	Calibration levels required: 20x range: 3 calibration standards. 50x range: 4 calibration standards. 100x range: 5 calibration standards. Lowest near but above EDL. If %RSD ≤ 20%, linearity assumed and average RF used. Alternatively use a calibration curve.	Minimum of 3 levels, lowest near but above MDL. If %RSD <10, linearity assumed and average RF used. Alternatively, use a calibration curve.	Minimum of 3 levels, lowest near but above MDL. If %RSD <10, linearity assumed and average RF used. Alternatively, use a calibration curve. Calibrate system "daily."	Three levels (except multi-components at one level) the lowest at the CRQL. The %RSD must be ≤20% for target compounds, except α-BHC and Δ-BHC which must be ≤25%, ≤30% for surrogates. Maximum of 2 target compounds may exceed 20% RSD per column, but must be ≤30% RSD. Analyze Resolution Check and PEM standards before initial calibration.

(7) Methods 6630B and 6630C are GC methods, Method 6630D is a GC/MS method cross referenced as Method 6410B.

Organic

Determination of Organochlorine Pesticides and PCBs by GC (Continued)

Method ⇒ Parameter ⇓	SW-846 Method 8082 Congener	SW-846 Method 8082 Aroclor	SW-846 Method 8081A	EPA 500 Series Method 508	EPA 600 Series Method 608	Std. Methods Method 6630 (7)	CLP PEST (Organic SOW)
Continuing Calibration	Verify calibration at the beginning of each 12 hour shift. Calibration standards must be injected after every 20 samples (10 recommended). If response factor of all analytes is not within ±15% of mean response factor, recalibrate. Samples containing target compounds above the reportable limit must be bracketed by an acceptable calibration verification standard. Alternate the use of high and low standards. Standard RT must fall within daily RT window or system is out of control. Samples injected after criteria was exceeded must be reanalyzed.	Verify calibration at the beginning of each 12 hour shift. Calibration standards must be injected after every 20 samples (10 recommended). If the cal. factor of all analytes is not within ±15% of mean cal. factor, recalibrate. Samples containing target compounds above the reportable limit must be bracketed by an acceptable calibration verification standard. Alternate the use of high and low standards. Standard RT must fall within daily RT window or system is out of control. Samples injected after criteria was exceeded must be reanalyzed.	Verify calibration at the beginning of each 12 hour shift. Calibration standards must be injected after every 20 samples (10 recommended). If average response of all analytes is not within ±15% of predicted response, recalibrate. Samples containing target compounds above the reportable limit must be bracketed by an acceptable calibration verification standard. Alternate the use of high and low standards. Standard RT must fall within daily RT window or system is out of control. Samples injected after criteria was exceeded must be reanalyzed.	A calibration check standard must be analyzed at the beginning and end of analysis day (8 hrs.; each at a different level). If not within ±20% of predicted response, recalibrate.	One or more calibration standards analyzed daily. If not within ±15% of predicted response, recalibrate.	Inject standards frequently to insure optimum operating conditions. Verify calibration each working day. If not within ±15% of predicted response, recalibrate.	Instrument Blank, Performance Evaluation Mixture and midpoint calibration standard mixtures A and B are run once per 12 hour period. Standard must be within ±25% of predicted response.

(7) Methods 6630B and 6630C are GC methods, Method 6630D is a GC/MS method cross referenced as Method 6410B.

GENIUM PUBLISHING CORPORATION

Organic

Determination of Organochlorine Pesticides and PCBs by GC (Continued)

Method ⇒ Parameter ⇓	SW-846 Method 8082 Congener	SW-846 Method 8082 Aroclor	SW-846 Method 8081A	EPA 500 Series Method 508	EPA 600 Series Method 608	Std. Methods Method 6630 (7)	CLP PEST (Organic SOW)
Surrogate Standards	2,4,5,6-Tetrachloro-m-xylene. Results must fall within laboratory established control limits. Laboratory can use 70-130% as interim limits until in-house limits are developed.	Decachloro-biphenyl Results must fall within laboratory established control limits. Laboratory can use 70-130% as interim limits until in-house limits are developed.	Decachloro-biphenyl and 2,4,5,6-Tetrachloro-m-xylene. Dual column analysis use 4-chloro-3-nitrobenzo-trifluoride. Results must fall within laboratory established control limits. Laboratory can use 70-130% as interim limits until in-house limits are developed.	4,4'-Dichlorobiphenyl (DCB) at 5.0 µg/ml in extract. %R = 70-130.	Not specified.	Not specified.	Two surrogates, Decachlorobi-phenyl and 2,4,5,6-Tetra chloro-m-xylene. Advisory limits = 30-150%. Note: Limits must be met for method blank.
Internal Standards	Highly Recommended Decachloro-biphenyl Sample IS response must be ±50% of average area calculated during calibration.	Not required	Optional. If used penta-chloronitro-benzene or 1-Bromo-2-nitroben-zene at 50 µg/ml Sample IS response must be ±50% of average area calculated during calibration.	Optional. If used, penta-chloroni-trobenzene at 0.1µg/ml in extract. Others can be used if acceptance criteria met. Sample IS response must be ±30% of daily standard response.	Optional; no standards specified.	Optional; not specified.	Not required.

(7) Methods 6630B and 6630C are GC methods, Method 6630D is a GC/MS method cross referenced as Method 6410B.

Determination of Organochlorine Pesticides and PCBs by GC (Continued)

Method ⇒ Parameter ⇓	SW-846 Method 8082 Congener	SW-846 Method 8082 Aroclor	SW-846 Method 8081A	EPA 500 Series Method 508	EPA 600 Series Method 608	Std. Methods Method 6630 (7)	CLP PEST (Organic SOW)
Accuracy/ Precision	One MS/ MSD per 20 samples or each batch of samples, whichever is more frequent. Compare results to laboratory established limits. (See also: **QC Check Standards/ Samples**). Laboratory can use 70-130% as interim limits until in house limits are developed.	One MS/ MSD per 20 samples or each batch of samples, whichever is more frequent. Compare results to laboratory established limits. (See also: **QC Check Standards/ Samples**). Laboratory can use 70-130% as interim limits until in house limits are developed.	One MS/ MSD per 20 samples or each batch of samples, whichever is more frequent. Compare results to laboratory established limits. (See also: **QC Check Standards/ Samples**). Laboratory can use 70-130% as interim limits until in house limits are developed.	One MS (Laboratory Fortified Sample Matrix) per 10 samples or each set of samples, whichever is more frequent. % Recovery for analyte must be with-in Table 2 (See pg. A-29) %R values ±30%. Analyze a QC sample from an external source at least quarterly. (See also: **QC Check Standards/ Samples**).	One MS per 10 samples from each site or 1 MS per month, whichever is more frequent. Compare %R to Table 3 (see page A-72). If MS results fall outside acceptance criteria, a QC check standard must be analyzed. (See also: **QC Check Standards/ Samples**).	One MS per 10 samples from each site or 1 MS per month, whichever is more frequent. Compare %R to Table 6630:V (see page A-102). (See also: **QC Check Standards/ Samples**).	One MS/ MSD per sample delivery group or 1 per 20 samples, whichever is greater. Six compounds are used for spiking; advisory limits in Exhibit D, Section III, Paragraph 16.4 (see page A-2).
Blanks	One method blank per extraction batch (up to 20 samples) or when there is a change in reagents, whichever is more frequent.	One method blank per extraction batch (up to 20 samples) or when there is a change in reagents, whichever is more frequent.	One method blank per extraction batch (up to 20 samples) or when there is a change in reagents, whichever is more frequent.	One method blank with each batch of samples extracted or when new reagents used.	One method blank with each batch of samples extracted or when new reagents used.	One method blank with each batch of samples extracted.	One method blank per each case, or 20 samples (including MS and reana- lyses) or each extraction batch or 14 day period samples re- ceived from case, which- ever is more frequent. Concentration <CRQL for all target compounds. Instrument Blanks at start of 12 hr sequence, then every 12 hours. All tar- get com- pounds <0.5 times CRQL.

(7) Methods 6630B and 6630C are GC methods, Method 6630D is a GC/MS method cross referenced as Method 6410B.

Organic

Determination of Organochlorine Pesticides and PCBs by GC (Continued)

Method ⇒ / Parameter ⇓	SW-846 Method 8082 Congener	SW-846 Method 8082 Aroclor	SW-846 Method 8081A	EPA 500 Series Method 508	EPA 600 Series Method 608	Std. Methods Method 6630 (7)	CLP PEST (Organic SOW)
Preservation/ Storage Conditions	If residual chlorine present, add sodium thiosulfate. Store at 4 °C. Protect from light.	If residual chlorine present, add sodium thiosulfate. Store at 4 °C. Protect from light.	If residual chlorine present, add sodium thiosulfate. Store at 4 °C. Protect from light.	Mercuric chloride as biocide. Sodium thiosulfate if residual chlorine present. Store at 4 °C. Protect from light.	Adjust pH to 5-9 if extraction not to be done within 72 hr of sampling. Add sodium thiosulfate if residual chlorine present and Aldrin is being determined. Store at 4 °C.	Adjust pH to 5-9 if extraction not to be done within 72 hr of sampling. Add sodium thiosulfate if residual chlorine present and Aldrin is being determined. Store at 4 °C.	Protect from light. Store at 4 °C.
Holding Time (4)	Extraction: 7 days (aqueous) 14 days (solids) Analysis: 40 days after extraction	Extraction: 7 days (aqueous) 14 days (solids) Analysis: 40 days after extraction	Extraction: 7 days (aqueous) 14 days (solids) Analysis: 40 days after extraction	Extraction: 7 days Analysis: 14 days after extraction	Extraction: 7 days Analysis: 40 days after extraction	Extraction: 7 days Analysis: 40 days after extraction	Extraction: (separatory funnel) 5 days of sample receipt (aqueous) (Continuous liquid-liquid) started within 5 days of sample receipt (aqueous) 10 days of sample receipt (solids) Analysis: 40 days after extraction
Field Sample Amount Required (5)	1 liter amber (aqueous) 4 oz. (solid) glass container Teflon lined top.	1 liter amber (aqueous) 4 oz. (solid) glass container Teflon lined top.	1 liter amber (aqueous) 4 oz. (solid) glass container Teflon lined top.	1 liter amber glass container with graduations if possible Teflon lined top.	1 liter amber glass container Teflon lined top.	1 liter amber glass container Teflon lined top.	1 liter amber (aqueous) 4 oz. (solid) glass container Teflon lined top.
Amount for Extraction	1 liter (aqueous) 30 grams (low level solid) 2 grams (medium level solid).	1 liter (aqueous) 30 grams (low level solid) 2 grams (medium level solid).	1 liter (aqueous) 30 grams (low level solid) 2 grams (medium level solid).	1 liter	1 liter	1 liter	1 liter (aqueous) 30 grams (low level solid) 2 grams (medium level solid).

Notes:

(4) Unless otherwise indicated, holding times are from the date of sample collection.

(5) Approximate volumes to be gathered for analysis. Additional volumes are required for the generation of QC data.

(7) Methods 6630B and 6630C are GC methods, Method 6630D is a GC/MS method cross referenced as Method 6410B.

Organic

Organic

Determination of Organochlorine Pesticides and PCBs by GC (Continued)

Method ⇒ Parameter ⇓	SW-846 Method 8082 Congener	SW-846 Method 8082 Aroclor	SW-846 Method 8081A	EPA 500 Series Method 508	EPA 600 Series Method 608	Std. Methods Method 6630 (7)	CLP PEST (Organic SOW)
Other Criteria (Method Specific)	Each tentative identification must be confirmed using the following options: 1) Second GC column. 2) GC/MS analysis. When using a second GC column for confirmation the second column must meet the same calibration acceptance criteria as the first column.	Each tentative identification must be confirmed using the following options: 1) Second GC column. 2) Identification based on a clearly identifiable Aroclor pattern 3) GC/MS analysis. When using a second GC column for confirmation the second column must meet the same calibration acceptance criteria as the first column.	Compounds confirmed by two columns should be confirmed by GC/MS if concentration is sufficient. Check for dichlorodiphenyltrichloroethane (DDT) and endrin degradation. % breakdown must be ≤15%.	Laboratory Performance Check (LPC) sample analyzed daily to monitor instrument sensitivity, column performance, and chromatographic performance. Compare results to Table 4 (see page A-31) Check for dichlorodiphenyltrichloroethane (DDT) and endrin degradation daily. % breakdown must be < 20%. At least quarterly analyze independent reference standard. Results must be within calibration check limits. When doubt exists in compound identification, second column confirmation should be used.	When doubt exists in compound identification, second column confirmation or GC/MS should be used.	When doubt exists in compound identification, second column confirmation or GC/MS should be used.	Second column confirmation is mandatory. GC/MS confirmation mandatory for positive samples of sufficient concentration. Check for dichlorodiphenyltrichloroethane (DDT) and endrin degradation every 12 hours. % breakdown must be <20% for each compound and <30% combined. Resolution criteria between adjacent peaks: mid-level individual standards ≥90%; PEM = 100%; Resolution Check Mixture ≥60%. A standard of an identified Aroclor must be run within 72 hrs of detection in a sample.

(7) Methods 6630B and 6630C are GC methods, Method 6630D is a GC/MS method cross referenced as Method 6410B.

(8) In situations where only a few Aroclors are of interest for a specific project, the analyst may employ a five point initial calibration of each of the Aroclors of interest and not use the 1016/1260 mixture.

Determination of Organochlorine Pesticides and PCBs by GC

Chemical Name	CAS Number	SW-846 Method 8082 Congener	SW-846 Method 8082 Aroclor	SW-846 Method 8081A	EPA 500 Series Method 508	EPA 600 Series Method 608	Std. Methods Method 6630	CLP PEST (Organic SOW)
Alachlor	15972-60-8			•				
Aldrin	309-00-2			•	•	•	•	•
Aroclor-1016	12674-11-2		•		•	•	•	•
Aroclor-1221	11104-28-2		•		•	•	•	•
Aroclor-1232	11141-16-5		•		•	•	•	•
Aroclor-1242	53469-21-9		•		•	•	•	•
Aroclor-1248	12672-29-6		•		•	•	•	•
Aroclor-1254	11097-69-1		•		•	•	•	•
Aroclor-1260	11096-82-5		•		•	•	•	•
2-Chlorobiphenyl	2051-60-7	•						
2,3-Dichlorobiphenyl	16605-91-7	•						
2,2',5-Trichlorobiphenyl	37680-65-2	•						
2,4',5-Trichlorobiphenyl	16606-02-3	•						
2,2',3,5'-Tetrachlorobiphenyl	41464-39-5	•						
2,2',5,5'-Tetrachlorobiphenyl	35693-99-3	•						
2,3',4,4'-Tetrachlorobiphenyl	32598-10-0	•						
2,2',3,4,5'-Pentachlorobiphenyl	38380-02-8	•						
2,2',4,5,5'-Pentachlorobiphenyl	37680-73-2	•						
2,3,3',4',6-Pentachlorobiphenyl	38380-03-9	•						
2,2',3,4,4',5'-Hexachlorobiphenyl	35065-28-2	•						
2,2',3,4,5,5'-Hexachlorobiphenyl	52712-04-6	•						
2,2',3,5,5',6-Hexachlorobiphenyl	52663-63-5	•						
2,2',4,4',5,5'-Hexachlorobiphenyl	35065-27-1	•						
2,2',3,3',4,4',5-Heptachlorobiphenyl	35065-30-6	•						
2,2',3,4,4',5,5'-Heptachlorobiphenyl	35065-29-3	•						
2,2',3,4,4',5',6-Heptachlorobiphenyl	52663-69-1	•						
2,2',3,4',5,5',6-Heptachlorobiphenyl	52663-68-0	•						
2,2',3,3',4,4',5,5',6-Nonachlorobiphenyl	40186-72-9	•						
α-BHC	319-84-6			•	•	•	•	•
β-BHC	319-85-7			•	•	•	•	•
δ-BHC [1]	319-86-8			•	•	•	•	•

Determination of Organochlorine Pesticides and PCBs by GC (Continued)

Chemical Name	CAS Number	SW-846 Method 8082 Congener	SW-846 Method 8082 Aroclor	SW-846 Method 8081A	EPA 500 Series Method 508	EPA 600 Series Method 608	Std. Methods Method 6630	CLP PEST (Organic SOW)
γ-BHC (Lindane)	58-89-9			•	•	•	•	•
Captafol	2425-06-1			•				
Captan	133-06-2						•	
α-Chlordane	5103-71-9			•	•			•
Chlordane	57-74-9			•	•	•	•	
γ-Chlordane	5103-74-2			•	•			•
Chloroneb	2675-77-6			•	•			
Chlorobenzilate [1]	510-15-6			•	•			
Chloropropylate	99516-95-7			•				
Chlorothalonil	1897-45-6			•	•			
DBCP	96-12-8			•				
DCPA	1861-32-1			•	•			
4,4'-DDD	72-54-8			•		•	•	•
4,4'-DDE	72-55-9			•		•	•	•
4,4'-DDT	50-29-3			•		•	•	•
Diallate	2302-16-4			•				
Dichlone	117-80-6			•				
Dichloran	N/A						•	
Dicofol	115-32-2			•				
Dieldrin	60-57-1			•	•	•	•	•
Endosulfan I	959-98-8			•	•	•	•	•
Endosulfan II	33213-65-9			•	•	•	•	•
Endosulfan sulfate	1031-07-8			•	•	•	•	•
Endrin	72-20-8			•	•	•	•	•
Endrin aldehyde	7421-93-4			•	•	•	•	•
Endrin ketone	53494-70-5			•				•
Etridiazole	2593-15-9			•	•			
Hallowax-1000	58718-67-5			•				
Halowax-1001	58718-67-5			•				
Halowax-1013	12616-35-2			•				
Halowax-1014	12616-36-3			•				
Halowax-1051	2234-13-1			•				
Halowax-1099	39450-05-0			•				
Heptachlor	76-44-8			•	•	•	•	•
Heptachlor epoxide	1024-57-3			•	•	•	•	•
Hexachlorobenzene	118-74-1			•	•			
Hexachloro-cyclopentadiene	77-47-4			•				
Isodrin	465-73-6			•				
Kepone	143-50-0							

Determination of Organochlorine Pesticides and PCBs by GC (Continued)

Chemical Name	CAS Number	SW-846 Method 8082 Congener	SW-846 Method 8082 Aroclor	SW-846 Method 8081A	EPA 500 Series Method 508	EPA 600 Series Method 608	Std. Methods Method 6630	CLP PEST (Organic SOW)
Malathion [2]	121-75-5						•	
Methoxychlor	72-43-5			•	•		•	•
Methyl parathion [2]	298-00-0						•	
Mirex	2385-85-5			•			•	
Nitrofen	1836-75-5			•				
trans-Nonachlor	39765-80-5			•				
Parathion [2]	56-38-2						•	
Pentachloronitro-benzene (PCNB)	82-68-8			•			•	
cis-Permethrin	52645-53-1				•			
trans-Permethrin	51877-74-8			•	•			
Perthane	72-56-0			•				
Propachlor	1918-16-7			•	•			
Strobane	8001-50-1			•			•	
Toxaphene	8001-35-2			•	•	•	•	•
Trifluralin [2]	1582-09-8			•	•		•	

Notes:

[1] δ-BHC and Chlorobenzilate listed as tentatively measurable by Method 508.

[2] Trifluralin and organophosphorus pesticides including Parathion, Methyl parathion and Malathion listed as tentatively measurable by Method 6630.

Determination of Semivolatile Organic Compounds by GC/MS

Method ⇒ Parameter ⇓	SW-846 Method 8270C	EPA 500 Series Method 525.2	EPA 600 Series Method 625	Std. Methods Method 6410	CLP SVOA (Organic SOW)
Applicability	Groundwaters, soils, sediments, sludges, and non-water miscible wastes	Drinking water and raw source water	Wastewater	Domestic, industrial wastewaters, natural and potable waters	Water, soil, and sediment
Number of Compounds (1)	241 total	118 total	81 total	81 total	64 total
Method Validation (2)	Extract and analyze 4 replicates of QC check standard. Compare accuracy and precision results to Table 6 (see page A-144).	Extract and analyze 4 to 7 replicates of QC check sample at 2-5 µg/L (Laboratory Fortified Blank). %R = 70-130 %RSD < 30 Extract and analyze 7 replicates at 0.5 µg/L over 3 to 4 days. Use results to calculate MDLs. MDLs must be sufficient to detect analytes at regulatory levels.	Extract and analyze 4 replicates of QC check standard. Compare accuracy and precision results to Table 6 (see page A-79).	Extract and analyze 4 replicates of QC check standard. Compare accuracy and precision results to Table 6410:V (see page A-99). Determine MDL (3).	Not specified.

(1) Analyte lists may vary among methods; a smaller list in one method is not necessarily a subset of a larger list in another method.

(2) Initial, one-time demonstration of ability to generate acceptable accuracy and precision. Procedure may need to be repeated if changes in instrumentation or methodology occur.

(3) MDL determination for Standard Methods 6000 Series: Analyze a minimum of seven check samples (concentration = 0.2 times MCL or 10 times estimated MDL.). Average percent recovery should be 80% to 120% of true value with %RSD ≤ 35%. Use results to determine MDLs. Broader acceptance ranges exist for some compounds with lower extraction efficiency and are indicated in the specific method.

Organic

Determination of Semivolatile Organic Compounds by GC/MS (Continued)

Method ⇒ Parameter ⇓	SW-846 Method 8270C	EPA 500 Series Method 525.2	EPA 600 Series Method 625	Std. Methods Method 6410	CLP SVOA (Organic SOW)
QC Check Standards/ Samples	If MS/MSD results fall outside ranges designated in Table 6 (see page A-144) or preferably lab designated limits, a QC check standard or LCS must be analyzed and fall within those ranges.	Analyze one QC check sample (Laboratory Fortified Blank) per 20 samples or each batch of samples processed together within a 12 hour work shift, whichever is more frequent. (see footnote (4) for further details). %R = 70-130 %RSD < 30 Analyze replicate QC check samples and a QC sample from an external source at least quarterly.	If MS results fall outside ranges in Table 6 (see page A-79), a QC check standard must be analyzed and fall within those ranges.	If MS results fall outside ranges in Table 6410:V (see page A-99), a QC check standard must be analyzed and fall within those ranges.	Not specified.
Method Detection Limit	EQLs listed in Table 2 (see page A-138).	MDLs listed in Tables 3 (see page A-48), 4 (see page A-52), 5 (see page A-56), 6 (see page A-60).	MDLs listed in Tables 4 (see page A-76) and 5 (see page A-78).	MDLs listed in Tables 6410:I (see page A-95) and 6410:II (see page A-97).	CRQLs for TCL listed in Exhibit C (see page A-3).
Standard Solution Expiration (5)	Stock standards: 1 year. Store between -10 °C to -20 °C in Teflon sealed bottles, protect from light. Calibration standards: 1 year. Store between -10 °C to -20 °C. Daily continuing calibration standard: 1 week. Store at 4 °C.	Stock standards: Expiration not specified. Store in dark, cool place in amber vials at 4°C or less. Calibration standards: Expiration not specified. Store in dark, cool place in amber vials.	Stock standards: 6 months. Store at 4 °C in Teflon sealed bottles, protect from light. Calibration standards: not specified.	Stock standards: 6 months. Store at 4 °C in Teflon sealed bottles, protect from light. Calibration standards: not specified.	Stock standards: 1 year. Store at 4 °C. (± 2 °C) in Teflon sealed amber bottles. Calibration standards: 1 year. Store at 4 °C (± 2 °C) in Teflon sealed amber bottles. Daily continuing calibration standard: 1 week. Store at 4 °C (± 2 °C) in Teflon sealed amber bottles.

(4) If the LFB for each batch of samples contains the individual PCB congeners listed in the analyte cross reference section, then an LFB for each aroclor is not required. At least one LFB containing toxaphene should be extracted for each 24 hour period during which extractions are performed. Toxaphene should be fortified in a separate LFB from other method analytes.

(5) Indicates maximum usage time. If comparisons to QC check standards indicate a problem, more frequent preparation may be necessary.

Determination of Semivolatile Organic Compounds by GC/MS (Continued)

Method ⇒ / Parameter ⇓	SW-846 Method 8270C	EPA 500 Series Method 525.2	EPA 600 Series Method 625	Std. Methods Method 6410	CLP SVOA (Organic SOW)
Initial Calibration	Minimum of 5 levels, lowest near but above MDL. %RSD for CCCs < 30. RF for SPCCs >0.050. **Option 1**: If CF or RF %RSD ≤ 15 or if mean %RSD of all analytes ≤ 15 use avg. CF or RF. Alternatively use a cal. curve (see options below). **Option 2**: If CF or RF % RSD >15 or if mean % RSD of all analytes > 15, then use cal. curve: a) linear cal. using least squares regression; b) linear cal. using weighted least squares regression. **Requirements:** r ≥ 0.99, not forced through origin, do not use origin as cal. point. **Options 3**: Where instrument response is non-linear over wide working range or above procedures fail acceptance criteria, non-linear cal. may be employed. **Requirements:** COD ≥ 0.99, not forced through origin, do not use origin as a cal. point, other requirements exist pending curve fitting model chosen.	Six levels. If %RSD < 30, average RF is used. Alternatively, use a calibration curve. If all analytes are to be determined 2 or 3 sets of cal. standards will probably be required.	Minimum of 3 levels, lowest near but above MDL. If %RSD < 35, linearity assumed and average RF used. Alternatively, use a calibration curve.	Minimum of three levels, one near but above MDL. If %RSD < 35, linearity assumed and average RF used. Alternatively, use a calibration curve.	Five levels except for 8 compounds listed in **Other Criteria** which require 4 levels. RRF and %RSD criteria listed in Table 5 of Exhibit D/SV (see page A-6). Maximum of 4 compounds listed in Table 5 permitted to exceed criteria if RRF ≥ 0.010 and %RSD ≤40%. All others RRF ≥0.010.

GENIUM PUBLISHING CORPORATION

Determination of Semivolatile Organic Compounds by GC/MS (Continued)

Method ⇒ Parameter ⇓	SW-846 Method 8270C	EPA 500 Series Method 525.2	EPA 600 Series Method 625	Std. Methods Method 6410	CLP SVOA (Organic SOW)
Continuing Calibration/ Calibration Verification	Mid-level calibration standard run every 12 hours. RF for SPCCs > 0.050. RF of CCCs must be < 20% difference from initial calibration. Internal Standards: RT must be within ±30 seconds from last mid-point standard level of the initial calibration; area must be -50 to +100%.	Mid-level calibration standard analyzed at the beginning of each 12 hour shift. RF must be within ±30% of initial calibration. See footnote (8) for further details. Absolute areas of quant. ions for IS and surrogates must not decrease more than 30% from previous check standard, or by more than 50% from initial calibration.	One or more calibration standards analyzed each day. If not within ±20% of predicted response, recalibrate.	One or more calibration standards analyzed each day. If not within ±20% of predicted response, recalibrate.	A 50 µg/mL calibration standard is run every 12 hours. The % difference must be ≤25% from the initial calibration. As with initial calibration up to 4 compounds in Table 5 of Exhibit D/SV (see page A-6) permitted to fail criteria if RRF >0.010 and %D ≤40%.
Surrogate Standards	Nitrobenzene-d_5, 2-Fluorobiphenyl, p-Terphenyl-d_{14}, Phenol-d_6, 2-Fluorophenol, and 2,4,6-Tribromophenol. Compare %R to laboratory established limits.	Perylene-d_{12}, 1,3-dimethyl-2-nitrobenzene, triphenylphosphate Optional: Pyrene-d_{10}.	Minimum of three from Table 8 (see page A-81). % recovery limits not specified.	Minimum of 3 from Table 6410:IV (see page A-98). % recovery limits not specified.	Same as 8270C plus 1,2-Dichlorobenzene-d_4 and Chlorophenol-d_4. Recovery limits listed in Table 7 of Exhibit D/SV (see page A-8). If 2 or more acid surrogates or 2 or more BN surrogates fail criteria or %R < 10% for any surrogate, reanalyze/re-extract. RRT must be ± 0.06 RRT units of the continuing calibration standard.

(8)   Minimum volume for analysis. Additional volumes are required for the generation of QC data.

Organic

Determination of Semivolatile Organic Compounds by GC/MS (Continued)

Method ⇒ Parameter ⇓	SW-846 Method 8270C	EPA 500 Series Method 525.2	EPA 600 Series Method 625	Std. Methods Method 6410	CLP SVOA (Organic SOW)
Internal Standards	1,4-Dichloroben-zene-d$_4$, Naphth-alene-d$_8$, Ace-naphthene-d$_{10}$, Phenan-threne-d$_{10}$, Chrysene-d$_{12}$, and Perylene-d$_{12}$. Recommended: RT must be with-in ±30 seconds from last calibra-tion; area must be -50 to +100%.	Acenaph-thene-d$_{10}$, Phenan-threne-d$_{10}$, and Chrysene-d$_{12}$. Sample IS re-sponse must be >70% of standard response. Terphenyl d$_{14}$ is added to each extract to monitor IS recovery from extraction.	Optional. If used, minimum of 3 from Table 8 (see page A-81).	Minimum of 3 [some recom-mended in Table 6410:IV (see page A-98)].	1,4-Dichloro-benzene-d$_4$, Naphthalene-d$_8$, Acenaphthene -d$_{10}$, Phenan-threne-d$_{10}$, Chrysene-d$_{12}$, and Pery-lene-d$_{12}$. IS RT must be within ±30 seconds from last calibration; area must be -50 to +100%.
Accuracy/ Precision	One MS/MSD and one LCS per 20 samples or each batch of samples, which-ever is more frequent. Com-pare results to laboratory estab-lished limits. See Table 6 for Multilaboratory performance data (see page A-144). (See also: **QC Check Standards/ Samples**).	Analyze MS repli-cates (Laboratory Fortified Sample Matrix) to deter-mine effect of matrix. Analyze one LFM per 20 samples. If a variety of matrices are encountered, LFM should be documented for all sample sources. (See also: **QC Check Standards/ Samples**).	One MS per 20 samples from each site or one MS per month, whichever is more frequent. Compare %R to Table 6 (see page A-79). (See also: **QC Check Standards/ Samples**).	One MS per 20 samples from each site or one per month, whichever is more frequent. Compare results to Table 6410:V (see page A-99). (See also: **QC Check Standards/ Samples**).	One MS/MSD per each case or 20 samples or each extraction batch or each 14 day period samples received from case, which-ever is more frequent. 11 compounds used for spiking. Compare results to Table 6 of Exhibit D/SV (see page A-8).
Blanks	One method blank per extraction batch (up to 20 samples) or when there is a change in reagents, whichever is more frequent.	One method blank with each batch of samples extracted as a group within a 2 hour work shift. Field Reagent Blank (FRB) is recommended with each sample set.	One method blank with each batch of samples extracted or when new reagents used.	One method blank with each batch of samples.	One method blank per each case, or 20 samples (in-cluding MS and reanalyses) or each extraction batch or grouped samples from case over 14 day period, whichever is more frequent. Concentration < CRQL for all compounds except phthalate esters < 5 times CRQL.

Determination of Semivolatile Organic Compounds by GC/MS (Continued)

Method ⇒ Parameter ⇓	SW-846 Method 8270C	EPA 500 Series Method 525.2	EPA 600 Series Method 625	Std. Methods Method 6410	CLP SVOA (Organic SOW)
Preservation/ Storage Conditions	Sodium thiosulfate if residual chlorine (aqueous). Store at 4 °C.	See footnote (9) for residual chlorine. Adjust pH < 2 with HCl for unchlorinated water. Store at 4 °C.	Sodium thiosulfate if residual chlorine. Store at 4 °C.	Sodium thiosulfate if residual chlorine. Store at 4 °C.	Protect from light. Store at 4 °C.
Holding Time (7)	Extraction: 7 days (aqueous). 14 days (solids). Analysis: 40 days after extraction.	Extraction: 14 days. Analysis: 30 days after extraction. (If analysis is for carboxin, diazinon, disulfoton, disulfoton sulfoxide, fenamiphos, or terbufos extract immediately).	Extraction: 7 days Analysis: 40 days after extraction.	Extraction: 7 days Analysis: 40 days after extraction.	Extraction: (Continuous liquid-liquid) started within 5 days of sample receipt (aqueous);10 days of sample receipt (solid). Analysis: 40 days after extraction.
Field Sample Amount (8)	1 liter (aqueous) 4 oz. (solid) glass container Teflon lined top	1 liter glass container Teflon lined top	1 liter glass container Teflon lined top	1 liter glass container Teflon lined top	1 liter (aqueous) 4 oz. (solid) glass container Teflon lined top
Amount for Extraction	1 liter (aqueous) 30 grams (low level solid) 2 grams (medium level solid)	1 liter	1 liter	1 liter	1 liter (aqueous) 30 grams (low level solid) 1 gram (medium level solid)

(7) Unless otherwise indicated, holding times are from the date of sample collection.

(8) Minimum volume for analysis. Additional volumes are required for the generation of QC data.

(9) If residual chlorine present, dechlorinate with sodium sulfate prior to adjusting pH to <2 with 6N HCL. Store at 4 °C in a dark place. If Cyanizine is to be determined, collect a separate sample. Do not dechlorinate or acidify until extraction. If atraton and/or prometon are to be determined, collect a separate sample, dechlorinate and do not acidify.

Organic

Determination of Semivolatile Organic Compounds by GC/MS (Continued)

Method ⇒ Parameter ⇓	SW-846 Method 8270C	EPA 500 Series Method 525.2	EPA 600 Series Method 625	Std. Methods Method 6410	CLP SVOA (Organic SOW)
Other Criteria (Method Specific)	Tuning: 50 ng decafluorotriphenylphosphine (DFTPP) initially and every 12 hours; acceptance criteria in Table 3 (see page A-143). Qualitative ID: All ions > 30% intensity must be ±30% of standard; ±0.06 RRT units of standard RRT. Library searches may be made for the purpose of tentative identification.	Tuning: 5 ng decafluorotriphenylphosphine (DFTPP), Breakdown check: endrin and 4,4′-DDT, both at the beginning of each 12 hour shift; acceptance criteria for DFTPP in Table 1 (see page A-43). Endrin and 4,4′-DDT breakdown ≤20%. Qualitative ID: All ions > 10% intensity must be ±20% of standard ion; RT ±10 seconds of standard RT. GC performance: Anthracene and phenanthrene should be separated by baseline. Benzo[a]anthracene and Chrysene should be separated by a valley 25% of the average peak height. Greater than 99% of the compounds in the calibration standards must be recognized by peak identification software. The abundance of m/z 67 at the RT of endrin aldehyde must be <10% of the abundance of m/z 67 produced by endrin. LSE cartridges must be checked for contamination.	Tuning: 50 ng decafluorotriphenylphosphine (DFTPP) at the beginning of each day; acceptance criteria in Table 9 (see page A-81). Qualitative ID: Characteristic ions for each analyte must maximize in the same scan or within 1 scan; relative peak heights of 3 characteristic ions must be ±20% of standard; RT (characteristic ions) within ±30 seconds RT of analyte. Check column performance daily with benzidine and pentachlorophenol.	Tuning: 50 ng decafluorotriphenylphosphine (DFTPP) at the beginning of each day; acceptance criteria in Table 6410:III (see page A-97). Qualitative ID: Characteristic masses [Table 6410:I (see page A-95) and Table 6410:II (see page A-97)] - each compound maximizes in same scan or within 1 scan; RT within ±30 seconds of RT of standard; relative peak heights of 3 characteristic masses in EICPs within ±20% of relative intensities of masses in reference spectrum. Check column performance daily with benzidine and pentachlorophenol.	Tuning: 50 ng decafluorotriphenylphosphine (DFTPP) initially and every 12 hours; acceptance criteria in Table 1 of Exhibit D (see page A-5). Up to 30 compounds with responses >10% of nearest internal standard are tentatively identified via a library search. Qualitative ID: All ions > 10% intensity must be present and ±20% of standard area; ±0.06 RRT units of standard RRT. GPC cleanup of soils mandatory. Eight compounds, 2,4-Dinitrophenol, 2,4,5-Trichlorophenol, 2-Nitroaniline, 3-Nitroaniline, 4-Nitroaniline, 4-Nitrophenol, 4,6-Dinitro-2-methylphenol and Pentachlorophenol require only a 4 point calibration.

GENIUM PUBLISHING CORPORATION

Organic

Determination of Semivolatile Organic Compounds by GC/MS

Chemical Name	CAS Number	SW-846 Method 8270C	EPA 500 Series Method 525.2	EPA 600 Series Method 625	Std. Methods Method 6410	CLP SVOA (Organic SOW)
Acenaphthene	83-32-9	•		•	•	•
Acenaphthylene	208-96-8	•	•	•	•	•
Acetophenone	98-86-2	•				
1-Acetyl-2-thiourea	591-08-2	•				
2-Acetylaminofluorene	53-96-3	•				
Alachlor	15972-60-8		•			
Aldrin	309-00-2	•	•	•	•	
Ametryn	834-12-8		•			
2-Aminoanthraquinone	117-79-3	•				
Aminoazobenzene	60-09-3	•				
4-Aminobiphenyl	92-67-1	•				
3-Amino-9-ethylcarbazole	132-32-1	•				
Anilazine	101-05-3	•				
Aniline	62-53-3	•				
o-Anisidine	90-04-0	•				
Anthracene	120-12-7	•	•	•	•	•
Aramite	140-57-8	•				
Aroclor-1016	12674-11-2	•	•	•	•	
Aroclor-1221	11104-28-2	•	•	•	•	
Aroclor-1232	11141-16-5	•	•	•	•	
Aroclor-1242	53469-21-9	•	•	•	•	
Aroclor-1248	12672-29-6	•	•	•	•	
Aroclor-1254	11097-69-1	•	•	•	•	
Aroclor-1260	11096-82-5	•	•	•	•	
Atraton	1610-17-9		•			
Atrazine	1912-24-9		•			
Azinphos methyl	86-50-0	•				
Barban	101-27-9	•				
Benzidine	92-87-5	•		•	•	
Benzoic acid	65-85-0	•				
p-Benzoquinone	106-51-4	•				
Benzo(a)anthracene	56-55-3	•	•	•	•	•
Benzo(a)pyrene	50-32-8	•	•	•	•	•
Benzo(b)fluoranthene	205-99-2	•	•	•	•	•
Benzo(g,h,i)perylene	191-24-2	•	•	•	•	•
Benzo(k)fluoranthene	207-08-9	•	•	•	•	•
Benzyl alcohol	100-51-6	•				
α-BHC	319-84-6	•		•	•	
β-BHC	319-85-7	•		•	•	
δ-BHC	319-86-8	•		•	•	

Determination of Semivolatile Organic Compounds by GC/MS (Continued)

Chemical Name	CAS Number	SW-846 Method 8270C	EPA 500 Series Method 525.2	EPA 600 Series Method 625	Std. Methods Method 6410	CLP SVOA (Organic SOW)
γ-BHC (Lindane)	58-89-9	•	•	•	•	
4-Bromophenyl-phenylether	101-55-3	•		•	•	•
Bromacil	314-40-9		•			
Bromoxynil	1689-84-5	•				
Butachlor	23184-66-9		•			
Butylate	2008-41-5		•			
Butylbenzylphthalate[1]	85-68-7	•	•	•	•	•
2-sec-Butyl-4,6-Dinitrophenol	2425-06-1	•				
Captafol	2425-06-1	•				
Captan	133-06-2	•				
Carbaryl	63-25-2	•				
Carbazole	86-74-8					•
Carbofuran	1563-66-2	•				
Carbophenothion	786-19-6	•				
α-Chlordane	5103-71-9		•			
Chlordane	57-74-9	•		•	•	
γ-Chlordane	5103-74-2		•			
trans-Nonachlor-Chlordane	39765-80-5		•			
Chlorfenvinphos	470-90-6	•				
Chlorneb	2675-77-6		•			
5-Chloro-2-methylaniline	95-79-4	•				
4-Chloro-3-methylphenol	59-50-7	•		•	•	•
4-Chloroaniline	106-47-8	•				•
Chlorobenzilate	510-15-6	•	•			
2-Chlorobiphenyl	2051-60-7		•			
bis(2-Chloroethoxy)methane	111-91-1	•		•	•	•
bis(2-Chloroethyl)ether	111-44-4	•		•	•	•
bis(2-Chloroisopropyl)ether	108-60-1	•		•	•	•
3-(Chloromethyl) pyridine hydrochloride	6959-48-4	•				
1-Chloronaphthalene	90-13-1	•				
2-Chloronaphthalene	91-58-7	•		•	•	•
2-Chlorophenol	95-57-8	•		•	•	•
4-Chloro-1,2-phenylenediamine	95-83-0	•				

(1) Some methods list as Benzyl butyl phthalate.

Organic

Determination of Semivolatile Organic Compounds by GC/MS (Continued)

Chemical Name	CAS Number	SW-846 Method 8270C	EPA 500 Series Method 525.2	EPA 600 Series Method 625	Std. Methods Method 6410	CLP SVOA (Organic SOW)
4-Chloro-1,3-phenylenediamine	5131-60-2	•				
4-Chlorophenyl-phenylether	7005-72-3	•		•	•	•
Chlorothalonil	1897-45-6		•			
Chlorpropham	101-21-3		•			
Chlorpyrifos	2921-88-2		•			
Chrysene	218-01-9	•	•	•	•	•
Coumaphos	56-72-4	•				
p-Cresidine	120-71-8	•				
Crotoxyphos	7700-17-6	•				
Cyanazine	21725-46-2		•			
Cycloate	1134-23-2		•			
2-Cyclohexyl-4,6-dinitrophenol	131-89-5	•				
Dacthal(DCPA)	1861-32-1		•			
4,4'-DDD	72-54-8	•	•	•	•	
4,4'-DDE	72-55-9	•	•	•	•	
4,4'-DDT	50-29-3	•	•	•	•	
Demeton-o	298-03-3	•				
Demeton-s	126-75-0	•				
Diallate (trans or cis)	2303-16-4	•				
2,4-Diaminotoluene	95-80-7	•				
Diazinon	333-41-5		•			
Dibenzofuran	132-64-9	•				•
Dibenzo(a,e)pyrene	192-65-4	•				
Dibenz(a,h)anthracene	53-70-3	•	•	•	•	•
Dibenz(a,j)acridine	224-42-0	•				
1,2-Dibromo-3-Chloropropane	96-12-8	•				
Dichlone	117-80-6	•				
1,2-Dichlorobenzene	95-50-1	•		•	•	•
1,3-Dichlorobenzene	541-73-1	•		•	•	•
1,4-Dichlorobenzene	106-46-7	•		•	•	•
3,3'-Dichlorobenzidine	91-94-1	•		•	•	•
2,3-Dichlorobiphenyl	16605-91-7		•			
2,4-Dichlorophenol	120-83-2	•		•	•	•
2,6-Dichlorophenol	87-65-0	•				
Dichlorovos	62-73-7	•	•			
Dicrotophos	141-66-2	•				
Dieldrin	60-57-1	•	•	•	•	
Diethyl sulfate	64-67-5	•				

Organic

Determination of Semivolatile Organic Compounds by GC/MS (Continued)

Chemical Name	CAS Number	SW-846 Method 8270C	EPA 500 Series Method 525.2	EPA 600 Series Method 625	Std. Methods Method 6410	CLP SVOA (Organic SOW)
Diethylphthalate	84-66-2	•	•	•	•	•
Diethylstilbesterol	56-53-1	•				
Dihydrosaffrole	56312-13-1	•				
Dimethoate	60-51-5	•				
3,3'-Dimethoxybenzidine	119-90-4	•				
p-Dimethylaminoazo-benzene	60-11-7	•				
3,3'-Dimethylbenzidine	119-93-7	•				
7,12-Dimethylbenz[A]-anthracene	57-97-6	•				
2,4-Dimethylphenol	105-67-9	•		•	•	•
Dimethylphthalate	131-11-3	•	•	•	•	•
α,α-Dimethylphenethyl-amine	122-09-8	•				
Di-n-butylphthalate	84-74-2	•	•	•	•	•
4,6-Dinitro-2-methylphenol	534-52-1	•				•
1,3-Dinitrobenzene	99-65-0	•				
1,4-Dinitrobenzene	100-25-4	•				
1,2-Dinitrobenzene	528-29-0	•				
2,4-Dinitrophenol	51-28-5	•		•	•	•
2,4-Dinitrotoluene	121-14-2	•	•	•	•	•
2,6-Dinitrotoluene	606-20-2	•	•	•	•	•
Dinocap	6119-92-2	•				
Di-n-octylphthalate	117-84-0	•		•	•	•
Dinoseb	88-85-7	•				
Dioxathion	78-34-2	•				
Diphenamid	957-51-7		•			
Diphenylamine	122-39-4	•				
5,5-Diphenylhydantoin	57-41-0	•				
1,2 Diphenylhydrazine	122-66-7	•				
Disulfoton	298-04-4	•	•			
Disulfoton sulfoxide	2497-07-6		•			
Disulfoton sulfone	2497-06-5		•			
Di(2-ethylhexyl)adipate	103-23-1		•			
Endosulfan I	959-98-8	•	•	•	•	
Endosulfan II	33213-65-9	•	•	•	•	
Endosulfan sulfate	1031-07-8	•	•	•	•	

Organic

Determination of Semivolatile Organic Compounds by GC/MS (Continued)

Chemical Name	CAS Number	SW-846 Method 8270C	EPA 500 Series Method 525.2	EPA 600 Series Method 625	Std. Methods Method 6410	CLP SVOA (Organic SOW)
Endrin	72-20-8	•	•	•	•	
Endrin aldehyde	7421-93-4	•	•	•	•	
Endrin ketone	53494-70-5	•				
EPN	2104-64-5	•				
EPTC	759-94-4		•			
Ethion	563-12-2	•				
Ethoprop	13194-48-4		•			
Ethyl carbamate	51-79-6	•				
Ethyl methanesulfonate	62-50-0	•				
Ethyl parathion	56-38-2	•				
bis(2-Ethylhexyl)phthalate	117-81-7	•	•	•	•	•
Etridiazole	2593-15-9		•			
Famphur	52-85-7	•				
Fenamiphos	22224-92-6		•			
Fenarimol	60168-88-9		•			
Fensulfothion	115-90-2	•				
Fenthion	55-38-9	•				
Fluchloralin	33245-39-5	•				
Fluoranthene	206-44-0	•		•	•	•
Fluorene	86-73-7	•	•	•	•	•
Fluridone	59756-60-4		•			
Heptachlor	76-44-8	•	•	•	•	
Heptachlor epoxide	1024-57-3	•	•	•	•	
2,2',3,3',4,4',6-Heptachlorobiphenyl	52663-71-5		•			
Hexachlorobenzene	118-74-1	•	•	•	•	•
2,2',4,4',5,6'-Hexachlorobiphenyl	60145-22-4		•			
Hexachlorobutadiene	87-68-3	•				•
Hexachlorocyclo-hexane, alpha	319-84-6		•			
Hexachlorocyclo-hexane, beta	319-85-7		•			
Hexachlorocyclo-hexane, delta	319-86-8		•			
Hexachlorocyclo-pentadiene	77-47-4	•	•	•	•	•
Hexachloroethane	67-72-1	•		•	•	•
Hexachlorophene	70-30-4	•				
Hexachloropropene	1888-71-7	•				

Determination of Semivolatile Organic Compounds by GC/MS (Continued)

Chemical Name	CAS Number	SW-846 Method 8270C	EPA 500 Series Method 525.2	EPA 600 Series Method 625	Std. Methods Method 6410	CLP SVOA (Organic SOW)
Hexamethyl phosphoramide	680-31-9	•				
Hexazinone	51235-04-2		•			
Hydroquinone	123-31-9	•				
Indeno(1,2,3-cd)pyrene	193-39-5	•	•	•	•	•
Isodrin	465-73-6	•				
Isophorone	78-59-1	•	•	•	•	•
Isosafrole	120-58-1	•				
Kepone	143-50-0	•				
Leptophos	21609-90-5	•				
Maleic anhydride	108-31-6	•				
Malathion	121-75-5	•				
Merphos	150-50-5		•			
Mevinphos	7786-34-7	•	•			
Mestranol	72-33-3	•				
Methapyrilene	91-80-5	•				
Methoxychlor	72-43-5	•	•			
Methyl paraoxon	950-35-6		•			
Methyl parathion	298-00-0	•				
2-Methyl-4,6-dinitrophenol	534-52-1	•		•	•	
3-Methylcholanthrene	56-49-5	•				
4,4'-Methylenebis(2-chloroaniline)	101-14-4	•				
4,4'-Methylenebis(N,N-dimethylaniline)	101-61-1	•				
Methylmethanesulfonate	66-27-3	•				
2-Methylnaphthalene	91-57-6	•				•
2-Methyl-5-Nitroaniline	99-55-8	•				
2-Methylphenol	95-48-7	•				•
3-Methylphenol	108-39-4	•				
4-Methylphenol	106-44-5	•				•
2-Methylpyridine	109-06-8	•				
Metolachlor	51218-45-2		•			
Metribuzin	21087-64-9		•			
Mexacarbate	315-18-4	•				
MGK 264	113-48-4		•			
Mirex	2385-85-5	•				
Molinate	2212-67-1		•			
Monocrotophos	6923-22-4	•				
Naled	300-76-5	•				
Naphthalene	91-20-3	•		•	•	•

GENIUM PUBLISHING CORPORATION

Determination of Semivolatile Organic Compounds by GC/MS (Continued)

Chemical Name	CAS Number	SW-846 Method 8270C	EPA 500 Series Method 525.2	EPA 600 Series Method 625	Std. Methods Method 6410	CLP SVOA (Organic SOW)
1,4-Naphthoquinone	130-15-4	•				
2-Naphthylamine	91-59-8	•				
1-Naphtylamine	134-32-7	•				
Napropamide	15299-99-7		•			
Nicotine	54-11-5	•				
5-Nitroacenaphthene	602-87-9	•				
2-Nitroaniline	88-74-4	•				•
3-Nitroaniline	99-09-2	•				•
4-Nitroaniline	100-01-6	•				•
Nitrobenzene	98-95-3	•		•	•	•
4-Nitrobiphenyl	92-93-3	•				
Nitrofen	1836-75-5	•				
5-Nitro-o-anisidine	99-59-2	•				
5-Nitro-o-toluidine	99-55-8	•				
2-Nitrophenol	88-75-5	•		•	•	•
4-Nitrophenol	100-02-7	•		•	•	•
Nitroquinoline-1-oxide	56-57-5	•				
N-Nitrosodibutylamine	924-16-3	•				
N-Nitrosodiethylamine	55-18-5	•				
N-Nitroso-di-n-propylamine	621-64-7	•		•	•	•
N-Nitrosodiphenylamine	86-30-6	•		•	•	•
N-Nitrosodimethylamine	62-75-9	•		•	•	
N-Nitrosomethylethyl-amine	10595-95-6	•				
N-Nitrosomorpholine	59-89-2	•				
N-Nitrosopiperidine	100-75-4	•				
N-Nitrosopyrrolidine	930-55-2	•				
Norflurazon	27314-13-2		•			
2,2',3,3',4,5',6,6'-Octachlorobiphenyl	40186-71-8		•			
Octamethyl pyrophosphoramide	152-16-9	•				
4,4'-Oxydianiline	101-80-4	•				
Parathion	56-38-2	•				
Pebulate	1114-71-2		•			
Pentachlorobenzene	608-93-5	•				
2,2',3',4,6-Pentachlorobiphenyl	60233-25-2		•			
Pentachloronitro-benzene (PCNB)	82-68-8	•				
Pentachlorophenol	87-86-5	•	•	•	•	•

Determination of Semivolatile Organic Compounds by GC/MS (Continued)

Chemical Name	CAS Number	SW-846 Method 8270C	EPA 500 Series Method 525.2	EPA 600 Series Method 625	Std. Methods Method 6410	CLP SVOA (Organic SOW)
Permethrin, *cis*	54774-45-7		•			
Permethrin, *trans*	51877-74-8		•			
Phenacetin	62-44-2	•				
Phenanthrene	85-01-8	•	•	•	•	•
Phenobarbital	50-06-6	•				
Phenol	108-95-2	•		•	•	•
1,4-Phenylenediamine	106-50-3	•				
Phorate	298-02-2	•				
Phosalone	2310-17-0	•				
Phosmet	732-11-6	•				
Phosphamidon	13171-21-6	•				
Phthalic anhydride	85-44-9	•				
2-Picoline	109-06-8	•				
Piperonyl sulfoxide	120-62-7	•				
Prometon	1610-18-0		•			
Prometryn	7287-19-6		•			
Pronamide	23950-58-5	•	•			
Propachlor	1918-16-7		•			
Propazine	139-40-2		•			
Propylthiouracil	51-52-5	•				
Pyrene	129-00-0	•	•	•	•	•
Pyridine	110-86-1	•				
Resorcinol	108-46-3	•				
Safrole	94-59-7	•				
Simazine	122-34-9		•			
Simetryn	1014-70-6		•			
Stirofos	22248-79-9		•			
Strychnine	57-24-9	•				
Sulfallate	95-06-7	•				
Tebuthiuron	34014-18-1		•			
Terbacil	5902-51-2		•			
Terbufos	13071-79-9	•	•			
Terbutryn	886-50-0		•			
1,2,4,5-Tetrachlorobenzene	95-94-3	•				
2,2',4,4'-Tetrachlorobiphenyl	2437-79-8		•			
2,3,4,6-Tetrachlorophenol	58-90-2	•				
Tetrachlorvinphos	961-11-5	•				
Tetraethyl dithiopyrophosphate	3689-24-5	•				

Determination of Semivolatile Organic Compounds by GC/MS (Continued)

Chemical Name	CAS Number	SW-846 Method 8270C	EPA 500 Series Method 525.2	EPA 600 Series Method 625	Std. Methods Method 6410	CLP SVOA (Organic SOW)
Tetraethyl pyrophosphate	107-49-3	•				
Thionazine	297-97-2	•				
Thiophenol (Benzenethiol)	108-98-5	•				
Toluene diisocyanate	584-84-9	•				
o-Toluidine	95-53-4	•				
Toxaphene	8001-35-2	•	•	•	•	
Triademefon	43121-43-3		•			
1,2,4-Trichlorobenzene	120-82-1	•		•	•	•
2,4,5-Trichlorobiphenyl	15862-07-4		•			
2,4,5-Trichlorophenol	95-95-4	•				•
2,4,6-Trichlorophenol	88-06-2	•		•	•	•
Tricyclazole	41814-78-2		•			
0,0,0-Triethylphosphor-othioate	126-68-1	•				
Trifluralin	1582-09-8	•	•			
Trimethyl phosphate	512-56-1	•				
2,4,5-Trimethylaniline	137-17-7	•				
1,3,5-Trinitrobenzene	99-35-4	•				
Tri-p-tolyl phosphate(h)	78-32-0	•				
Tris(2,3-dibromopropyl) phosphate	126-72-7	•				
Vernolate	1929-77-7		•			

Organic

Determination of Volatile Organic

Method ⇒ Parameter ⇓	SW-846 Method 8260B	EPA 500 Series Method 524.2
Applicability	Groundwaters. soils, sediments, sludges, non-water miscible wastes & others	Drinking water, raw source water
Number of Analytes (1)	121 total	84 total
Method Validation (2)	Analyze 4 replicates of QC check standard. Compare to average recovery (R), standard deviation of recovery (S) and relative standard deviation (RSD) values given in Tables 6 (see page A-128) and 7 (see page A-130). Results for RSD must not exceed 2.6 times the single laboratory RSD or 20%, whichever is greater, and the mean recovery must lie within the interval R ± 3 times S or R ± 30%, whichever is greater.	Analyze 5-7 replicates of QC check sample (Laboratory Fortified Blank) at a concentration range of 2-5 µg/L. %R = 80-120. %RSD < 20.
QC Check Standards/ Samples	The MS/MSD percent recovery (R_i) should lie within the QC acceptance criteria determined from the analysis of laboratory control samples during method validation. If the MS recovery is out of criteria, a QC check standard must be analyzed and fall within those ranges. Express the accuracy assessment as a percent recovery interval of ±2 standard deviations.	Analyze one QC check sample (Laboratory Fortified Blank) per 20 samples or each batch of samples processed together within a work shift, whichever is more frequent. %R = 70-130. %RSD = ± 30.
Method Detection Limit	MDLs listed in Tables 1 (see page A-119) and 2 (see page A-121). EQLs listed in Table 3 (see page A-123).	Analyze a minimum of 7 LFBs at a low concentration, MDLs listed in Tables 4 through 6 (see page A-37 through A-41) as a guide. Alternatively, use calibration data to estimate a concentration for each analyte that will yield a peak with a 3 to 5 signal to noise response.

Notes:

(1) Analyte list may vary among methods; a smaller list in one method is not necessarily a subset of a larger list in another method.

(2) Initial, one-time, demonstration of ability to generate acceptable accuracy and precision. Procedure may need to be repeated if changes in instrumentation or methodology occur.

Compounds by GC/MS

EPA 600 Series Method 624	Standard Methods Method 6210A	CLP VOA (Organic SOW)	⇐ Method ⇓ Parameter
Municipal and industrial discharge	Most environmental water samples: 6210B municipal and industrial wastewaters; 6210C and 6210D drinking water and raw source.	Water, soil, and sediment	**Applicability**
31 total	6210B: 34 total 6210C: 51 total 6210D: 58 total	33 total	**Number of Analytes (1)**
Analyze 4 replicates of QC check standard. Compare accuracy and precision results to Table 5 (see page A-75).	6210B: Analyze 4 replicates of QC check standard. Compare accuracy and precision results to Table 6210:IV (see page A-87). Determine MDL (6) 6210C and 6210D: Analyze 7 replicates of a QC check standard (recommended spiking level is 2 µg/L). Average recovery must be 90-110% of true value and standard deviation of recovery must be = 35% Determine MDL (6).	Not specified.	**Method Validation (2)**
Analyze one QC check standard each working day. (Frequency may be reduced if MS recoveries meet QC criteria). Compare %R to Table 5 (See page A-75). If MS results fall outside acceptance criteria, a QC check standard must be analyzed and fall within acceptance criteria.	Analyze a QC check standard with every 10 samples (a minimum of 2 per month for 6210C and 6210D, recommended spiking level 10 µg/L). (Frequency may be reduced if MS recoveries meet QC criteria). Compare %R to Table 6210:IV (see page A-87). For methods 6210C and 6210D acceptable recoveries are 60-140%. If MS results fall outside acceptance criteria, a QC check standard must be analyzed and fall within acceptance criteria. If QC check standard then fails, the analyst must identify and correct the problem, results for the compounds failing in the unspiked sample are suspect. 6210C and 6210D: analyze a low level QC check standard weekly.	Not specified.	**QC Check Standards**
MDLs listed in Table 1 (see page A-73).	MDLs listed in Table 6210:I (see page A-85).	CRQLs for TCL listed in Exhibit C (see page A-9).	**Method Detection Limit**

Organic

Organic

Determination of Volatile Organic

Method ⇒ Parameter ⇓	SW-846 Method 8260B	EPA 500 Series Method 524.2
Standard Solution Expiration (3)	Stock standards (except gases): 6 months. Store between -10 °C to -20 °C in Teflon sealed amber bottles with minimal headspace, protect from light. Stock gas standards: 2 months. Storage conditions same as stock standards. Secondary dilution standards: 1 week. Storage conditions same as stock standards. Calibration standards: Daily.	Stock and primary dilution standards (except gases): 4 weeks. Store at 4 °C in Teflon sealed bottles. For acrylonitrile, methyl iodide, and methyl acrylate, 1 week under these conditions. Stock gas standards: 1 week. Store at < 0 °C or 1 day at room temperature. Aqueous calibration standards: 1 hour, unless sealed in a sample bottle immediately.
Initial Calibration	Minimum of 5 levels, lowest near but above MDL. %RSD should be <15 for every compound, however for CCCs < 30. RF for SPCCs = 0.300 for chlorobenzene and 1,1,2,2-tetrachloroethane; = 0.100 for chloromethane and 1,1-dichloroethane, (> 0.100 for bromoform) Option 1: If CF or RF %RSD <15 or if mean %RSD of all analytes <15 then use averageCF or RF. Alternatively use a cal. curve (see option below). Option 2: If CF or RF %RSD >15 or if the mean %RSD of all analytes >15, then use a cal. curve: a) linear cal. using least squares regression; b) linear cal. using weighted least squares regression. Requirement: r ≥0.99, not forced through the origin, do not use origin as a cal. point. Option 3: Where instrument response is non-linear over wide working range or above procedures fail acceptance criteria, non-linear cal. may be employed. Requirement: COD ≥ 0.99, not forced through origin, do not use origin as a cal. point, other requirements exist pending curve fitting model chosen.	3, 4, or 5 levels (depending on calibration range), lowest 2-10 times MDL. If %RSD = 20, linearity assumed and average RF used. Alternatively, use a calibration curve.

Notes:

(3) Indicates maximum usage time. If comparisons of QC check standards indicate a problem, more frequent preparation may be necessary.

GENIUM PUBLISHING CORPORATION

Compounds by GC/MS (Continued)

EPA 600 Series Method 624	Standard Methods Method 6210A	CLP VOA (Organic SOW)	⇐ Method ⇓ Parameter
Stock standards (except gases): 1 month. Store between -10 °C and -20 °C in Teflon sealed bottles. Stock gas and 2-chloroethylvinyl ether standards: 1 week. Stored between -10 °C and -20 °C in Teflon sealed bottles away from light. Aqueous calibration standards: 1 hour; 24 hours if stored at 4 °C in Teflon sealed bottles with zero headspace.	Stock standards (except gases and 2-chloroethylvinyl ether): 1 month. Store between -10 °C and -20 °C in Teflon sealed bottles away from light. Stock gas and 2-chloroethylvinyl ether standards: 1 week. Store between -10 °C and -20 °C. Aqueous calibration standards: 1 hour; 24 hours if stored in sealed vials with zero headspace.	Stock standards (except gases and reactive compounds): 6 months. Store between -10 °C and -20 °C in Teflon sealed vials with minimum headspace. Stock gas and reactive compounds standards: 2 months; store between -10 °C and -20 °C in Teflon sealed vials with minimum headspace. Ampulated standard extracts store as per manufacture recommendations or 2 years from prep date (once opened use for no longer than one week). Aqueous calibration standards: 1 hour; 24 hours if stored at 4 °C in Teflon sealed vials with zero head space.	**Standard Solution Expiration (3)**
Minimum of 3 levels, lowest near but above MDL. If % RSD < 35, linearity assumed and average RF used. Alternatively, use a calibration curve.	Minimum of three levels for method 6210B, a minimum of 5 levels for 6210C and 6210D, one near but above MDL. If %RSD < 35, linearity assumed and average RF used. Alternatively, use a calibration curve.	Five levels (10,20,50,100 and 200). RRF and RSD criteria listed in Table 5 of Exhibit D/VOA (see page A-13). Primary characteristic ions used for quantitation are listed in Tables 2 (see page A-11) and 4 (see page A-12) of Exhibit D/VOA. Up to 2 compounds in table 5 of exhibit D/VOA are permitted to fail RRF and RSD criteria provided RRF ≥0.010 and % RSD ≥40.0%.	**Initial Calibration**

Determination of Volatile Organic

Organic

Method ⇒ Parameter ⇓	SW-846 Method 8260B	EPA 500 Series Method 524.2
Continuing Calibration/ Calibration Verification	Mid-level calibration standard run every 12 hours. RF for SPCCs same as for initial calibration. RF of CCCs must be < 20% difference from initial calibration. Internal Standards: RT must be within ±30 seconds from last mid-point standard level of the initial calibration; area must be -50 to +100%.	Mid-level calibration standard analyzed at the beginning of each 8 hour shift. RF must be within ± 30% of initial calibration. Alternatively, if a linear or second order regression is used, the concentration measured must be within 30% of the true value. Absolute areas of quant. ions for IS and surrogates must not decrease more than 30% from previous check standard or by more than 50% from initial calibration.
Surrogate Standards	4-Bromofluorobenzene, Dibromofluoromethane, Toluene-d_8, and 1,2-Dichloroethane-d_4 are recommended. Compare %R to laboratory established limits or to Table 8 (see page A-131) if laboratory limits not available. Laboratory limits must fall within limits specified in Table 8. (see page A-131)	4-Bromofluorobenzene and 1,2-Di-chlorobenzene-d_4. Additional surrogates optional. Recovery limits not specified. Absolute area of quant. ions should not vary more than 50%.
Internal Standards	Fluorobenzene, Chlorobenzene-d_5, 1,4-Dichlorobenzene-d_4 are recommended. RT must be ±30 seconds from last calibration check (12 hours); area must be -50 to +100%.	Fluorobenzene. Additional internal standards optional. Absolute area of quant. ions should not vary more than 50%.
Accuracy/ Precision	One MS/MSD per 20 samples or each batch of samples, whichever is more frequent. The MS/MSD percent recovery (R_i) should lie within the QC acceptance criteria determined from the analysis of laboratory control samples during method validation. (See also: **QC Check Standards/Samples.**)	No MS required, unless matrix effects are suspected. Sample results should be flagged and MS results reported with sample. Replicate LFB check standards and independent quality control sample (QCS) analyzed at least quarterly. (See also: **QC Check Standards/Samples.**)
Method Blanks	One method blank per extraction batch (up to 20 samples) or when there is a change in reagents, whichever is more frequent.	One method blank with each batch of samples processed together, within a work shift. One field reagent blank (FRB) per batch of samples.

GENIUM PUBLISHING CORPORATION

Compounds by GC/MS (Continued)

EPA 600 Series Method 624	Standard Methods Method 6210A	CLP VOA (Organic SOW)	⟸ Method ⇓ Parameter
QC check standard analyzed each working day. Compare results to Table 5 (see page A-75).	QC check sample analyzed each working day and at a minimum frequency of 10% of all samples. For method 6210B compare results to Table 6210:IV (see page A-87). For 6210C and 6210D recoveries must be 60-140% of expected value.	A mid-level calibration standard is analyzed every 12 hours. The % difference criteria is listed in Table 5 of Exhibit D/VOA. Up to 2 compounds in Table 5 of Exhibit D/VOA (see page A-13) permitted to fail criteria if RRF > 0.010 and %D = 40%.	**Continuing Calibration/ Calibration Verification**
Minimum of 3 from Table 3 (see page A-74). % Recovery limits not specified.	Minimum of 3 from Table 6210:III (see page A-86). % Recovery limits not specified. For methods 6210C and 6210D recoveries should be 60-140%.	(System Monitoring Compounds) 4-Bromofluorobenzene, 1,2-Dichloroethane-d_4, and Toluene-d_8. Recovery limits in Table 7 of Exhibit D/VOA (see page A-14). RRT must be within 0.06 RRT units of continuing calibration standard.	**Surrogate Standards**
Optional. If used, minimum of 3 from Table 3 (see page A-74).	Minimum of 3 from Table 6210:III (see page A-86). Internal standards must not be affected by method or matrix.	Bromochloromethane, 1,4-Di-fluorobenzene, and Chlorobenzene-d_5. RT must be ±30 seconds from last calibration; area must be 50 to 100%. See Table 3 of Exhibit D/VOA (see page A-12) for target compound assignments.	**Internal Standards**
One MS per 20 samples from each site or 1 per month, whichever is more frequent. Compare %R to Table 5 (see page A-75). (See also: **QC Check Standards/Samples.**)	One MS per 10 samples analyzed and one MS per 20 samples from each site or one per month, whichever is more frequent. Spike samples at 20 µg/L or 1:5 times the background concentration. Compare results to Table 6210:IV (see page A-87). MS not required by 6210C and 6210D (See also: **QC Check Standards/Samples.**)	One MS/MSD per each case or 20 samples or each extraction batch or each 14 day period samples received from case, whichever is more frequent (only 5 compounds spiked). Compare results to Table 8 of Exhibit D/VOA (see page A-14).	**Accuracy/ Precision**
One method blank per day.	One method blank per day. For 6210C and 6210D blank should be <0.1 µg/L.	One method blank every 12 hours. Concentration = CRQL for all but methylene chloride = 2.5 times CRQL; acetone; 2-butanone = 5 times CRQL.	**Method Blanks**

Organic

Determination of Volatile Organic

Method ⇒ Parameter ⇓	SW-846 Method 8260B	EPA 500 Series Method 524.2
Preservation/ Storage Conditions	pH <2 with HCl or H_2SO_4 (aqueous). Sodium thiosulfate if residual chlorine (aqueous). Store at 4 °C.	pH < 2 with HCl. Ascorbic acid if residual chlorine (use sodium thiosulfate for dechlorination if gases are not to be determined). Acidification may be omitted if analysis is for Trihalomethanes (THMs) only sodium thiosulfate was used to dechlorinate, or if sample foams upon addition of HCl. Store at 4 °C. **Note:** Dechlorination must be performed prior to acidification. Do not mix dechlorination agents with HCl prior to sampling.
Holding Time (4)	14 days	14 days, 24 hours if unpreserved (except for THMs)
Field Sample Amount Required (5)	40 mL VOA vial in duplicate without headspace (air bubbles) (aqueous) 4 oz. (solid) glass container Teflon lined septa	40 mL VOA vial in duplicate without headspace (air bubbles) glass container Teflon lined septa
Amount for Extraction	5 mL (aqueous) or 25 mL (aqueous) 5 grams (solid)	25 mL (5 mL may be used if MDLs can be achieved)
Other Criteria (Method Specific)	Tuning: 50 ng bromofluorobenzene (BFB) initially and every 12 hours; acceptance criteria in Table 4 (see page A-123). Qualitative ID: The intensities of the characteristic ions of a compound maximize in the same scan or within one scan of each other. RRT of the sample component is within 0.06 RRT units of the RRT of the standard component. The relative intensities of the characteristic ions are within 30% of those ions in the reference spectrum. Tentative ID: All ions >10% intensity must be ±20% of standard; ±0.06 RRT units of standard RRT. Molecular ion must be present. Library searches may be made for the purpose of tentative identification.	Tuning: 25 ng bromofluorobenzene (BFB) at the beginning of each 12 hour shift; acceptance criteria in Table 3 (see page A-36). Qualitative ID: All ions >10% intensity must be ±20% of standard ion; sample RT within 3 times SD of RT in calibration.

Notes:
(4) Unless otherwise indicated, holding times are from the date of sample collection.
(5) Approximate volumes to be gathered for analysis. Additional volumes are required for the generation of QC data.

GENIUM PUBLISHING CORPORATION

Compounds by GC/MS (Continued)

EPA 600 Series Method 624	Standard Methods Method 6210A	CLP VOA (Organic SOW)	⇐ Method ⇓ Parameter
pH <2 with HCl. Sodium thiosulfate if residual chlorine. Store at 4 °C.	Store at 4 °C. 4 drops 6N Hcl/40ml If residual chlorine present: ascorbic acid (25 mg/40ml), or sodium thiosulfate (3 mg/40ml), or sodium sulfite (3 mg/40ml).	Protect from light. Store at 4 °C. Aqueous: pH of 2 All samples in group should be stored together. Storage blanks required. Samples should be stored for 60 days after the data package has reached the EPA.	**Preservation/ Storage Conditions**
14 days	14 days	10 days from sample receipt	**Holding Time (4)**
40 mL VOA vial in duplicate without headspace (air bubbles) glass container Teflon lined septa	40 mL VOA vial in duplicate without headspace (air bubbles) glass container Teflon lined septa	40 mL VOA vial in duplicate without headspace (air bubbles) (aqueous) 4 oz. (solid) glass container Teflon lined septa	**Field Sample Amount Required (5)**
5 mL	5 mL for 6210B 25 mL for 6210C and 6210D	5 mL (aqueous) 5 grams (solid) 4 grams (medium-level solid)	**Amount for Extraction**
Tuning: 50 ng bromo-fluorobenzene (BFB) initially and every 24 hours; acceptance criteria in Table 2 (see page A-74). Qualitative ID: Three characteristic ions must be ± 20% of standard and must maximize within 1 scan.	Tuning: 50 ng bromofluorobenzene (BFB) at the beginning of each day; acceptance criteria in Table 6210:II (see page A-86). Qualitative ID: Characteristic masses of each compound must reach the maximization point in same scan or within 1 scan; RT within 30 seconds of RT of authentic compound; relative peak height of 3 characteristic masses in EICP within 20% of relative intensities of masses in reference spectrum. Structural isomers that meet qualitative identification criteria must also be acceptably resolved (baseline to valley height between isomers <25% of the sum of the two peak heights), otherwise report as isomeric pairs.	Tuning: 50 ng bromofluorobenzene (BFB) initially and every 12 hours; acceptance criteria in Table 1 of Exhibit D/VOA (see page A-10). Qualitative ID: All ions >10% intensity must be present and ± 20% of standard; ± 0.06 RRT units of standard RRT. Up to 30 compounds not listed in Exhibit C for the volatile or semi-volatile organic fraction, with responses >10% of nearest IS are tentatively identified via a library search. Relative intensities of reference spectrum must be present in sample spectrum and within 20%. Molecular ion must be present.	**Other Criteria (Method Specific)**

(6) MDL determination for Standard Methods 6000 Series: Analyze a minimum of seven check samples (concentration = 0.2 times MCL or 10 times estimated MDL.). Average percent recovery should be 80% to 120% of true value with %RSD ≤35%. Use results to determine MDLs. Broader acceptance ranges exist for some compounds with lower extraction efficiency and are indicated in the specific method.

Determination of Volatile Organic Compounds by GC/MS

Chemical Name	CAS Number	SW-846 Method 8260B	EPA 500 Series Method 524.2	EPA 600 Series Method 624	Standard Methods Method 6210A			CLP VOA (Organic SOW)
					6210 B	6210 C	6210 D	
Acetone	67-64-1	•	•					•
Acetonitrile	75-05-8	•						
Acrolein	107-02-8	•						
Acrylonitrile	107-13-1	•	•					
Allyl alcohol	107-18-6	•						
Allyl chloride	107-05-1	•	•					
Benzene	71-43-2	•	•	•	•	•	•	•
Benzyl chloride (α-chlorotoluene)	100-44-7	•						
Bromoacetone	598-31-2	•						
Bromobenzene	108-86-1	•	•			•	•	
Bromochloromethane	74-97-5	•	•			•	•	
Bromodichloromethane	75-27-4	•	•	•	•	•	•	•
Bromoform	75-25-2	•	•	•	•	•	•	•
Bromomethane	74-83-9	•	•	•	•	•	•	•
n-Butanol	71-36-3	•						
2-Butanone	78-93-3	•	•					•
t-Butyl alcohol	75-65-0	•						
n-Butylbenzene	104-51-8	•	•				•	
sec-Butylbenzene	135-98-8	•	•			•	•	
tert-Butylbenzene	98-06-6	•	•			•	•	
Carbon disulfide	75-15-0	•	•					•
Carbon tetrachloride	56-23-5	•	•	•	•	•	•	•
Chloroacetonitrile	107-14-2	•	•					
Chloral hydrate	302-17-0	•						
Chlorobenzene	108-90-7	•	•	•	•	•	•	•
1-Chlorobutane	109-69-3	•	•					
2-Chloro-1,3-butadiene	126-99-8	•						
Chloroethane	75-00-3	•	•	•	•	•	•	•
2-Chloroethanol	107-07-3	•						
bis-(2-Chloroethyl) sulfide	505-60-2	•						
2-Chloroethylvinyl ether	110-75-8	•		•	•			
Chloroform	67-66-3	•	•	•	•	•	•	•
Chloromethane	74-87-3	•	•	•	•	•	•	•
Chloroprene	126-99-8	•						
3-Chloropropene	107-05-1	•						
3-Chloropropionitrile	542-76-7	•						
1-Chlorohexane	544-10-5	•						
2-Chlorotoluene	95-49-8	•	•			•	•	

GENIUM PUBLISHING CORPORATION

Determination of Volatile Organic Compounds by GC/MS (Continued)

Chemical Name	CAS Number	SW-846 Method 8260B	EPA 500 Series Method 524.2	EPA 600 Series Method 624	Standard Methods Method 6210A			CLP VOA (Organic SOW)
					6210 B	6210 C	6210 D	
4-Chlorotoluene	106-43-4	•	•			•	•	
Crotonaldehyde	4170-30-3	•						
1,2-Dibromo-3-chloropropane	96-12-8	•	•			•	•	
Dibromochloromethane	124-48-1	•	•	•	•	•	•	•
1,2-Dibromoethane	106-93-4	•	•			•	•	
Dibromomethane	74-95-3	•	•			•	•	
trans-1,4-Dichloro-2-butene	110-57-6	•	•					
cis-1,4-Dichloro-2-butene	1476-11-5	•						
1,4-Dichloro-2-butene	764-41-0							
1,2-Dichlorobenzene	95-50-1	•	•	•	•	•	•	
1,3-Dichlorobenzene	541-73-1	•	•	•	•	•	•	
1,4-Dichlorobenzene	106-46-7	•	•	•	•	•	•	
Dichlorodifluoromethane	75-71-8	•	•					
1,1-Dichloroethane	75-34-3	•	•	•	•	•	•	•
1,2-Dichloroethane	107-06-2	•	•	•	•	•	•	•
1,1-Dichloroethene	75-35-4	•	•	•	•	•	•	•
cis-1,2-Dichloroethene	156-59-4	•	•			•	•	
trans-1,2-Dichloroethene	156-60-5	•	•		•	•	•	
1,2-Dichloroethene (total)	540-59-0							•
1,2-Dichloropropane	78-87-5	•	•	•	•	•	•	•
1,3-Dichloropropane	142-28-9	•	•			•	•	
2,2-Dichloropropane	590-20-7	•	•			•	•	
1,3 Dichloro-2-propanol	96-23-1	•						
1,1-Dichloropropanone	513-88-2		•					
1,1-Dichloropropene	563-58-6	•	•			•	•	
cis-1,3-Dichloropropene	10061-01-5	•	•	•	•			•
trans-1,3-Dichloropropene	10061-02-6	•	•	•	•			•
1,2,3,4 Diepoxybutane	298-18-0	•						
Diethyl ether	60-29-7	•	•					
1,4 Dioxane	123-91-1	•						
Epichlorohydrin	106-89-8	•						
Ethanol	64-17-5	•						
Ethyl acetate	141-78-6	•						
Ethyl benzene	100-41-4	•	•	•	•	•	•	•
Ethyl methacrylate	97-63-2	•	•					
Ethylene oxide	75-21-8	•						
Hexachlorobutadiene	87-68-3	•	•			•	•	

Organic

Determination of Volatile Organic Compounds by GC/MS (Continued)

Chemical Name	CAS Number	SW-846 Method 8260B	EPA 500 Series Method 524.2	EPA 600 Series Method 624	Standard Methods Method 6210A			CLP VOA (Organic SOW)
					6210 B	6210 C	6210 D	
Hexachloroethane	67-72-1	•	•					
2-Hexanone	591-78-6	•	•					•
2-Hydroxypropionitrile	78-97-7	•						
Iodomethane	74-88-4	•						
Isobutyl alcohol	78-83-1	•						
Isopropylbenzene	98-82-8	•	•			•	•	
4-Isopropyl toluene	99-87-6	•	•				•	
Malononitrile	109-77-3	•						
Methacrylonitrile	126-98-7	•	•					
Methanol	67-56-1	•						
Methylacrylate	96-33-3	•	•					
Methyl iodide	74-88-4		•					
Methyl methacrylate	80-62-6	•	•					
4-Methyl-2-pentanone	108-10-1	•	•					•
Methyl-t-butyl ether	1634-04-4	•	•					
Methylene chloride	75-09-2	•	•	•	•	•	•	•
Naphthalene	91-20-3	•	•				•	
Nitrobenzene	98-95-3	•						
2-Nitropropane	79-46-9	•	•					
N-Nitroso-di-n-butylamine	924-16-3	•						
Paraldehyde	123-63-7	•						
Pentachloroethane	76-01-7	•	•					
2-Pentanone	107-87-9	•						
2-Picoline	109-06-8	•						
1-Propanol	71-23-8	•						
2-Propanol	67-63-0	•						
Propargyl alcohol	107-19-7	•						
6-Propiolactone	57-57-8	•						
Propionitrile	107-12-0	•	•					
n-Propylamine	107-10-8	•						
n-Propylbenzene	103-65-1	•	•			•	•	
Pyridine	110-86-1	•						
Styrene	100-42-5	•	•			•	•	•
1,1,1,2-Tetrachloroethane	630-20-6	•	•			•	•	
1,1,2,2-Tetrachloroethane	79-34-5	•	•	•	•	•	•	•
Tetrachloroethene	127-18-4	•	•	•	•	•	•	•

GENIUM PUBLISHING CORPORATION

Organic

Determination of Volatile Organic Compounds by GC/MS (Continued)

Chemical Name	CAS Number	SW-846 Method 8260B	EPA 500 Series Method 524.2	EPA 600 Series Method 624	Standard Methods Method 6210A			CLP VOA (Organic SOW)
					6210 B	6210 C	6210 D	
Tetrahydrofuran	109-99-9		•					
Toluene	108-88-3	•	•	•	•	•	•	•
1,2,3-Trichlorobenzene	87-61-6	•	•				•	
1,2,4-Trichlorobenzene	120-82-1	•	•				•	
1,1,1-Trichloroethane	71-55-6	•	•	•	•	•	•	•
1,1,2-Trichloroethane	79-00-5	•	•	•	•	•	•	•
Trichloroethene	79-01-6	•	•	•	•	•	•	•
Trichlorofluoromethane	75-69-4	•	•	•	•	•	•	
1,2,3-Trichloropropane	96-18-4	•	•			•	•	
1,2,4-Trimethylbenzene	95-63-6	•	•				•	
1,3,5-Trimethylbenzene	108-67-8	•	•				•	
Vinyl acetate	108-05-4	•						
Vinyl chloride	75-01-4	•	•	•	•	•	•	•
m-Xylene	108-38-3	•	•		•	•	•	
o-Xylene	95-47-6	•	•		•	•	•	
p-Xylene	106-42-3	•	•		•	•	•	
Xylene (total)	1330-20-7							•

Organic

Determination of Volatile

Method ⇒ Parameter ⇓	EPA 600 Series Method 602	Standard Methods Method 6220A	SW-846 Method 8021B
Applicability	Municipal and industrial discharge water	6220B: municipal and industrial wastewaters 6220C: raw source and drinking waters	Groundwater, soils, sludge, water miscible liquid waste and non-water miscible waste & others.
Number of Analytes (1)	7 total	6220B: 7 total (analyte list may be extended) 6220C: 28 total (3)	70 total
Method Validation (2)	Analyze 4 replicates of QC check standard spiked at 20 µg/L. Compare accuracy and precision results to Table 2 (see page A-69).	6220B: Analyze 4 replicates of QC check standard. Compare accuracy and precision results to Table 6220:II (see page A-91). 6220C: Analyze 7 replicate QC check standards spiked at 0.4 µg/L. Average recoveries must be 90-110%, standard deviation must be ≤ 35%. Determine MDL (4)	Analyze 4 replicates of QC check standard (parameters of interest at 20 µg/L). Compare accuracy and precision results to Table 2 (see page A-112).
QC Check Standards/ Samples	If MS results fall outside ranges designated in Table 2 (see page A-69), a QC check standard must be analyzed and fall within those ranges. Analyze one QC check standard with every 10 samples or one per month whichever is more frequent. Compare results to Table 2 (see page A-69)	6220B: If MS results fall outside ranges designated in Table 6220:II (see page A-91), a QC check standard must be analyzed and fall within those ranges. If QC check standard fails then data is suspect. Analyze one QC check sample per 10 samples analyzed or one per month. 6220C: Analyze one QC check sample per 10 samples analyzed or two per month. Analyze one low level QC check standard weekly.	If MS results fall outside the ranges designated, a QC check standard must be analyzed and fall within those ranges.
Method Detection Limit	MDLs listed in Table 1 (see page A-68).	MDLs listed in Table 6220:I (see page A-91).	MDLs listed in Table 1 (see page A-110). EQLs listed in Table 3 (see page A-114).

Notes:

(1) Analyte lists may vary among methods; a smaller list in one method is not necessarily a subset of a larger list in another method.

(2) Initial, one-time demonstration of ability to generate acceptable accuracy and precision. Procedure may need to be repeated if changes in instrumentation or methodology occur.

(3) Method 6220C is not applicable to the analysis of styrene in chlorinated water.

(4) MDL determination for Standard Methods 6000 Series: Analyze a minimum of seven check samples (concentration = 0.2 times MCL or 10 times estimated MDL.). Average percent recovery should be 80% to 120% of true value with %RSD < 35%. Use results to determine MDLs. Broader acceptance ranges exist for some compounds with lower extraction efficiency and are indicated in the specific method.

Organic

Organic Compounds By GC

EPA 500 Series Method 502.2	EPA 600 Series Method 601	Standard Methods Method 6230A	⇐ Method Parameter ⇓
Drinking water, raw source water	Municipal and industrial discharge water	6230B: Municipal and industrial wastewater. 6230C and 6230D: Raw source and drinking waters.	**Applicability**
60 total	29 total	6230B: 29 total 6230C: 38 total 6230D: 58 total	**Number of Analytes (1)**
Analyze 4-7 replicates of QC check sample with a concentration range of 0.1-5 µg/L. (Laboratory Fortified Blank) %R = 80-120. %RSD < 20.	Analyze 4 replicates of QC check standard spiked at 20 µg/L. Compare accuracy and precision results to Table 2 (see page A-69).	6230B: Analyze 4 replicates of QC check standard. Compare accuracy and precision results to Table 6230:II (see page A-93) or optionally Table 6230:III (see page A-94). 6230C and 6230D: Analyze 7 replicate QC check standards spiked at 0.4 µg/L. Average recovery should be 90-110%, standard deviation ≤ 35%. Determine MDL (6)	**Method Validation (2)**
Analyze one QC check sample (Laboratory Fortified Blank) per batch of samples processed together within a 12-hour work shift. %R = 80-120 %RSD < 20 Analyze replicate QC check samples and a QC sample from an external source at least quarterly.	If MS results fall outside the ranges designated in Table 2 (see page A-69), a QC check standard must be analyzed and fall within those ranges. Analyze one QC check standard for every 10 samples analyzed or one per month if only 1 to 10 samples a month. Compare results with Table 2 (see page A-69)	6230B: If MS results fall outside the ranges designated in Table 6230:II (see page A-93), a QC check standard must be analyzed and fall within those ranges. If the QC check fails, the data from the unspiked sample is suspect. 6230C and 6230D: No MS required. Analyze one QC check standard per 10 samples analyzed with a minimum of two per month. Analyze one low level QC check standard weekly. Acceptable recoveries are 60-140%.	**QC Check Standards/ Samples**
MDLs listed in Table 2 (see page A-25) and Table 4 (see page A-27).	MDLs listed in Table 1 (see page A-68).	6230B: MDLs listed in Table 6230:I (see page A-92)	**Method Detection Limit**

Organic

Determination of Volatile

Method ⇒ Parameter ⇓	EPA 600 Series Method 602	Standard Methods Method 6220A	SW-846 Method 8021B
Standard Solution Expiration (5)	Stock standards: 1 month. Store at 4 °C in Teflon sealed vials away from light. Aqueous calibration standards: 1 day.	Stock standards: 1 month. Store between -10 °C and -20 °C in Teflon sealed bottles away from light. Aqueous calibration standards: 1 hour; 24 hours if stored in sealed vials with zero headspace.	Stock standards (except gases and reactive compounds): 6 months. Store between -10 ˚C to -20 ˚C with minimum headspace in Teflon sealed bottles away from light. Stock gas and reactive compound standards: 1 week or manufacturer's expiration date. Same conditions as stock standards. Aqueous calibration standards: 1 hour; 24 hours if stored with zero headspace.
Initial Calibration	Minimum of 3 levels, lowest near but above MDL. If %RSD < 10, assume linearity and average RF used. Alternatively use a calibration curve.	6220B: Minimum of 3 levels. If %RSD <35 linearity assumed and average RF used. 6220C: Minimum of 5 levels. If %RSD < 10, linearity assumed and average RF used. Alternatively use a calibration curve or single point calibration. Single point calibration standard must produce response within ±20% of unknowns.	Minimum of 5 levels, lowest near but above MDL. **Option 1**: If CF or RF %RSD <20 or if mean %RSD of all analytes < 20 use avg. CF or RF. Alternatively use a cal. curve (see options below). **Option 2**: If CF or RF %RSD >20 or if mean %RSD of all analytes > 20, then use cal. curve: a linear cal. using least squares regression, **Requirements:** r ≥ 0.99, not forced through origin; do not use origin as cal. point. **Option 3**: Where instrument response is non-linear over wide working range or above procedures fail acceptance criteria, non-linear cal. may be employed. **Requirements:** COD ≥ 0.99, not forced through origin; do not use origin as cal. point., other requirements exist pending curve fitting model chosen.

Notes:

(5) Indicates maximum usage time. If comparisons to QC check standards indicate a problem, more frequent preparation may be necessary.

Organic

Organic Compounds By GC (Continued)

EPA 500 Series Method 502.2	EPA 600 Series Method 601	Standard Methods Method 6230A	⇐ Method Parameter ⇓
Stock and primary dilution standards (except gases): 4 weeks if stored at 4 °C in PTFE sealed bottles. Stock and primary dilution gases: 1 week. Store at < 0 °C; 1 day at room temperature. Calibration standards: 1 hour unless stored at 4 °C with zero headspace.	Stock standards, (except gases and reactive compounds, i.e.: 2-chloroethyl- vinyl ether): 1 month. Store between -10 °C to -20 °C with minimum head-space in Teflon sealed bottles away from light. Stock gas and reactive compound standards: 1 week. Same conditions as stock standards. Aqueous calibration standards: 1 hour; 24 hours if stored with zero headspace.	Stock standards, (except gases and 2-chloroethyl-vinyl ether): 1 month. Store between -10 °C and -20 °C in Teflon sealed bottles away from light. Stock gas and 2-chloroethylvinyl ether standards: 1 week. Store between -10 °C and -20 °C away from light. Aqueous calibration standards: 1 hour; 24 hours if stored in sealed vials with zero headspace.	**Standard Solution Expiration (5)**
3, 4, or 5 levels (depending on calibration range), lowest 2-10 times MDL. If %RSD < 10, linearity assumed and average RF used. Alternatively use a calibration curve.	Minimum of 3 levels, lowest near but above MDL. If %RSD < 10, linearity assumed and average RF used. Alternatively use a calibration curve.	Minimum of 3 levels for 6230B, minimum of 5 levels for 6230C and 6230D, lowest near but above MDL. If %RSD < 10, linearity assumed and average RF used. Alternatively use a calibration curve.	**Initial Calibration**

Organic

Organic

Method ⇒ Parameter ⇓	EPA 600 Series Method 602	Standard Methods Method 6220A	SW-846 Method 8021B
Continuing Calibration/ Calibration Verification	QC check standard analyzed each working day. Compare results to Table 2 (see page A-69).	6220B: Analyze one or more calibration standards daily. Compare results for each compound to Table 6220:II (see page A-91). Recalibrate for any failed compounds. 6220C: Analyze one or more calibration standards daily. If not within ±20% of predicted response, recalibrate.	Mid-level calibration standard run at the beginning and end of each analysis sequence (12 hours) or more frequently if warranted. Average response must not exceed ± 15% of mean response of initial calibration. Notify data user if response for individual analyte exceeds ± 15%. Samples which contain target compounds above the reportable limit must be bracketed by acceptable standards. Standard RT must fall within 12 hour shift RT window or window must be re-established or instrument maintenance performed.
Surrogate Standards	Add surrogates to encompass range of temperature program. (α,α,α-Trifluorotoluene is recommended). % Recovery limits not specified	α,α,α-Trifluorotoluene % Recovery limits not specified.	At least two surrogate compounds. Recommended: Bromochlorobenzene, and 1,4-Dichlorobutane. Results must fall within lab established control limits.
Internal Standards	Optional. If used, α,α,α-Trifluorotoluene is recommended.	Optional. If used, α,α,α-Trifluorotoluene is recommended.	Optional. If used, Fluorobenzene and 2-Bromo-1-chloropropane. Area must be ± 3 SD of calibration standards.
Accuracy/ Precision	One MS per 10 samples from each site or 1 per month, whichever is more frequent. Compare %R to Table 2 (see page A-69). (See also: **QC Check Standard/ Samples.**)	6220B: One MS per 10 samples analyzed from each site or one per month, whichever is more frequent. Compare %R to Table 6220:II (see page A-91). (See also: **QC Check Standard/ Samples.**)	One MS/MSD per 20 samples or each batch of samples, whichever is more frequent. Compare results to Table 2 (see page A-112). (See also: **QC Check Standards/ Samples.**)
Blanks	One method blank per day.	One method blank per day.	One method blank per extraction batch (up to 20 samples) or when there is a change in reagents, whichever is more frequent.

Organic Compounds By GC (Continued)

EPA 500 Series Method 502.2	EPA 600 Series Method 601	Standard Methods Method 6230A	⇐ Method Parameter ⇓
Analyze one or more calibration standards once every 12 hours. If not within ±20% of predicted response, repeat with a fresh cal. standard. If still outside limits, re-calibrate.	QC check standard analyzed each working day. Compare results to Table 2 (see page A-69).	QC check standard analyzed each working day. For 6230B compare results to Table 6230:II (see page A-93). For 6230C and 6230D ± 20%.	**Continuing Calibration/ Calibration Verification**
%R=80-120.	Bromochloromethane, 2-Bromo-1-chloropropane, and 1,4-Dichlorobutane	Bromochloromethane, 2-Bromo-1-chloropropane, and 1,4-Dichlorobutane	**Surrogate Standards**
Optional. 1-Chloro-2-fluoro-benzene, or 2-Bromo-1-chloro-propane and Fluorobenzene recommended. Area must be ±3 SD or ± 20% of the mean response of calibration standards.	Optional. If used, surrogate compounds have been successfully used as internal standards.	Optional. 2-Bromo-1-chloro-propane or 1,4-Dichlorobutane recommended. Select one or more of the compounds with similar analytical behavior to compounds of interest.	**Internal Standards**
MS not required. Laboratory Fortified Blanks (LFB) should be run at least quarterly to determine precision. (See also: **QC Check Standards/ Samples**.)	One MS per 10 samples from each site or 1 per month, whichever is more frequent. Compare %R to Table 2 (see page A-69). (See also: **QC Check Standards/ Samples**.)	6230B: One MS per 10 samples from each site or 1 per month, whichever is more frequent. Compare %R to Table 6230:II (see page A-93). 6230C and 6230D: No MS required. (See also: **QC Check Standards/ Samples**.)	**Accuracy/ Precision**
One method blank per batch of samples processed at the same time. Field Reagent Blank (FRB) with each sample set.	One method blank per day.	One method blank per day.	**Blanks**

Organic

Determination of Volatile Organic

Method ⇒ Parameter ⇓	EPA 600 Series Method 602	Standard Methods Method 6220A	SW-846 Method 8021B
Preserva-tion/ Storage Conditions	pH < 2 with HCl. If residual chlorine, add sodium thiosulfate. Store at 4 °C.	pH < 2 with HCl. Add ascorbic acid, sodium thiosulfate or sodium sulfite if residual chlorine. Store at 4 °C.	Sodium thiosulfate if residual chlorine, adjust pH to <2 (aqueous). Store at 4 °C.
Holding Time (6)	14 days	14 days	14 days
Field Sample Amount Required (7)	40 mL VOA vial in duplicate without headspace (air bubbles) glass container Teflon lined septa	40 mL VOA vial in duplicate without headspace (air bubbles) glass container Teflon lined septa	40 mL VOA vial in duplicate without headspace (air bubbles) (aqueous) 4 oz. (solid) glass container Teflon lined top
Amount for Extraction	5 mL	5 mL	5 mL (aqueous) 5 grams (solid)
Other Criteria (Method Specific)	When analyzing unfamiliar samples, support identifications by at least one additional qualitative technique, such as second dissimilar column confirmation or GC/MS.	When analyzing unfamiliar samples, support identifications by at least one additional qualitative technique, such as second dissimilar column confirmation or GC/MS. 6220C: Retention times must vary by <10% over 8 hour period.	When doubt exists in compound identification, second column or GC/MS confirmation should be used.

Notes:

(6)   Unless otherwise indicated, holding times are from the date of sample collection.

(7)   Approximate volumes to be gathered for analysis. Additional volumes are required for the generation of QC data.

Organic

Organic Compounds By GC (Continued)

EPA 500 Series Method 502.2	EPA 600 Series Method 601	Standard Methods Method 6230A	⇐ Method Parameter ⇓
pH < 2 with HCl. Ascorbic acid if residual chlorine (use sodium thiosulfate for dechlorination if gases are not to be determined). Acidification may be omitted if analysis is for Trihalomethanes (THMs) only sodium thiosulfate was used to dechlorinate, or if sample foams upon addition of HCl. Store at 4 °C. **Note:** Dechlorination must be performed prior to acidification. Do not mix dechlorination agents with HCl prior to sampling.	Sodium thiosulfate if residual chlorine. Store at 4 °C.	pH < 2 with HCl. Add ascorbic acid, sodium thiosulfate or sodium sulfite if residual chlorine is present. Store at 4 °C.	**Preservation/ Storage Conditions**
14 days	14 days	14 days	**Holding Time (6)**
40 mL VOA vial in duplicate without headspace (air bubbles) glass container Teflon lined top	40 mL VOA vial in duplicate without headspace (air bubbles) glass container Teflon lined top	40 mL VOA vial in duplicate without headspace (air bubbles) glass container Teflon lined top	**Field Sample Amount Required (7)**
5 mL	5 mL	5 mL	**Amount for Extraction**
When doubt exists in compound identification, second column or GC/MS confirmation recommended. Retention times of suspect peaks must agree within ± 3 SD of the standards for identification.	When doubt exists in compound identification, second dissimilar column confirmation recommended.	When analyzing unfamiliar samples, support identifications by at least one additional qualitative technique. Retention times should not vary by >10% over 8 hours.	**Other Criteria (Method Specific)**

Determination of Volatile Organic Compounds By GC

Chemical Name	CAS Number	SW-846 Method	EPA 500 Method	EPA 600 Method	Standard Methods Method 6220A		EPA 600 Method	Standard Methods Method 6230A		
		8021B	502.2	602	6220B	6220C	601	6230B	6230C	6230D
Allyl chloride	107-05-1	•								
Benzene	71-43-2	•	•	•	•	•				•
Benzyl chloride (α-chlorotoluene)	100-44-7	•								
Bis(2-chloroethoxy)-methane	111-91-1									
Bis(2-chloroisopropyl) ether	108-60-1	•								
Bromoacetone	598-31-2	•								
Bromobenzene	108-86-1	•	•			•			•	•
Bromochloromethane	74-97-5	•	•						•	•
Bromodichloromethane	75-27-4	•	•				•	•	•	•
Bromoform	75-25-2	•	•				•	•	•	•
Bromomethane	74-83-9	•	•				•	•	•	•
n-Butylbenzene	104-51-8	•	•			•				•
sec-Butylbenzene	135-98-8	•	•			•				•
tert-Butylbenzene	98-06-6	•	•			•				•
Carbon tetrachloride	56-23-5	•	•				•	•	•	•
Chlorobenzene	108-90-7	•	•	•	•	•	•	•	•	•
Chloroethane	75-00-3	•	•				•	•	•	•
2-Chloroethanol	107-07-3	•								
2-Chloroethylvinyl ether	110-75-8	•					•	•		
Chloroform	67-66-3	•	•				•	•	•	•
1-Chlorohexane	544-10-5									
Chloromethane	74-87-3	•	•				•	•	•	•
Chloromethyl methyl ether	107-30-2	•								
Chloropropene	126-99-8	•								
2-Chlorotoluene	95-49-8	•	•			•			•	•
4-Chlorotoluene	106-43-4	•	•			•			•	•
1,2-Dibromo-3-chloropropane	96-12-8	•	•							•
Dibromochloromethane	124-48-1	•	•				•	•		
1,2-Dibromoethane	106-93-4	•	•						•	•
Dibromomethane	74-95-3	•	•						•	•
1,2-Dichlorobenzene	95-50-1	•	•	•	•	•	•	•	•	•
1,3-Dichlorobenzene	541-73-1	•	•	•	•	•	•	•	•	•
1,4-Dichlorobenzene	106-46-7	•	•	•	•	•	•	•	•	•
1.4-Dichloro-2-butene	764-41-0									
Dichlorodifluoromethane	75-71-8	•	•				•	•	•	•
1,1-Dichloroethane	75-34-3	•	•				•	•	•	•

Organic

Determination of Volatile Organic Compounds By GC

Chemical Name	CAS Number	SW-846 Method 8021B	EPA 500 Method 502.2	EPA 600 Method 602	Standard Methods Method 6220A		EPA 600 Method 601	Standard Methods Method 6230A		
					6220B	6220C		6230B	6230C	6230D
1,2-Dichloroethane	107-06-2	•	•				•	•	•	•
1,1-Dichloroethene	75-35-4	•	•				•	•	•	•
cis-1,2-Dichloroethene	156-59-2	•	•						•	•
trans-1,2-Dichloroethene	156-60-5	•	•				•	•	•	•
1,2-Dichloropropane	78-87-5	•	•				•	•	•	•
1,3-Dichloropropane	142-28-9	•	•						•	•
2,2-Dichloropropane	590-20-7	•	•						•	•
1,1-Dichloropropene	563-58-6	•	•						•	•
1,3-Dichloro-2-propanol	96-23-1	•								
cis-1,3-Dichloropropene	10061-01-5	•	•				•	•		
trans-1,3-Dichloropropene	10061-02-6	•	•				•	•		
Epichlorhydrin	106-89-8	•								
Ethyl benzene	100-41-4	•	•	•	•	•				•
Hexachlorobutadiene	87-68-3	•	•			•				•
Isopropylbenzene	98-82-8	•	•							•
4-Isopropyltoluene	99-87-6	•	•			•				•
Methylene chloride	75-09-2	•	•				•	•	•	•
Methyl iodide	74-88-4									
Naphthalene	91-20-3	•	•			•				•
n-Propylbenzene	103-65-1	•	•			•				•
Styrene	100-42-5	•	•							•
1,1,1,2-Tetrachloroethane	630-20-6	•	•						•	
1,1,2,2-Tetrachloroethane	79-34-5	•	•				•	•	•	•
Tetrachloroethene	127-18-4	•	•			•	•	•	•	•
Toluene	108-88-3	•	•	•	•	•				•
1,2,3-Trichlorobenzene	87-61-6	•	•			•				•
1,2,4-Trichlorobenzene	120-82-1	•	•			•				•
1,1,1-Trichloroethane	71-55-6	•	•				•	•	•	•
1,1,2-Trichloroethane	79-00-5	•	•				•	•	•	•
Trichloroethene	79-01-6	•	•			•	•	•	•	•
Trichlorofluoromethane	75-69-4	•	•				•	•	•	•
1,2,3-Trichloropropane	96-18-4	•	•						•	•
1,2,4-Trimethylbenzene	95-63-6	•	•			•				•
1,3,5-Trimethylbenzene	108-67-8	•	•			•				•
Vinyl chloride	75-01-4	•	•				•	•	•	•
m-Xylene	108-38-3	•	•			•				•
o-Xylene	95-47-6	•	•			•				•
p-Xylene	106-42-3	•	•			•				•
Xylene (total)	1330-20-7									

Organic

Analytical Methods - Inorganic Constituents

The numerous analytical methods for inorganic constituent testing have been organized into six major categories:

- Cyanide (total and amenable)
- Organic carbon (total)
- Mercury identified by cold vapor atomic absorption
- Trace metals identified by flame and graphite furnace atomic absorption spectroscopy
- Trace metals identified by inductively coupled plasma technique
- Trace metals identified by inductively coupled plasma/mass spectroscopy

Within each category there is a section called *Method Comparison* that summarizes the key method requirements and components for each testing method. The methods are compared on a side-by-side basis, which allows the user to quickly compare key aspects of the test methods. Where appropriate the user is referred to a table in Appendix A that provides details about an aspect of the test method. The tables in Appendix A are taken directly from the regulations.

For determining if a specific inorganic analyte is analyzed by a test method, each category that identifies more than one specific analyte has an *Analyte Listing* of all analytes addressed by the test methods within the category. This is also presented on a side-by-side basis so that the user can quickly see which method(s) cover the specific inorganic analyte in question.

If you want to find a test method for a specific inorganic analyte but don't know what category(ies) it belongs to, refer to Appendix C. Appendix C is an alphabetical listing of all inorganic analytes specified in the method comparison sections. Use this appendix to identify the category to turn to for additional information.

Inorganic

Determination of Total and Amenable Cyanide

Method ⇒ / Parameter ⇓	SW-846 Methods 9010B, 9012A, 9014	EPA Series Methods 335.1, 335.2	EPA Series Method 335.4	Standard Methods Methods 4500-CN C, D, E, F and G	CLP CN (Inorganic SOW) Method 335.2 CLP-M
Applicability	Drinking water, natural surface waters, domestic and industrial wastewaters and soil extracts.	Drinking, surface, saline water, domestic and industrial wastes.	Drinking, surface, saline water, domestic and industrial wastes.	Wastewater, groundwaters, and solid waste.	Water, soil, and sediments.
Number of Analytes	Total CN, free CN and CN amenable to chlorination	Total CN and CN amenable to chlorination	Total CN only	Total CN and CN amenable to chlorination	Total CN only
Method Validation (1)	Determine Method Detection Limit (MDL) by analyzing at least three portions of a solution at a concentration near 3 to 5 times the estimated MDL.	Not specified.	Initial demonstration of performance: 1. The Linear Calibration Range (LCR) must be determined initially and verified every 6 months. The verification of linearity must use a minimum of a blank and 3 standards. 2. A Quality Control sample (QCS), an independent standard, is prepared and analyzed at least quarterly to verify the calibration standards and instrument performance. If not within ±10% of stated value, determine source of problem and correct before continuing with analyses. 3. Determine MDLs by analyzing seven replicates of Laboratory fortified blanks. MDLs must be determined every six months.	1. Determine Method Detection Limit (MDL) by analyzing seven portions of a solution at a concentration near the estimated MDL. 2. Analyze independently prepared standard in replicate. The mean amount recovered should be ± 3 standard deviations of the mean value.	1. Determine Instrument detection Limits (IDL) by making seven consecutive measurements of a standard solution at a concentration 3x-5x the instrument manufacturer's suggested IDL on three nonconsecutive days. Determine IDLs quarterly. IDLs must meet specified CRDLs.

Notes:
(1)   Initial, one-time demonstration of ability to generate acceptable accuracy and precision.

Inorganic

Determination of Total and Amenable Cyanide (Continued)

Method ⇒ Parameter ⇓	SW-846 Methods 9010B, 9012A, 9014	EPA Series Methods 335.1, 335.2	EPA Series Method 335.4	Standard Methods Methods 4500-CN C, D, E, F and G	CLP CN (Inorganic SOW) Method 335.2 CLP-M
QC Check Standards/ Samples	9010B & 9014: Analyze check standards with each sample batch. If not within ±15% of stated value, reanalyze samples. 9012A: 1. After calibration is completed, verify by analyzing Independent check standard. If not within ±15% of stated value, a new calibration curve is required. 2. Verify calibration curve with every sample batch by analyzing a mid-range standard.	Not specified.	After calibration is completed, verify by analyzing QCS. If not within ±10% of stated value, terminate analysis and recalibrate instrument. Prepare and analyze a Laboratory Fortified Blank (LFB) with each batch of samples by fortifying laboratory reagent water with the QCS. If the recovery of the analyte is not within 90-110%, the analyte is judged out of control. Determine source of problem and correct before continuing with analyses.	Analyze externally supplied standards whenever analysis of known additions does not result in acceptable recovery or once each day, whichever is more frequent.	Verify initial calibration with a distilled independent standard (ICV). If not within ±15% of stated value, terminate analysis, correct problem, recalibrate instrument and verify calibration. Aqueous and solid Laboratory Control Samples (LCS) shall be prepared and analyzed for every analyte for every batch of aqueous or solid samples digested in a Sample Delivery Group (SDG) or per sample process batch, whichever is more frequent. Aqueous samples: A distilled ICV is used as the LCS. Limits: 85-115% Solid samples: Use certified material with supplied limits. If LCS is not within the limits specified, terminate analysis, correct problem, and redigest and reanalyze all samples associated with that LCS.
Method Detection Limit	Titration: 0.10 mg/L Colorimetric: 0.02 mg/L	Colorimetric 0.02 mg/L	0.005 mg/L	Titration: 1.0 mg/L Colorimetric: 0.02 mg/L Electrode: 0.05 mg/L	Titration (option A): 1.0 mg/L Manual and Semi-automated calorimetric (option B and C): 0.01 mg/L

Inorganic

Determination of Total and Amenable Cyanide (Continued)

Method ⇒ Parameter ⇓	SW-846 Methods 9010B, 9012A, 9014	EPA Series Methods 335.1, 335.2	EPA Series Method 335.4	Standard Methods Methods 4500-CN C, D, E, F and G	CLP CN (Inorganic SOW) Method 335.2 CLP-M
Standard Solution Expiration (2)	Stock standard: Not specified. Working Standard solution: Prepare daily.	Stock standard: Not specified. Working Standard solution: Prepare daily.	Stock standard: Not specified. Working Standard solution: Prepare daily.	Stock cyanide: Check weekly. Standard cyanide solution: Prepare daily.	Stock cyanide: Not specified. Standard cyanide: Not specified. Working Standard: Prepare daily.
Initial Calibration	Colorimetric: 6 levels plus blank Range: 40-800 µg/L	Colorimetric 6 levels plus blank Range: 40-800 µg/L	Colorimetric: Minimum of 3 levels and a blank Range: 5-500 µg/L.	Colorimetric: 1.Minimum of 3 levels plus a blank. 2.Range: 0.02 – 0.20 µg/ml. Electrode: 1.Minimum of four levels. 2.Range: 0.025-2.5 µg/ml.	Manual and Semi-automated Colorimetric: Minimum of 3 levels plus a blank (one standard at CRDL).
Continuing Calibration	Verify calibration with an independently prepared check standard every 15 samples. No criteria specified.	Not specified	Analyze Instrument performance check (IPC) solution (mid-range check standard) immediately following calibration, after every 10 samples and at the end of the run. 1. If not within ±10% of stated value, reanalyze IPC. 2. If second analysis of IPC is not within ±10% of stated value, discontinue analysis, determine the cause and/or in the case of drift recalibrate instrument. Reanalyze all samples since last compliant IPC.	Colorimetric: Recheck calibration curve periodically, and each time new reagents used. Periodically is not defined in the method.	Colorimetric: Analyze a Continuing calibration standard (CCV) solution (mid-range check standard) at the beginning, end and after every 10 samples. If not within ±15% of stated value, discontinue analysis, correct problem, recalibrate instrument, verify calibration, and reanalyze all samples since last compliant CCV.
Surrogate Standards	Not applicable.	Not applicable.	Not applicable.	Not applicable.	Not applicable.
Internal Standards	Not applicable.	Not applicable.	Not applicable.	Not applicable.	Not applicable.

Notes:
(2) Indicates maximum usage time. If comparisons to QC check standards indicate a problem, more frequent preparation may be necessary.

Determination of Total and Amenable Cyanide (Continued)

Method ⇒ Parameter ⇓	SW-846 Methods 9010B, 9012A, 9014	EPA Series Methods 335.1, 335.2	EPA Series Method 335.4	Standard Methods Methods 4500-CN C, D, E, F and G	CLP CN (Inorganic SOW) Method 335.2 CLP-M
Accuracy/ Precision	9010B & 9014: One MS per 20 samples. Add standard to ensure a level of 40 µg/L. No recovery criteria specified. One duplicate per 20 samples. If CV >20%, re-analyze samples. 9012A: One MS and MSD per 10 samples: Add standard to ensure a level of 40 µg/L. No recovery criteria specified.	Spike and analyze one sample: Add standard to ensure a level of 20 µg/L. No frequency or recovery criteria specified.	Spike and analyze one sample out of every 10 before sample distillation (Laboratory fortified sample matrix: LFM). The added analyte concentration should be the same as that used in the LFB. %R = 90-110	One MS per 10 samples. %R not specified. One duplicate per 20 samples. %RPD not specified.	One MS per sample delivery group or per matrix type: Add standard to ensure a level of 100 µg/L. If recovery not within 75-125%, analyze a post digestion spike unless the sample result exceeds the spike concentration by a factor of 4. One duplicate per SDG or per matrix type: Use control limit of 20% for %RPD if the sample and duplicate are both greater than 5x the CRDL. Else, use a control limit of ± CRDL.

Inorganic

Determination of Total and Amenable Cyanide (Continued)

Method ⇒ Parameter ⇓	SW-846 Methods 9010B, 9012A, 9014	EPA Series Methods 335.1, 335.2	EPA Series Method 335.4	Standard Methods Methods 4500-CN C, D, E, F and G	CLP CN (Inorganic SOW) Method 335.2 CLP-M
Blanks	9010B & 9014: One reagent blank per batch or every 20 samples. 9014 (Titration): One reagent blank per sample batch. 9012A & 9014 (Colorimetric): One calibration blank per calibration. Process and analyze a method blank with each batch of samples.	Colorimetric One calibration blank. No distillation blank specified.	A Laboratory reagent blank (LRB) is carried through the entire sample preparation and analysis scheme with each batch of samples. Values that exceed the MDL indicate contamination should be suspected and corrective actions must be taken before continuing analysis. A Calibration blank is to be analyzed after each IPC solution.	Analyze reagent blank with new reagents and every 20 samples.	Process and analyze one preparation blank per sample delivery group or per sample digestion batch, whichever is more frequent. 1. If the analyte conc. > CRDL, the lowest conc. of that analyte in the associated samples must be ≥10x the blank conc., else all samples associated with the blank with the analyte conc. <10x the blank conc. and >CRDL, shall be redigested and reanalyzed for that analyte. 2. If the analyte conc. is less than (- CRDL), then all samples <10x CRDL associated with that batch shall be redigested and reanalyzed for that analyte. Analyze calibration blanks after ICV and CCV solutions. 3. If absolute value of blank for analyte >CRDL, terminate analysis, correct problem, recalibrate instrument, verify calibration and reanalyze all samples since last compliant CCB.

GENIUM PUBLISHING CORPORATION

Inorganic

Determination of Total and Amenable Cyanide (Continued)

Method ⇒ Parameter ⇓	SW-846 Methods 9010B, 9012A, 9014	EPA Series Methods 335.1, 335.2	EPA Series Method 335.4	Standard Methods Methods 4500-CN C, D, E, F and G	CLP CN (Inorganic SOW) Method 335.2 CLP-M
Preservation/ Storage Conditions	Aqueous: pH ≥12 with NaOH. Sodium arsenite or ascorbic acid if oxidizing agents present. Store at 4 °C (aqueous). Store at 4 °C (solid).	pH > 12 with NaOH. Ascorbic acid if residual chlorine present. Store at 4 °C.	pH ≥ 12 with NaOH. Sodium arsenite or ascorbic acid if oxidizing agents present. Store at 4 °C.	Preserve with sodium arsenite or <0.1 g/L sodium thiosulphate if oxidizing agents present. Preserve with lead acetate or lead carbonate if sulfide ion present. Add 2 ml of 3.5% ethylenediamine solution per 100 ml of sample if aldehydes present. Total CN: Adjust pH > 12 to 12.5 and store in cool dark place if samples not immediately analyzed. Amenable to chlorination: Add 100 mg sodium thiosulphate per liter of sample.	Aqueous: pH > 12 with NaOH. Ascorbic acid if residual chlorine present. Store at 2-6 °C (aqueous). Store at 2-6 °C (solid).
Holding Times (3)	14 days	14 days (24 hours when sulfide present)	14 days	14 days (24 hours when sulfide present)	12 days from sample receipt
Field Sample Amount Required (4)	1 liter (aqueous) 200 grams (solid) Glass or polyethylene container	1 liter Glass or plastic container	250 ml Glass or plastic container	1 liter Glass or polyethylene container	Not specified Glass or polyethylene container
Amount for Extraction	500 ml (1000 ml if both total and amenable CN) (aqueous) 1-5 grams (2-10 grams if both total and amenable CN) (solid)	500 ml (1000 ml if both total and amenable CN)	50 ml	Total CN: 500 ML Amenable to chlorination: 1000 ml. Solid: 0.50 grams	500 ml (aqueous) 1-5 grams (solid)

Notes:

(3) Unless otherwise indicated, holding times are from the date of sample collection.

(4) Approximate volumes to be gathered for analysis. Additional volumes are required for the generation of QC data.

Inorganic

Determination of Total and Amenable Cyanide (Continued)

Method ⇒ Parameter ⇓	SW-846 Methods 9010B, 9012A, 9014	EPA Series Methods 335.1, 335.2	EPA Series Method 335.4	Standard Methods Methods 4500-CN C, D, E, F and G	CLP CN (Inorganic SOW) Method 335.2 CLP-M
Other Criteria (Method Specific)	Distill a high and a low standard; If recoveries are not within ± 10% of undistilled standards, the source of the problem should be identified and resolved before continuing analyses. If sulfides present, all standards must be distilled in the same manner as the samples. Samples suffering from matrix interferences shall use MSA.	Distill a high and a low standard; If recoveries are not within ±10% of undistilled standards, the source of the problem should be identified and resolved before continuing analyses. If sulfides present, all standards must be distilled in the same manner as the samples.	Distill a high and a low standard; If recoveries are not within ± 10% of undistilled standards, the source of the problem should be identified and resolved before continuing analyses.	Check distillation apparatus by analyzing a 1.0 ppm distilled standard. %R = 96-104 No corrective action specified.	A distilled ICV is used as the Laboratory control sample (LCS). %R = 85-115. A separate Laboratory control sample (LCS) is required for all soil samples. Control limits supplied with material.

Inorganic

GENIUM PUBLISHING CORPORATION

Determination of Total Organic Carbon

Method ⇒ Parameter ⇓	SW-846 Method 9060	EPA Series Method 415.1	Standard Methods Method 5310B
Applicability	Groundwater, surface and saline waters, domestic and industrial wastes	Groundwater, surface and saline waters, domestic and industrial wastes	Ground and surface waters, wastewater
Number of Analytes	No specific compounds; includes natural sugars, mercaptans, alkanes, low molecular weight alcohols and oils, cellulose fibers, oily matter adsorbed on silt.	No specific compounds; includes natural sugars, mercaptans, alkanes, low molecular weight alcohols and oils, cellulose fibers, oily matter adsorbed on silt	No specific compounds; classes of organic compounds not specified.
Method Validation (1)	Determine Method Detection Limit (MDL) by analyzing at least three portions of a solution at a concentration near 3 to 5 times the estimated MDL.	Not specified.	Determine Method Detection Limit (MDL) by analyzing seven portions of a solution at a concentration near the estimated MDL. Analyze independently prepared standard in replicate. The mean amount recovered should be within 3 standard deviations of the mean value.
QC Check Standards/ Samples	Verify calibration with an independently prepared check standard. No criteria specified. Analyze laboratory control sample when appropriate.	Not specified.	Analyze externally supplied standards whenever analysis of known additions does not result in acceptable recovery or once each day, whichever is more frequent.
Method Detection Limit	1 mg/L	1 mg/L	1 mg/L
Standard Solution Expiration (2)	Not specified.	Not specified.	Not specified.
Initial Calibration	Per instrument manufacturer's specifications.	Per instrument manufacturer's specifications.	Minimum of 3 standards plus blank. Range not specified.
Continuing Calibration	Verify calibration with an independently prepared check standard every 15 samples. No criteria specified.	Not specified.	Not specified.
Surrogate Standards	Not applicable.	Not applicable.	Not applicable.
Internal Standards	Not applicable.	Not applicable.	Not applicable.
Accuracy/ Precision	One MS per 20 samples. %R not specified. One duplicate every 10 samples. No criteria specified.	Not specified.	One MS per 10 samples. %R not specified. One duplicate per 20 samples. %RPD not specified.

Notes:
(1) Initial, one-time demonstration of ability to generate acceptable accuracy and precision.
(2) Indicates maximum usage time. If comparisons to QC check standards indicate a problem, more frequent preparation may be necessary.

Inorganic

Determination of Total Organic Carbon (Continued)

Method ⇒ Parameter ⇓	SW-846 Method 9060	EPA Series Method 415.1	Standard Methods Method 5310B
Blanks	Process and analyze a method blank with each batch of samples. One calibration blank per calibration.	Method blanks analyzed; frequency not specified.	Analyze reagent blank with new reagents and every 20 samples.
Preservation/ Storage Conditions	pH < 2 with HCl or H_2SO_4. Protect from light & atmospheric O_2. Store at 4 °C.	pH < 2 with HCl or H_2SO_4. Protect from light & atmospheric O_2. Store at 4 °C.	pH < 2 with H_2SO_4 or H_3PO_4 only if inorganic carbon is later purged. Protect from light. Store at 4 °C.
Holding Time (3)	28 days	28 days	28 days
Field Sample Amount Required (4)	Not specified. Glass or polyethylene container	25 ml Glass or polyethylene container	100 ml Amber glass with TFE-lined cap.
Amount for Extraction	Not specified.	Not specified.	10 - 15 ml.
Other Criteria (Method Specific)	Quadruplicate analysis is required. Report the average and the range.	None.	Subtract procedural blank from each instrument response. Repeat injections until consecutive peaks are obtained that are reproducible to within ± 10%

Notes:

(3) Unless otherwise indicated, holding times are from the date of sample collection.

(4) Approximate volumes to be gathered for analysis. Additional volumes are required for generation of QC data.

Inorganic

GENIUM PUBLISHING CORPORATION

Determination of Mercury by Cold Vapor Atomic Absorption Spectroscopy

Method ⇒ Parameter ⇓	SW-846 Methods 7470A/7471A	EPA Series Methods 245.1/245.5	Standard Methods Method 3500-Hg B	CLP (Inorganic SOW) Method 245.1/245.2/245.5 CLP-M
Applicability	Mobility-procedure extracts, aqueous wastes, ground waters, soils, sediments, bottom deposits, and sludge-type materials.	Drinking, surface, ground, sea, brackish, industrial, domestic wastewaters, soils, sediments, and sludges.	Drinking, surface, saline, groundwaters, and wastewaters.	Water, soil, and sediment.
Number of Analytes	Hg only	Hg only	Hg only	Hg only
Method Validation (1)	Determine Method Detection Limit (MDL) by analyzing at least three portions of a solution at a concentration near 3 to 5 times the estimated MDL.	Initial demonstration of performance: 1. Linear Dynamic Range: Determined from a linear calibration prepared from a minimum of three different concentration standards, determined by analyzing succeedingly higher standard concentrations of analyte(s) until analyte concentration is no more than 10% below the stated value of the standard. Verified annually. 2. Quality Control sample (QCS): An independent standard, is prepared and analyzed at least quarterly to verify the calibration standards and instrument performance. If not within ±10% of stated value, determine source of problem and correct before continuing with analyses. 3. Determine MDLs by analyzing seven replicates of Laboratory fortified blanks. MDLs must be determined annually.	1. Determine Method Detection Limit (MDL) by analyzing seven portions of a solution at a concentration near the estimated MDL. 2. Analyze independently prepared standard in replicate. The mean amount recovered should be within ±3 standard deviations of the mean value.	Determine Instrument Detection Limits (IDL) by making seven consecutive measurements of a standard solution at a concentration 3x-5x the instrument manufacturer's suggested IDL on three nonconsecutive days. Determine IDLs quarterly. IDLs must meet specified CRDLs.

Notes:
(1) Initial, one-time demonstration of ability to generate acceptable accuracy and precision.

Determination of Mercury by Cold Vapor Atomic Absorption Spectroscopy (Continued)

Method ⇒ Parameter ⇓	SW-846 Methods 7470A/7471A	EPA Series Methods 245.1/245.5	Standard Methods Method 3500-Hg B	CLP (Inorganic SOW) Method 245.1/245.2/245.5 CLP-M
QC Check Standards/ Samples	Verify each calibration with an independent calibration check standard. If not within ±10% of stated value, calibration is invalid. Process and analyze one laboratory control sample with each batch.	After calibration is completed, verify by analyzing QCS. If not within ±10% of stated value, terminate analysis and recalibrate instrument. Prepare and analyze a Laboratory Fortified Blank (LFB) with each batch of samples by fortifying laboratory reagent water with the QCS. If not within ±15% of stated value, determine source of problem and correct before continuing with analyses.	Verify calibration with independent check standard. %R = 95-105.	Initial calibration verified with independent standard (ICV). If not within ±20% of stated value, terminate analysis, correct problem, recalibrate instrument and verify calibration. One solid Laboratory Control Sample (LCS) shall be prepared and analyzed for each analyte for every group of solid samples in a Sample Delivery Group (SDG) or per sample process batch, whichever is more frequent. If not within the limits specified, terminate analysis, correct problem, and redigest and reanalyze all samples associated with that LCS.
Method Detection Limit	0.0002 mg/L	0.0002 mg/L	0.001 mg/L	CRDL 0.0002 mg/L
Standard Solution Expiration (2)	Stock standards: Not specified. Working Standard solution: Prepare daily. Calibration standards: Prepare daily	Stock standards: Not specified. Calibration standards: Prepare daily.	Stock standards: Not specified. Calibration standards: Prepare daily.	Stock standards: Not specified. Calibration standards: Prepare daily.
Initial Calibration	5 levels and a blank. Range 0.5 -10 µg/L.	Five levels plus a blank. Range 0.5 - 10 µg/L	Three levels plus a blank. Range 1 - 5 µg/L.	245.1 and 245.5: 5 levels plus a blank. 245.2: 8 levels plus a blank. One calibration standard at the CRDL.

Notes:
(2) Indicates maximum usage time. If comparisons to QC check standards indicate a problem, more frequent preparation may be necessary.

Determination of Mercury by Cold Vapor Atomic Absorption Spectroscopy (Continued)

Method ⇒ Parameter ⇓	SW-846 Methods 7470A/7471A	EPA Series Methods 245.1/245.5	Standard Methods Method 3500-Hg B	CLP (Inorganic SOW) Method 245.1/245.2/245.5 CLP-M
Continuing Calibration	Analyze a mid-level calibration standard or calibration check standard after every 10 samples. If not within ±20% of stated value, Reanalyze previous ten samples.	Instrument performance check (IPC) solution. Analyze a mid-level calibration standard immediately after calibration, every 10 samples and at the end of the run. 1. For initial analysis, %R = 95-105. 2. For subsequent analyses of standard, %R = 90-110. 3. If analysis of IPC is not within specified limits, discontinue analysis, determine the cause and/or in the case of drift recalibrate instrument. Reanalyze all samples since last compliant IPC	Analyze a midpoint check standard at the beginning, end and after every 9 samples. 1. If not within ±5% of stated value, a potential problem exists. 2. If not within ±10% of stated value, take corrective action.	Analyze a Continuing calibration standard (CCV) solution (mid-range check standard) at the beginning, end and after every 10 samples. If not within ±20% of stated value, discontinue analysis, correct problem, recalibrate instrument, verify calibration, and reanalyze all samples since last compliant CCV.
Surrogate Standards	Not applicable.	Not applicable.	Not applicable.	Not applicable.
Internal Standards	Not applicable.	Not applicable.	Not applicable.	Not applicable.
Accuracy/ Precision	One MS/MSD, and duplicate for each batch of samples No criteria specified.	Spike and analyze one sample with low Hg background out of every 10 samples before sample digestion. (Laboratory fortified sample matrix: LFM). The added analyte concentration should be the same as that used in the LFB. %R = 70-130 Duplicates: No frequency or recovery criteria specified.	One MS per 10 sample. %R = 85-115 One duplicate per 20 samples. %RPD not specified.	One MS per SDG or per matrix type. %R = 75-125 One duplicate per SDG or per matrix type. Use control limit of 20% for %RPD if the sample and duplicate are both greater than 5x the CRDL. Else, use a control limit of ± the CRDL.

Inorganic

Determination of Mercury by Cold Vapor Atomic Absorption Spectroscopy (Continued)

Method ⇒ Parameter ⇓	SW-846 Methods 7470A/7471A	EPA Series Methods 245.1/245.5	Standard Methods Method 3500-Hg B	CLP (Inorganic SOW) Method 245.1/245.2/245.5 CLP-M
Blanks	Process and analyze a method blank with each batch of samples. After calibration, verify the calibration curve by analyzing a calibration blank.	A Laboratory reagent blank (LRB) is carried through the entire sample preparation scheme with each batch of 20 or fewer samples of the same matrix. When LRB values constitute 10% or more of the analyte level or is 2.2x the analyte MDL, whichever is greater, fresh aliquots of the sample must be prepared and analyzed again for the affected analytes after the source of the contamination has been corrected. A Calibration blank (CCB) is to be analyzed after each IPC solution. Value(s) of the CCB should be < MDL. A field blank should be prepared and analyzed as required by the data user.	Process and analyze a procedural blank with each set of samples. If blank values for analytes are greater than MDL, repeat sample preparation and analysis with cleaner reagents or glassware. Analyze reagent blank whenever new reagents are used and every 20 samples. Analyze a calibration blank at the beginning, end and after every 9 samples.	Process and analyze one preparation blank per sample delivery group or per sample digestion batch, whichever is more frequent. 1. If the analyte conc. > CRDL, the lowest conc. of that analyte in the associated samples must be ≥10x the blank conc., else all samples associated with the blank with that analyte conc. <10x the blank conc. and >CRDL, shall be redigested and reanalyzed for that analyte. 2. If the analyte conc. is less than (- CRDL) then all samples <10x CRDL associated with that batch shall be redigested and reanalyzed for that analyte. Analyze calibration blanks after ICV and CCV solutions. 3. If absolute value of blank for analyte >CRDL, terminate analysis, correct problem, recalibrate instrument, verify calibration and reanalyze all samples since last compliant CCB.
Preservation/ Storage Conditions	Aqueous: pH < 2 with HNO_3 Solid: Store at 4 °C.	Aqueous: pH < 2 with HNO_3 Solid: Not specified.	pH < 2 with HNO_3 Store at 4 °C.	Aqueous: pH < 2 with HNO_3 Solid: Store at 2-6 °C.
Holding Time (3)	28 days	28 days	28 days	26 days from sample receipt

Notes:

(3) Unless otherwise indicated, holding times are from the date of sample collection.

Inorganic

Determination of Mercury by Cold Vapor Atomic Absorption Spectroscopy (Continued)

Method ⇒ Parameter ⇓	SW-846 Methods 7470A/7471A	EPA Series Methods 245.1/245.5	Standard Methods Method 3500-Hg B	CLP (Inorganic SOW) Method 245.1/245.2/245.5 CLP-M
Field Sample Amount Required (4)	400 ml (aqueous) 200 grams (solid) Glass or polyethylene container	1 liter (aqueous) Glass or polyethylene container	500 ml Glass or polyethylene container	Not specified (aqueous) Not specified (solid) Glass or polyethylene container
Amount for Extraction	100 ml (aqueous) 0.2 grams (solids)	100 ml (aqueous) 0.2 grams (solids)	100 ml	100 ml (aqueous) 0.2 grams (solids)
Other Criteria (Method-Specific)	Interferences tests Dilution test: Analyze one diluted sample per batch. The concentration of the analyte should be at least 25x the estimated detection limit. If all the samples in batch are <10x MDL or if the corrected results for the diluted sample are not within 10% of the undiluted sample, perform Recovery test: Add a known amount of the analyte to bring the concentration of the analyte to 2-5x the original concentration. If the recovery is not within 85-115%, the Method of Standard Additions shall be used for all samples in the batch.	For solid samples, process and analyze samples in triplicate.	Verify the absence of interferences by performing Dilution test: Dilute one sample 1:10. Analyze the diluted and undiluted samples. Results should be comparable.	To verify calibration linearity near CRDL, analyze a CRA standard at the CRDL or IDL, whichever is greater at the beginning of each run. No limits specified.

Notes:
(4) Approximate volumes to be gathered for analysis. Additional volumes are required for the generation of QC data.

Determination of Trace Metals by Flame and Graphite Furnace Atomic Absorption Spectroscopy

Method ⇒ Parameter ⇓	SW-846 Method 7000 Series	EPA Series Method 200 Series	Standard Methods Method 3000 Series	CLP (Inorganic SOW) Method 200 series CLP-M
Applicability	Drinking, surface, saline waters, domestic and industrial wastes, extracts, sludges, soils, sediments, industrial and other solid wastes	Drinking, surface, and saline waters; domestic and industrial wastes	Drinking, surface, saline, groundwaters, and wastewaters	Water, soil, and sediment
Number of Analytes (1)	27 total	33 total	39 total	22 total (including alternate methods)
Method Validation (2)	Determine Method Detection Limit (MDL) by analyzing at least three portions of a solution at a concentration near 3 to 5 times the estimated MDL.	Not specified.	1. Determine Method Detection Limit (MDL) by analyzing seven portions of a solution at a concentration near the estimated MDL. 2. Analyze independently prepared standard in replicate. The mean recovery should be within 3 standard deviations of the mean value.	Determine Instrument Detection Limits (IDL) by making seven consecutive measurements of a standard solution at a concentration 3x-5x the instrument manufacturer's suggested IDL on three nonconsecutive days. Determine IDLs quarterly. IDLs must meet specified CRDLs.
QC Check Standards/ Samples	Verify each calibration with an independent calibration check standard. If not within ±10% of stated value, calibration is invalid. Process and analyze one laboratory control sample with each batch.	Performance Sample: One (blind) sample analyzed per year for metals being determined. Results must be within EPA control limits. OPTIONAL: Known reference standards analyzed once per quarter for metals being determined	Verify calibration with independent check standard. %R = 95-105.	Verify initial calibration with an independent standard (ICV). If not within ±10% of stated value, terminate analysis, correct problem, recalibrate instrument and verify calibration. Aqueous and solid Laboratory Control Samples (LCS) shall be prepared and analyzed for every analyte for every batch of aqueous or solid samples digested in a Sample Delivery Group (SDG) or per sample process batch, whichever is more frequent. Aqueous samples: A distilled ICV is used as the LCS. Limits: 85-115% Solid samples: Use certified material with supplied limits. If LCS is not within the limits specified, terminate analysis, correct problem, and redigest and reanalyze all samples associated with that LCS.

Notes:

(1) Analyte lists may vary among methods; a smaller list in one method is not necessarily a subset of a larger list in another method.

(2) Initial, one-time demonstration of ability to generate acceptable accuracy and precision.

Inorganic

Determination of Trace Metals by Flame and Graphite Furnace Atomic Absorption Spectroscopy (Continued)

Method ⇒ Parameter ⇓	SW-846 Method 7000 Series	EPA Series Method 200 Series	Standard Methods Method 3000 Series	CLP (Inorganic SOW) Method 200 series CLP-M
Method Detection Limit	Detection limits listed in Table 1 of Method 7000 (see page A-109).	Detection limits listed in Table 1 of Metals Introductory chapter (see page A-16).	FLAA: IDLs listed in Table 3111:I of Method 3111 (see page A-82). GFAA: IDLs listed in Table 3113:II of Method 3113 (see page A-83).	CRDLs listed in Exhibit C (see page A-15).
Standard Solution Expiration (3)	Stock standards: Not specified. Calibration standards: Prepare daily.	Stock standards: Not specified. Calibration standards: Prepare daily.	Stock standards: Not specified. Calibration standards: Prepare daily.	Stock standards: Not specified. Calibration standards: Not specified.
Initial Calibration	Minimum of 3 levels plus a blank.	Minimum of 4 levels plus a blank.	Minimum of 3 levels plus a blank.	Minimum of 3 levels plus a blank (one standard at CRDL).
Continuing Calibration	Analyze a mid-level calibration standard or calibration check standard after every 10 samples (GFAA: every 10 injections). If not within ±20% of stated value, reanalyze previous ten samples.	Analyze a low-level standard daily or every 20 samples, whichever is more frequent. %R = 90-110.	Analyze a midpoint check standard at the beginning, end and after every 9 samples. 1. If not within ±5% of stated value, a potential problem exists. 2. If not within ±10% of stated value, take corrective action.	Analyze a Continuing calibration standard (CCV) solution (mid-range check standard) at the beginning, end and after every 10 samples. If not within ±10% of stated value, discontinue analysis, correct problem, recalibrate instrument, verify calibration, and re-analyze all samples since last compliant CCV.
Surrogate Standards	Not applicable.	Not applicable.	Not applicable.	Not applicable.
Internal Standards	Not applicable.	Not applicable.	Not applicable.	Not applicable.
Accuracy/ Precision	One MS/MSD, and duplicate for each batch of samples No criteria specified.	OPTIONAL: One duplicate sample per every 10 samples, or per each set of samples (if set contains < 10). Results within EPA control limits.	One MS per 10 sample. %R = 85-115 One duplicate per 20 samples. %RPD not specified.	One MS per sample delivery group or per matrix type. If recovery not within 75-125%, analyze a post digestion spike unless the sample result exceeds the spike concentration by a factor of 4. One duplicate per SDG or per matrix type. Use control limit of 20% for %RPD if the sample and duplicate are both greater than 5x the CRDL. Else, use a control limit of (±) the CRDL.

Notes:

(3) Indicates maximum usage time. If comparisons to QC check standards indicate a problem, more frequent preparation may be necessary.

Determination of Trace Metals by Flame and Graphite Furnace Atomic Absorption Spectroscopy (Continued)

Method ⇒ Parameter ⇓	SW-846 Method 7000 Series	EPA Series Method 200 Series	Standard Methods Method 3000 Series	CLP (Inorganic SOW) Method 200 series CLP-M
Blanks	Process and analyze a method blank with each batch of samples. After calibration, verify the calibration curve by analyzing a calibration blank.	Method blank not specified. Calibration blank with each calibration.	Process and analyze a procedural blank with each set of samples. If blank values for analytes are greater than MDL, repeat sample preparation and analysis with cleaner reagents or glassware. Analyze reagent blank with new reagents and every 20 samples. Analyze a calibration blank at the beginning, end and after every 9 samples.	Process and analyze one preparation blank per sample delivery group or per sample digestion batch, whichever is more frequent. 1. If the analyte conc. > CRDL, the lowest conc. of that analyte in the associated samples must be ≥ 10x the blank conc., else all samples associated with the blank with that analyte conc. <10x the blank conc. and >CRDL, shall be redigested and reanalyzed for that analyte. 2. If the analyte conc. is below the negative CRDL, then all samples <10x CRDL associated with that batch shall be redigested and reanalyzed for that analyte. Analyze calibration blanks after ICV and CCV solutions. 3. If absolute value of blank for analyte >CRDL, terminate analysis, correct problem, recalibrate instrument, verify calibration and reanalyze all samples since last compliant CCB.
Preservation/ Storage Conditions	Aqueous: pH < 2 with HNO$_3$ Solid: Store at 4 °C.	pH < 2 with HNO$_3$	pH < 2 with HNO$_3$	Aqueous: pH < 2 with HNO$_3$ Solid: Store at 2-6 °C.
Holding Time (4)	6 months	6 months	6 months	180 days from sample receipt
Field Sample Amount Required (5)	600 ml (aqueous) 200 grams (solid) Glass or polyethylene container	1 liter (aqueous) Glass or polyethylene container	500 ml Glass or polyethylene container	Aqueous: Not specified Solid: Not specified Glass or polyethylene container
Amount for Extraction	100 ml (aqueous) 2 grams (solid)	100 ml (aqueous)	100 ml	100 ml (aqueous) 1 gram (solid) Microwave digestion: 45 ml (aqueous) 0.5 grams (solid)

Notes:

(4) Unless otherwise indicated, holding times are from the date of sample collection.

(5) Approximate volumes to be gathered for analysis. Additional volumes are required for the generation of QC data.

Determination of Trace Metals by Flame and Graphite Furnace Atomic Absorption Spectroscopy (Continued)

Method ⇒ Parameter ⇓	SW-846 Method 7000 Series	EPA Series Method 200 Series	Standard Methods Method 3000 Series	CLP (Inorganic SOW) Method 200 series CLP-M
Other Criteria (Method Specific)	Interferences tests Dilution test: Analyze one diluted sample per batch. The concentration of the analyte should be at least 25x the estimated detection limit. If all the samples in batch are <10x MDL or if the corrected results for the diluted sample are not within 10% of the undiluted sample, perform Recovery test: Add a known amount of the analyte to bring the concentration of the analyte to 2-5x the original concentration. If the recovery is not within 85-115%, the Method of Standard Additions shall be used for all samples in the batch.	For GFAA, verify the absence of interferences by diluting and spiking each matrix. Use standard additions to compensate for matrix interferences.	Verify the absence of interferences by performing Dilution test Dilute one sample 1:10. Analyze the diluted and undiluted samples. Results should be comparable. GFAA: All samples require duplicate injections. 1. Variation of 10% is acceptable. 2. Average replicate values. Test every sample for recovery. If recovery not between 85 and 115%, perform MSA to compensate for matrix interferences.	To verify calibration linearity near CRDL analyze a CRA Standard at CRDL or IDL, whichever is greater, at the beginning of each run. GFAA: All samples require duplicate injection. 1.%RPD < 20. 2. Average replicate values. Test every sample for recovery. If recovery not between 85 and 115%, perform MSA to compensate for matrix interferences.

Inorganic

Trace Metals by Flame and Graphite Furnace Atomic Absorption Spectroscopy

Metal	Analytical Method Type[1]	SW-846 Method 7000 Series	EPA Method 200 Series	Std. Methods Method 3000	CLP Inorganic Method[2]
Aluminum	AAS	7020	202.1	3111	
Aluminum	GFAA		202.2	3113	
Antimony	AAS	7040	204.1	3111	
Antimony	GFAA	7041	204.2	3113	204.2
Arsenic	GFAA	7060A	206.2	3113	206.2
Arsenic	GHAA	7061A	206.3	3114	
Barium	AAS	7080A	208.1	3111	
Barium	GFAA	7081	208.2	3113	
Beryllium	AAS	7090	210.1	3111	
Beryllium	GFAA	7091	210.2	3113	210.2
Bismuth	AAS			3111	
Cadmium	AAS	7130	213.1	3111	
Cadmium	GFAA	7131A	213.2	3113	213.2
Calcium	AAS	7140	215.1	3111	215.1
Cesium	AAS			3111	
Chromium	AAS	7190	218.1	3111	
Chromium	GFAA	7191	218.2	3113	218.2
Cobalt	AAS	7200	219.1	3111	
Cobalt	GFAA	7201	219.2	3113	
Copper	AAS	7210	220.1	3111	
Copper	GFAA	7211	220.2	3113	
Gold	AAS		231.1	3111	
Gold	GFAA		231.2		
Iridium	AAS		235.1	3111	
Iridium	GFAA		235.2		
Iron	AAS	7380	236.1	3111	
Iron	GFAA	7381	236.2	3113	
Lead	AAS	7420	239.1	3111	
Lead	GFAA	7421	239.2	3113	239.2
Lithium	AAS	7430		3111	
Magnesium	AAS	7450	242.1	3111	242.1
Manganese	AAS	7460	243.1	3111	
Manganese	GFAA	7461	243.2	3113	
Molybdenum	AAS	7480	246.1	3111	
Molybdenum	GFAA	7481	246.2	3113	
Nickel	AAS	7520	249.1	3111	
Nickel	GFAA	7521	249.2	3113	
Osmium	AAS	7550	252.1	3111	
Osmium	GFAA		252.2		

Notes:

(1)   Analytical Method Type: AAS - Flame Atomic Absorption Spectroscopy (also called FLAA)
GFAA - Graphite Furnace Atomic Absorption Spectroscopy
GHAA - Gaseous Hydride Atomic Absorption Spectroscopy

(2)   All methods in this column have the suffix "CLP-M", indicating the method has been modified for the Contract Laboratory Program.

GENIUM PUBLISHING CORPORATION

Inorganic

Trace Metals by Flame and Graphite Furnace Atomic Absorption Spectroscopy (Continued)

Metal	Analytical Method Type[1]	SW-846 Method 7000 Series	EPA Method 200 Series	Std. Methods Method 3000	CLP Inorganic Method[2]
Palladium	AAS		253.1	3111	
Palladium	GFAA		253.2		
Platinum	AAS		255.1	3111	
Platinum	GFAA		255.2		
Potassium	AAS	7610	258.1	3111	258.1
Rhenium	AAS		264.1	3111	
Rhenium	GFAA		264.2		
Rhodium	AAS		265.1	3111	
Rhodium	GFAA		265.2		
Ruthenium	AAS		267.1	3111	
Ruthenium	GFAA		267.2		
Selenium	GFAA	7740	270.2	3113	270.2
Selenium	GHAA	7741A	270.3	3114	
Silicon/Silica (SiO_2)	AAS			3111	
Silver	AAS	7760A	272.1	3111	
Silver	GFAA	7761	272.2	3113	272.2
Sodium	AAS	7770	273.1	3111	273.1
Sodium	GFAA		273.2		
Strontium	AAS	7780		3111	
Thallium	AAS	7840	279.1	3111	
Thallium	GFAA	7841	279.2		279.2
Thorium	AAS			3111	
Tin	AAS	7870	282.1	3111	
Tin	GFAA		282.2	3113	
Titanium	AAS		283.1	3111	
Titanium	GFAA		283.2		
Vanadium	AAS	7910	286.1	3111	
Vanadium	GFAA	7911	286.2		
Zinc	AAS	7950	289.1	3111	
Zinc	GFAA	7951	289.2		

Notes:
(1) Analytical Method Type: AAS - Flame Atomic Absorption Spectroscopy (also called FLAA)
                                        GFAA - Graphite Furnace Atomic Absorption Spectroscopy
                                        GHAA - Gaseous Hydride Atomic Absorption Spectroscopy

(2) All methods in this column have the suffix "CLP-M", indicating the method has been modified for the Contract Laboratory Program.

Inorganic

Determination of Trace Metals by ICP

Method ⇒ / Parameter ⇓	SW-846 Method 6010B	EPA Series Method 200.7	Standard Methods Method 3120	CLP (Inorganic SOW) Method 200.7 CLP-M
Applicability	Nearly all matrices, including ground-water, extracts, industrial wastes, soils, sludges, sediments and solid wastes	Drinking water, surface, saline and industrial and domestic wastes	Water and waste-water	Water, soil, and sediment
Number of Analytes (1)	31 total	32 total	27 total	22 total
Method Validation (2)	Initial demonstration of performance: 1. Linear Dynamic Range (LDR): Established by determining the signal responses from a minimum of three different concentration standards across the range. The upper range limit should be within 10% of the highest measured level. Verified every six months. 2. Determine Method Detection Limit (MDL) by analyzing at least three portions of a solution at a concentration near 3 to 5 times the estimated MDL. 3. Interelement correction (IEC) factors must be determined.	Initial demonstration of performance: 1. Determination of the linear dynamic range. Determined by analyzing succeedingly higher standard concentrations of analyte(s) until analyte concentration is no more than 10% below the stated value of the standard. Verified annually. 2. A Quality Control sample (QCS), an independent standard, is prepared and analyzed three times at least quarterly to verify the calibration standards and instrument performance. The concentration of the analytes should be ≥ 1.0 mg/L except silver at a maximum of 0.5 mg/L. If the mean value of the three analyses is not within ± 5% of stated value, determine source of problem and correct before continuing with analyses. 3. Determine MDLs by analyzing seven replicates of Laboratory fortified blanks. MDLs must be determined annually.	1. Determine Method Detection Limit (MDL) by analyzing seven portions of a solution at a concentration near the estimated MDL. 2. Analyze independently prepared standard in replicate. The mean amount recovered should be within 3 standard deviations of the mean value.	1. Determine Instrument Detection Limits (IDL) by making seven consecutive measurements of a standard solution at a concentration 3x-5x the instrument manufacturer's suggested IDL on three nonconsecutive days. Determine IDLs quarterly. IDLs must meet specified CRDLs. 2. Interelement correction (IEC) factors must be determined at least annually. 3. A Linear Range verification check standard (LRA) shall be analyzed quarterly. %R=95-105%.

Notes:
(1) Analyte lists may vary among methods: a smaller list in one method is not necessarily a subset of a larger list in another method.
(2) Initial, one-time demonstration of ability to generate acceptable accuracy and precision.

GENIUM PUBLISHING CORPORATION

Determination of Trace Metals by ICP (Continued)

Method ⇒ Parameter ⇓	SW-846 Method 6010B	EPA Series Method 200.7	Standard Methods Method 3120	CLP (Inorganic SOW) Method 200.7 CLP-M
QC Check Standards/ Samples	Verify each element calibration with an Instrument Calibration Verification (ICV) solution prepared from a source different than the calibration standards. 1. If not within ±10% of stated value, stop analysis, correct problem, and recalibrate instrument. Process and analyze one laboratory control sample with each batch.	Prepare and analyze a Laboratory Fortified Blank (LFB) with each batch of samples by fortifying laboratory reagent water with the QCS. Spike levels are: Ag = 0.1 mg/L, K = 5.0 mg/L, all other analytes 0.2 mg/L or at a concentration approximately 100 times the MDL, whichever is greater. 1. If not within ±15% of stated value, determine source of problem and correct before continuing with analyses.	Verify each element calibration with an Instrument Quality Control (IQC) sample prepared from a source different than the calibration standards. 1. If not within ±5% of stated value, prepare new calibration std., and recalibrate the instrument. 2. If the recoveries from the reanalyses of IQC are not within ±5% of stated value, prepare new stock solution and new calibration std. and repeat calibration. Method Quality control sample: Carry IQC through the entire sample prep. procedure. 1. Analyze with every run. 2. If not within ±5% of stated value, loss of analyte or contamination during sample prep. procedure may have occurred.	Verify initial calibration with an independent standard (ICV). 1. If not within ±10% of stated value, terminate analysis, correct problem, recalibrate instrument and verify calibration. Aqueous and solid Laboratory Control Samples (LCS) shall be prepared and analyzed for every analyte for every batch of aqueous or solid samples digested in a Sample Delivery Group (SDG) or per sample process batch, whichever is more frequent. Aqueous samples: A distilled ICV is used as the LCS. Limits: 85-115% Solid samples: Use certified material with supplied limits. If LCS is not within the limits specified, terminate analysis, correct problem, and redigest and reanalyze all samples associated with that LCS.
Method Detection Limit	IDLs listed in Table 1 (see page A-106). MDLs vary by matrix.	IDLs listed in Table 1 (see page A-17). Determine MDLs annually. MDLs listed in Table.	IDLs listed in Table 3120:I (see page A-84).	CRDLs listed in Exhibit C (see page A-15).
Standard Solution Expiration (3)	Not specified.	Replace stock solutions when dilutions for calibration standards cannot be verified.	Not specified. Monitor calibration standards weekly by analyzing IQCS.	Not specified. Initially verify with QC standard and monitor weekly thereafter.
Initial Calibration	Per instrument manufacturer's specifications. Minimum of a blank and one standard.	Per instrument manufacturer's specifications. Minimum of a blank and one standard.	Per instrument manufacturer's specifications.	Per instrument manufacturer's specifications. Minimum of a blank and one standard.

Notes:

(3) Indicates maximum usage time. If comparisons to QC check standard indicate a problem, more frequent preparation may be necessary.

Inorganic

Determination of Trace Metals by ICP (Continued)

Method ⇒ / Parameter ⇓	SW-846 Method 6010B	EPA Series Method 200.7	Standard Methods Method 3120	CLP (Inorganic SOW) Method 200.7 CLP-M
Continuing Calibration	Analyze a Continuing Calibration Verification (CCV) Standard (mid-range check standard) at the beginning, end and after every 10 samples. 1. If not within ±10% of stated value, discontinue analysis, correct problem, recalibrate instrument, verify calibration, and reanalyze all samples since last compliant CCV. 2. The ICV may be used in lieu of the CCV.	Instrument performance check (IPC) solution. 1. Analyze a mid-level calibration standard immediately after calibration, every 10 samples and at the end of the run. 2. The element concentrations are as follows: Silver at a maximum of 0.5 mg/L, potassium, phosphorous, and silica at a concentration of 10 mg/L, the rest of the elements at 2.0 mg/L. 3. For initial analysis, %R = 95-105. 4. For subsequent analyses of standard, %R = 90-110. 5. If not within specified limits, reanalyze IPC. 6. If second analysis of IPC is not within specified limits, discontinue analysis, determine the cause and/or in the case of drift recalibrate instrument. Reanalyze all samples since last compliant IPC.	Analyze 2 mg/L Instrument Check Standard immediately after calibration and after every 10 samples. If not within ±5% of stated value, stop analysis, correct problem and recalibrate instrument.	Analyze a Continuing calibration standard (CCV) solution (mid-range check standard) at the beginning, end and after every 10 samples. 1. If not within ±10% of stated value, discontinue analysis, correct problem, recalibrate instrument, verify calibration, and reanalyze all samples since last compliant CCV.
Surrogate Standards	Not applicable.	Not applicable.	Not applicable.	Not applicable.
Internal Standards	Internal standards can be used as an alternative to MSA to correct for matrix interferences.	Internal standards can be used as an alternative to MSA to correct for matrix interferences.	Not applicable.	Not applicable.
Accuracy/ Precision	One MS per 20 samples or each batch of samples, whichever is more frequent. %R = 75-125. One duplicate per matrix batch. %RPD < 20% for each sample value > 10x IDL.	Spike and analyze one sample out of every 10 before sample digestion (Laboratory fortified sample matrix). 1. For water samples, the added analyte concentration should be the same as that used in the LFB. 2. %R = 70-130. Laboratory duplicates. Frequency of analysis and %RPD not specified.	One MS per 10 sample. %R = 85-115 One duplicate per 20 samples. %RPD not specified.	One MS per sample delivery group or per matrix type. If recovery not within 75-125%, analyze a post digestion spike unless the sample result exceeds the spike concentration by a factor of 4. One duplicate per SDG or per matrix type. Use control limit of 20% for %RPD if the sample and duplicate are both greater than 5x the CRDL. Else, use a control limit of (±) the CRDL.

Inorganic

GENIUM PUBLISHING CORPORATION

Determination of Trace Metals by ICP (Continued)

Method ⇒ Parameter ⇓	SW-846 Method 6010B	EPA Series Method 200.7	Standard Methods Method 3120	CLP (Inorganic SOW) Method 200.7 CLP-M
Blanks	Process and analyze one method blank per sample batch. Analyze calibration blank after each ICV and CCV. 1. If the results > ±3x IDL, then terminate the analysis, correct the problem, recalibrate, and reanalyze the previous 10 samples. Unless, the results are < 1/10th the concentration of the action level of interest, and no sample is within 10% of the action limit.	A field reagent blank (FRB) consisting of reagent water placed in a sample container in the laboratory and treated as a sample in all respects. Frequency of preparation and %R not specified. A Laboratory reagent blank (LRB) is carried through the entire sample preparation scheme with each batch of 20 or fewer samples of the same matrix. 1. When LRB values constitute 10% or more of the analyte level or is 2.2x the analyte MDL, whichever is greater, fresh aliquots of the sample must be prepared and analyzed again for the affected analytes after the source of the contamination has been corrected. A Calibration blank (CCB) is to be analyzed after each IPC solution. 1. Value(s) of the CCB should be < IDL. but > 3σ control limit of the calibration blank. 2. If not within specified limits, reanalyze calibration blank. 3. If second analysis of calibration blank is not within specified limits, discontinue analysis, determine the cause and/or in the case of drift recalibrate instrument. Reanalyze all samples since last compliant calibration blank.	Carry Method Blank through entire sample preparation procedure and analyze after initial analysis of calibration blank. 1. If blank values for analytes are greater than MDL, repeat sample preparation and analysis with cleaner reagents or glassware. Analyze reagent blank with new reagents and every 20 samples. Analyze calibration blank at the beginning of the run and between each sample or standard. If carryover is observed, repeat rinsing until blank values are obtained.	Process and analyze one preparation blank per sample delivery group or per sample digestion batch, whichever is more frequent. 1. If the analyte conc. >CRDL, the lowest conc. of that analyte in the associated samples must be ≥ 10x the blank conc., else all samples associated with the blank with that analyte conc. < 10x the blank conc. and >CRDL, shall be redigested and reanalyzed for that analyte. 2. If the analyte conc. is below the negative CRDL, then all samples < 10x CRDL associated with that batch shall be redigested and reanalyzed for that analyte. Analyze calibration blanks after ICV and CCV solutions. If absolute value of blank for analyte >CRDL, terminate analysis, correct problem, recalibrate instrument, verify calibration and reanalyze all samples since last compliant CCB.
Preservation/ Storage Conditions	Aqueous: pH < 2 with HNO₃ Solid: None.	pH < 2 with HNO₃	pH 2< with HNO₃	Aqueous: pH < 2 with HNO₃ Solid: Store at 4 °C.
Holding Time (4)	6 months	6 months	6 months	180 days from sample receipt.
Field Sample Amount Required (5)	600 ml (aqueous) 200 grams (solid) glass or polyethylene container	1 liter (aqueous) 200 grams (solid) glass or polyethylene container	500 ml glass or polyethylene container	Aqueous: Not specified Solid: Not specified Glass or polyethylene container

Notes:
(4) Unless otherwise indicated, holding times are from the date of sample collection.
(5) Approximate volumes to be gathered for analysis. Additional volumes are required for the generation of QC data.

Determination of Trace Metals by ICP (Continued)

Method ⇒ Parameter ⇓	SW-846 Method 6010B	EPA Series Method 200.7	Standard Methods Method 3120	CLP (Inorganic SOW) Method 200.7 CLP-M
Amount for Extraction	100 ml (aqueous) 2 grams (solid)	100 ml (aqueous) 2 grams (solid)	100 ml	100 ml (aqueous) 1 grams (solid) Microwave digestion: 45 ml (aqueous) 0.5 grams (solid)
Other Criteria (Method Specific)	An Interference Check Solution containing known concentrations of interfering elements and the elements of interest at a concentration between 0.5 and 1 mg/L is analyzed at the beginning of the run. %R = 80-120. Test each new matrix for interference: 1. Perform 1:5 dilution on sample containing analytes >10x IDL. Results of dilution should agree within ±10% of original measurement. 2. Perform post-digestion spike at 10 to 100x IDL. %R = 75-125. Use MSA if an interference is suspected or a new matrix is encountered.	Analyze SICS (Spectral Interference Check Solution) on a daily basis. 1. Concentration of SICS solutions are 300 mg/L Fe, 200 mg/L Al, 50 mg/L Ba, Be, Cd, Ce, Co, Cr, Cu, Mn, Mo, Ni, Sn, SiO_2, Ti, Tl, and V. 2. Apparent analyte concentration range = (conc. of interfering element times correction factor)/10. 3. If apparent analyte concentration is outside this range, correction factor must be updated. Test each new matrix for interference: 1. Perform 1:4 dilution on sample. If analyte concentration is >50x IDL., results of dilution should agree within ±10% of original measurement. 2. Perform post-digestion spike to produce a level at least 20x MDL to a maximum 100x MDL. %R = 75-125. If recoveries not within specified limits, use MSA.	Correct for spectral interferences by using computer software supplied by instrument manufacturer or use manual method based on interference correction factors. Test for matrix interference: For new or unusual samples. 1. If element in sample is > 1 mg/L, dilute sample 1:10 and analyze. 2. Alternately, or if the element concentration is < 1 mg/L, use a post digestion spike equal to 1 mg/L. If results of the serial dilution are not within ±5% or the %R for the analytical spike is not within 95-105%, complete the analysis by diluting the sample to eliminate the matrix effect or use MSA. If nonspectral interference correction is necessary, use MSA.	Analyze ICSA, ICSAB and CRI solutions at a frequency of not greater than 20 analytical samples. Must be followed immediately by CCV\CCB pair. 1. ICSA: For target analytes with CRDL ≤ 10 µg/L, results should fall within ± 2x CRDL of the analyte's true value (true value = 0 unless otherwise stated), otherwise use alternate method to quantify results for affected analytes. 2. CSAB: If results are not within ±20% of stated value, terminate analysis, correct problem, recalibrate instrument, and re-analyze all samples since last compliant ICS 3. To verify linearity near the CRDL, analyze CRI standard at 2x CRDL or 2x IDL, whichever is greater. No acceptance criteria. 4. One serial dilution per sample delivery group or per matrix type, whichever is more frequent. If analyte conc. > 50x the IDL in original sample for that analyte, the serial dilution (5x dilution) must agree within 10% of the original determination for that analyte.

Inorganic

Determination of Trace Metals by ICP

Chemical Name	SW-846 Method 6010B	EPA Series Method 200.7	Std. Methods Method 3120	CLP (Inorganic SOW) Method 200.7 CLP-M
Aluminum	•	•	•	•
Antimony	•	•	•	•
Arsenic	•	•	•	•
Barium	•	•	•	•
Beryllium	•	•	•	•
Boron		•	•	
Cadmium	•	•	•	•
Calcium	•	•	•	•
Cerium		•		
Chromium	•	•	•	•
Cobalt	•	•	•	•
Copper	•	•	•	•
Iron	•	•	•	•
Lead	•	•	•	•
Lithium	•	•	•	
Magnesium	•	•	•	•
Manganese	•	•	•	•
Mercury		•		
Molybdenum	•	•	•	
Nickel	•	•	•	•
Phosphorus	•	•		
Potassium	•	•	•	•
Selenium	•	•	•	•
Silica (SiO$_2$)		•	•	
Silver	•	•	•	•
Sodium	•	•	•	•
Strontium	•	•	•	
Thallium	•	•	•	•
Tin		•		
Titanium		•		
Vanadium	•	•	•	•
Zinc	•	•	•	•

Determination of Trace Metals by ICP-MS

Method ⇒ Parameter ⇓	SW-846 Method 6020	EPA Series Method 200.8
Applicability	Groundwater, aqueous samples, industrial wastes, soils, sludges, sediments and other solid wastes.	Groundwater, drinking water, surface water wastewater, sludges and soils.
Number of Analytes (1)	15 metals listed. Additional elements can be determined based upon precision and accuracy data.	21 total
Method Validation (2)	Initial demonstration of performance 1. Determine Method Detection Limit (MDL) by analyzing at least three portions of a solution at a concentration near 3 to 5 times the estimated MDL. 2. Determine Instrument Detection Limits(IDL) by calculating the average of the standard deviations of the three runs on three non-consecutive days from the analysis of a reagent blank with seven consecutive measurements per day. IDLs are determined quarterly.	Initial demonstration of performance: 1. Determination of the Linear Calibration Range. Verify linear calibration range limit by analyzing a minimum of three different concentration standards. Results must be within 10% of the true value. 2. A Quality Control sample (QCS), an independent standard, is prepared and analyzed three times at least quarterly to verify the calibration standards and instrument performance. The concentration of the analytes should be ≤ 100 µg/L except selenium at a concentration < 500 µg/L, and mercury at a concentration ≤ 5 µg/L. If the mean value of the three analyses is not within ± 10% of stated value or within the limits listed in Table 8 (see page A-24), reanalyze QCS. If the second analysis of the QCS is not within the specified limits, determine source of problem and correct before continuing with analyses. 3. Determine MDLs by analyzing seven replicates of Laboratory fortified blanks. MDLs must be determined annually.
QC Check Standards/ Samples	Verify each element calibration with an Instrument Calibration Verification (ICV) solution prepared from a source different than the calibration standards. If not within ±10% of stated value, stop analysis, correct problem, recalibrate instrument, verify calibration and reanalyze all samples since last compliant CCV. A Laboratory control sample (LCS) is analyzed for each analyte using the same sample preparation methods used for samples at a frequency of one per 20 samples or batch. No criteria specified.	Prepare and analyze a Laboratory Fortified Blank (LFB) with each batch of samples by fortifying laboratory reagent water with multielement stock standards. The final concentration of the spike added should range from 40 µg/L to 100 µg/L except for selenium and mercury. For selenium, the concentration should range from 200 µg/L to 500 µg/L. For mercury, the concentration should range from 2 µg/L to 5 µg/L. %R =85-115, if %R for analyte is outside control limits, correct problem before continuing analyses.
Method Detection Limit	IDLs, sensitivities, and linear ranges will vary with the matrices, instrumentation, and operating conditions.	Estimated IDLs listed in Table 1 (see page A-18). MDLs listed in Table 7 (see page A-23).
Standard Solution Expiration (3)	Not specified. Calibration standards are initially verified with QC standard and monitored weekly thereafter. Working ICS solutions are prepared weekly.	Expiration date for stock solutions not specified. Calibration standards should be prepared every two weeks or as needed and verified initially with QCS and monitored weekly thereafter.

Notes:

(1) Analyte lists may vary among methods: a smaller list in one method is not necessarily a subset of a larger list in another method.

(2) Initial, one-time demonstration of ability to generate acceptable accuracy and precision.

(3) Indicates maximum usage time. If comparisons to QC check standard indicate a problem, more frequent preparation may be necessary.

Inorganic

Determination of Trace Metals by ICP-MS (Continued)

Method ⇒ / Parameter ⇓	SW-846 Method 6020	EPA Series Method 200.8
Precalibration Routine	Analyze tuning solution (10 µg/L Li, Co, In, Tl) 4 times. RSD of absolute signals ≤ 5%. Analyze tuning solution and conduct mass calibration and resolution checks in the mass regions of interest. Adjust mass calibration if it has shifted by more than 0.1 AMU from the true value, mass calibration must be adjusted to the true value. The resolution must be verified to be less than 0.9 AMU full width at 10% peak height.	Analyze tuning solution (100 µg/L Be, Mg, Co, In, Pb) 5 times. RSD of absolute signals < 5%. Resolution at low mass is indicated by magnesium isotopes 24, 25, 26. Resolution at high mass is indicated by lead isotopes, 206, 207, 208. Adjust spectrometer resolution to produce a peak width of 0.75 AMU at 5% peak height. Adjust mass calibration if it has shifted by more than 0.1 AMU from unit mass
Initial Calibration	Per instrument manufacturer's specifications. Minimum of a blank and one standard for calibration. Use average of at least three integrations for calibration and sample analyses.	Per instrument manufacturer's specifications. Minimum of a blank and one standard for calibration. Use average of at least three integrations for calibration and sample analyses. Concentration of standards should range from 10 µg/L to 200 µg/L, except for mercury ≤ 5 µg/L, and selenium 50 µg/L to 500 µg/L.
Continuing Calibration	Analyze the ICV after calibrating instrument, after every 10 samples and after the last sample. If not within ±10% of stated value, stop analysis, correct problem, recalibrate instrument, verify calibration and reanalyze all samples since last compliant CCV.	Analyze calibration standards after calibrating instrument, after every 10 samples and at the end of the analytical run. If the recovery is not within ±10% of the calibration, recalibrate and continue analysis. If the recovery is not within ± 15% of the calibration, recalibrate and reanalyze all samples since last compliant continuing calibration standard.
Internal Standards	Recommended internal standards are ^{6}Li, ^{45}Sc, ^{89}Y, ^{103}Rh, ^{115}In, ^{159}Tb, ^{169}Ho, and ^{209}Bi. Concentrations not specified. The intensity levels of the internal standard of the calibration blanks and instrument check standards must agree within ±20% of the intensity level of the internal standard of the original calibration solution, otherwise stop the analysis, correct the problem, calibrate, and reanalyze the affected samples. If intensity levels of the internal standard for samples is not between 30 and 120% of the intensity level of the internal standard in the initial calibration standard, dilute fresh aliquot of sample 5x, add internal standard, and reanalyze. Repeat procedure until sample internal standard intensities fall within the prescribed window.	Recommended internal standards are Sc, Y, In, Te, and Bi. Concentration range of 20 µg/L to 200 µg/L. See Table 3 (see page A-20) for the mass numbers and limitations of use. If mercury is to be determined, add an aliquot of gold to provide a concentration of 100 µg/L in the final dilution of all blanks, standards, and samples. The absolute response of any internal standard must be within 60-125% of the original response in the calibration blank, otherwise flush instrument with rinse blank and monitor response of calibration blank. If response for calibration blank is within limits, dilute a fresh aliquot of sample 2x, add internal standard, and reanalyze. If response for calibration blank is not within limits, terminate analysis and correct problem.

Inorganic

Determination of Trace Metals by ICP-MS (Continued)

Method ⇒ Parameter ⇓	SW-846 Method 6020	EPA Series Method 200.8
Accuracy/ Precision	One MS/MSD, for each batch of samples No criteria specified. Analyze one duplicate sample for each matrix every 20 samples or analytical batch, whichever is more frequent. %RPD is < 20 for analytes values > 100x. If %RPD is not within specified limits, correct problem and reanalyze samples during out-of-control conditions.	Spike one sample (Laboratory Fortified Matrix) out of every 10 prior to sample digestion and analyze. For water samples, the added analyte concentration should be the same as that used in the LFB. %R = 70-130 %R is not calculated if analyte concentration of analyte added is < 30% of background concentration. If %R for analyte falls outside control limits and system is in control, notify user that analyte is suspect due to sample related effect. Laboratory duplicates. Frequency of analysis and %RPD not specified.
Blanks	One method blank per batch of samples processed at the same time. No criteria specified. Analyze calibration blank after each QCS and continuing calibration standard. 1. If the results > ±3x IDL, stop the analysis, correct the problem, and have the affected samples reanalyzed.	A calibration blank is analyzed after every continuing calibration standard. If the recovery is not within ±10% of the calibration, recalibrate and continue analysis. If the recovery is not within ± 15% of the calibration, recalibrate and reanalyze all samples since last compliant continuing calibration blank. A Field Reagent Blank (FRB) consisting of reagent water placed in a sample container in the laboratory and treated as a sample in all respects. Frequency of preparation and criteria not specified. A Laboratory reagent blank (LRB) is carried through the entire sample preparation scheme with each batch of 20 or fewer samples of the same matrix. 1. When LRB values constitute 10% or more of the analyte level or is 2.2x the analyte MDL, whichever is greater, fresh aliquots of the sample must be prepared and analyzed again for the affected analytes after the source of the contamination has been corrected.
Preservation/ Storage Conditions	Aqueous: pH < 2 with HNO_3 Solid: Store at 4 °C.	pH < 2 with HNO_3
Holding Time (4)	6 months	6 months
Field Sample Amount Required (5)	1 liter (aqueous) 200 grams (solid) Polyethylene or fluorocarbon container	1 liter (aqueous) 200 grams (solid) glass or polyethylene container
Amount for Extraction	100 ml (aqueous) 2 grams (solid)	100 ml (aqueous) 2 grams (solid)

Notes:

(4) Unless otherwise indicated, holding times are from the date of sample collection.

(5) Approximate volumes to be gathered for analysis. Additional volumes are required for the generation of QC data.

Inorganic

Determination of Trace Metals by ICP-MS (Continued)

Method ⇒ Parameter ⇓	SW-846 Method 6020	EPA Series Method 200.8
Other Criteria (Method Specific)	Analyze Interference Check Solution (ICS) containing known concentrations of interfering elements and the elements of interest to verify correction factors at the beginning of the run or every 12 hours, whichever is more frequent. Refer to Table 2 (see page A-107) for the recommended components and concentrations for solution A and AB. Test each new matrix for interference: 1. Dilution test: Perform 1+4 dilution on sample containing analytes >100x the reagent blank. Results of dilution should agree within ±10% of original measurement, if not, interference must be suspected. One per 20 samples per matrix per batch. 2. Post - digestion spike: Spike level based on the indigenous element concentration. If the recovery is not within 85-115%, perform Dilution test or MSA may be used to compensate for this effect. Results of dilution should agree within ±10% of original sample concentration. Sample calculations should include interference corrections, internal-standard normalization, and the summation of signals at 206, 207, and 208 m/z for lead. Dilute and reanalyze samples that are more concentrated than the linear range for analyte or measure an alternate less-abundant isotope.	All masses which might affect data quality must be monitored. See Table 4 (see page A-20) for recommended analytical isotopes and additional masses. Isobaric polyatomic ion interferences must be recognized and appropriate corrections made to the data. See Table 2 (see page A-19) for the more common interferences and the elements affected. Elemental equations for sample data calculations are listed in Table 5 (see page A-21). Dilute and reanalyze samples that are more concentrated than the linear range or measure an alternate less-abundant isotope.

Inorganic

Determination of Trace Metals by ICP-MS

Chemical Name	SW-846 Method 6020	EPA Series Method 200.8
Aluminum	•	•
Antimony	•	•
Arsenic	•	•
Barium	•	•
Beryllium	•	•
Cadmium	•	•
Chromium	•	•
Cobalt	•	•
Copper	•	•
Lead	•	•
Manganese	•	•
Mercury		•
Molybdenum		•
Nickel	•	•
Selenium		•
Silver	•	•
Thallium	•	•
Thorium		
Uranium		•
Vanadium		•
Zinc	•	•

Inorganic

Sampling Procedures and Preparation

This section provides a tabulation of environmental sample parameters including method references, sample amounts required for analysis, container specifications, sample preservation requirements, and extraction and analysis hold times. This information is useful in maintaining sample integrity as well as compliance with correct sampling procedures as specified in EPA regulation.

Also provided in this section is a summary of sample extraction and sample cleanup methods approved for various sample matrix types for each of the analytical methods listed. The user should note that for EPA SW-846 the extraction and clean-up methods are cited with their own method numbers.

Sampling

Method Reference and Sampling

PARAMETERS	SOLID SAMPLES Soil, Sediment, Sludge, Solid, Concentrated Liquids						
	METHOD REFERENCE	METHOD	MINIMUM SAMPLE AMOUNT FOR ANALYSIS	CONTAINER TYPE	PRESER-VATION	EXTRACTION HOLDING TIME	ANALYSIS HOLDING TIME
Acidity							
Acidity							
Acrolein & Acrylonitrile by GC							
Acetonitrile by GC/NPD							
Acrylonitrile by GC							
Acrylamide by GC							
Acrylamide, Acrolein & Acrylonitrile by HPLC							
Alkalinity							
Alkalinity							
Aniline & Selected Derivatives by GC/NPD	EPA SW846	8131	30 g	8 oz CWM	1	14 Days	40 Days
Bendiocarb by HPLC							
Bensulide by HPLC							
Bentazon by HPLC							
Benzidines by HPLC							
Biochemical Oxygen Demand (BOD)							
Biochemical Oxygen Demand (BOD)							
Bis(2-chloroethyl) Ether & Hydrolysis Products							
Biphenyl & Ortho-Phenylphenol by HPLC							
Boron							
Bromide							
Bromide							
Bromide by Ion Selective Electrode	EPA SW846	9211	50 g	4 oz CWM	1		28 Days
Benomyl & Carbendazim by HPLC							
Bromoxynil by HPLC							
Carbon (organic) TOC							
Carbon (organic) TOC							
Carbon (organic) TOC							

GENIUM PUBLISHING CORPORATION

Guidelines for Environmental Samples

METHOD REFERENCE	METHOD	MINIMUM SAMPLE AMOUNT FOR ANALYSIS	CONTAINER TYPE	PRESER-VATION	EXTRACTION HOLDING TIME	ANALYSIS HOLDING TIME	PARAMETERS
AQUEOUS SAMPLES Drinking, Surface, Ground, Saline, Domestic and Industrial Waste Water							
EPA WW	305.1, 305.2	100 mL	250 mL HDPE or Glass	1		14 Days	Acidity
SM 19th Edition	2310 A-B	100 mL	250 mL HDPE or Glass	1		14 Days	Acidity
40 CFR Pt. 136 App. A	603	5 mL	2 x 40 mL VOA Vial	8, pH 4-5, 1		14 Days	Acrolein & Acrylonitrile by GC
EPA SW846	8033	40 mL	2 x 40 mL VOA Vial	1, pH 4-5		14 Days	Acetonitrile by GC/NPD
EPA SW846	8031	40 mL	2 x 40 mL VOA Vial	1, pH 4-5		14 Days	Acrylonitrile by GC
EPA SW846	8032A	40 mL	2 x 40 mL VOA Vial	1, pH 4-5		14 Days	Acrylamide by GC
EPA SW846	8316	40 mL	2 x 40 mL VOA Vial	1		14 Days	Acrylamide, Acrolein & Acrylonitrile by HPLC
EPA WW	310.1, 310.2	100 mL	250 mL HDPE or Glass	1		14 Days	Alkalinity
SM 19th Edition	2320	200 mL	250 mL HDPE or Glass	1		14 Days	Alkalinity
EPA SW846	8131	1 L	1 L Amber Glass	8, 1	7 Days	14 Days	Aniline & Selected Derivatives by GC/NPD
40 CFR Pt. 136 App.A	639	1 L	1 L Amber Glass	1, pH 6-8	7 Days	40 Days	Bendiocarb by HPLC
40 CFR Pt. 136 App.A	636	1 L	1 L Amber Glass	1, pH 6-8	7 Days	40 Days	Bensulide by HPLC
40 CFR Pt. 136 App. A	643	1 L	1 L Amber Glass	1,8, pH 6-8	7 Days	30 Days	Betazon by HPLC
40 CFR Pt. 136 App.A	605	1 L	2 L Glass	8, pH~4, 1, Dark	7 Days	Analyze Immediately/7 Days under N$_2$	Benzidines by HPLC
EPA WW	405.1	1 L	1 L HDPE or Glass	1		48 Hours	Biochemical Oxygen Demand (BOD)
SM 19th Edition	5210 A-B	1 L	1 L HDPE or Glass	1		48 Hours	Biochemical Oxygen Demand (BOD)
40 CFR Pt. 136 App.A	642	1 L	1 L Amber Glass	1		A.S.A.P.	Biphenyl & Ortho-Phenylphenol by HPLC
EPA SW846	8430	1 L	1 L Amber Glass	1		14 Days	Bis(2-chloroethyl) Ether & Hydrolysis Products
SM 19th Edition	4500-B	100 mL	250 mL HDPE	None Required		6 Month	Boron
EPA WW	320.1	100 mL	250 mL HDPE or Glass	None Required		28 Days	Bromide
SM 19th Edition	4500-Br- A-C	100 mL	250 mL HDPE	None Required		28 Days	Bromide
EPA SW846	9211	100 mL	250 mL HDPE or Glass	1		28 Days	Bromide by Ion Selective Electrode
40 CFR Pt. 136 App.A	631	1 L	1 L Amber Glass	1	7 Days	40 Days	Benomyl & Carbendazim by HPLC
40 CFR Pt. 136 App.A	1661	10 mL	2x 40 mL VOA Vila	1,8, pH 3-7	7 Days	40 Days	Bromoxynil by HPLC
EPA SW846	9060	25 mL	60 mL HDPE	2 or 7, 1		28 Days	Carbon (organic) TOC
EPA WW	415.1, 415.2	25 mL	100 mL Glass	2, 1		28 Days	Carbon (organic) TOC
SM 19th Edition	5310 A-D	100 mL	125 mL Glass	1, 2		28 Days	Carbon (organic) TOC

Method Reference and Sampling

PARAMETERS	SOLID SAMPLES Soil, Sediment, Sludge, Solid, Concentrated Liquids						
	METHOD REFERENCE	METHOD	MINIMUM SAMPLE AMOUNT FOR ANALYSIS	CONTAINER TYPE	PRESER-VATION	EXTRACTION HOLDING TIME	ANALYSIS HOLDING TIME
Carbon Dioxide							
Carbonyl Compounds by HPLC	EPA SW846	8315A	25 g	8 oz CWM	1	3 Days	3 Days
Cation Exchange Capacity of Soil (Ammonium Acetate)	EPA SW846	9080	25 g	8 oz CWM		N.S.	N.S.
Cation Exchange Capacity of Soil (Sodium Acetate)	EPA SW846	9081	10 g	4 oz CWM		N.S.	N.S.
Chemical Oxygen Demand (COD)							
Chemical Oxygen Demand (COD)							
Chlordane by Immunoassay	EPA SW846	4041	Requires test kit for screening				N.S
Chloride							
Chloride							
Chloride							
Chloride by Ion Selective Electrode	EPA SW846	9212	50 g	4 oz CWM	1		28 Days
Chlorinated Herbicides by GC/ECD	EPA SW846	8151A	50 g	8 oz CWM	1	14 Days	40 Days
Chlorinated Herbicides by GC							
Chlorinated Herbicides							
Glyphosate Herbicides							
Phenoxy-Acid Herbicidees by GC	CFR 40, Pt. 136 App. A	1658	30 g	8 oz CWM	1	7 Days	40 Days
Chlorinated Hydrocarbons by GC	EPA SW846	8121	30 g	8 oz CWM	1	14 Days	40 Days
Chlorinated Hydrocarbons by GC							
Chlorine Dioxide							
Chlorine (Total Residual)							
Chlorine (Total Residual)							
Chlorine (Total in Petroleum Products)	EPA SW846	9075, 9076, 9077	1 g	N.S. Zero Headspace	1		N.S.
Chlorophyll							
Chromium VI	EPA SW846	7196A, 7197, 7198	2 g	4 oz CWM	1		24 Hours
Chromium VI							

Sampling

Guidelines for Environmental Samples

METHOD REFERENCE	METHOD	MINIMUM SAMPLE AMOUNT FOR ANALYSIS	CONTAINER TYPE	PRESER-VATION	EXTRACTION HOLDING TIME	ANALYSIS HOLDING TIME	PARAMETERS
AQUEOUS SAMPLES Drinking, Surface, Ground, Saline, Domestic and Industrial Waste Water							
SM 19th Edition	4500-CO$_2$ A-C	100 mL	250 mL HDPE or Glass	None Required		Analyze Immediately	Carbon Dioxide
EPA SW846	8315A	100 mL	500 mL Amber Glass	1	3 days	3 days	Carbonyl Compounds by HPLC
							Cation Exchange Capacity of Soil (Ammonium Acetate)
							Cation Exchange Capacity of Soil (Sodium Acetate)
EPA WW	410.1, 410.2, 410.3, 410.4	50 mL	1 L Glass	1, 10		28 Days	Chemical Oxygen Demand (COD)
SM 19th Edition	5220 A-D	100 mL	125 mL HDPE or Glass	1. 2		28 Days	Chemical Oxygen Demand (COD)
							Chlordane by Immunoassay
SM 19th Edition	4500-Cl- A-F	50 mL	125 mL HDPE	None Required		28 Days	Chloride
EPA WW	325.1, 325.2, 325.3	50 mL	125 mL HDPE or Glass	None Required		28 Days	Chloride
EPS SW846	9250, 9251, 9253	50 mL	125 mL HDPE or Glass	None Required		28 Days	Chloride
EPA SW846	9212	100 mL	250 mL HDPE or Glass	1		28 Days	Chloride by Ion Selective Electrode
EPA SW846	8151A	1 L	1 L Amber Glass	8, 1	7 Days	40 Days	Chlorinated Herbicides by GC/ECD
40 CFR Pt 136 App A	615	1 L	1 L Amber Glass	1	7 Days	40 Days	Chlorinated Herbicides by GC
SM 19th Edition	6640 A-B	40 mL	4 x 40 mL VOA Vial	8, 10, 1	14 Days	14 Days	Chlorinated Herbicides
SM 19th Edition	6651 A-B	500 mL	500 mL HDPE	8, 1 dark		14 Days	Glyphosate Herbicides
40 CFR Pt 136 App A	1658	1 L	1 L Amber Glass	1,8, pH 5-9	7 Days	40 Days	Phenoxy-Acid Herbicides, by GC
EPA SW846	8121	1 L	1 L Amber Glass	1	7 Days	40 Days	Chlorinated Hydrocarbons by GC
40 CFR Pt 136 App A	612	1 L	1 L Amber Glass	1	7 Days	40 Days	Chlorinated Hydrocarbons by GC
SM 19th Edition	4500-ClO$_2$ A-E	500 mL	1 L CWM or HDPE	None Required		Analyze Immediately	Chlorine Dioxide
EPA WW	330.1, 330.2, 330.3, 330.4, 330.5	200 mL	250 mL HDPE or Glass	None Required		Analyze Immediately	Chlorine (Total Residual)
SM 19th Edition	4500-CL A-I	500 mL	500 mL HDPE or Glass	None Required		Analyze Immediately	Chlorine (Total Residual)
							Chlorine (Total in Petroleum Products)
SM 19th Edition	10200 H	500 mL	1 L HDPE	Dark		30 Days	Chlorophyll
EPA SW846	7196A, 7197, 7198, 7199	100 mL	250 mL HDPE or Glass	1		24 Hours	Chromium VI
EPA WW	218.4, 218.5, 218.6	200 mL	250 mL HDPE or Glass	1		24 hours	Chromium VI

Sampling

Method Reference and Sampling

PARAMETERS	SOLID SAMPLES Soil, Sediment, Sludge, Solid, Concentrated Liquids						
	METHOD REFERENCE	METHOD	MINIMUM SAMPLE AMOUNT FOR ANALYSIS	CONTAINER TYPE	PRESER-VATION	EXTRACTION HOLDING TIME	ANALYSIS HOLDING TIME
Coliform, Fecal & Total							
Coliform, Fecal & Total							
Color							
Color							
Conductivity .							
Conductance, Specific							
Conductance							
Corrosivity	EPA SW846	7.2	10 g	N. S. Zero Headspace	Dark, 1		A.S.A.P.
Cyanide (Total)							
Cyanide (Total)							
Cyanide (Amenable to Chlorination)							
Cyanide (Amenable to Chlorination)							
Cyanide for Soil and Oil	EPA SW846	9013	50 g	4 oz CWM	1,4		14 Days
Cyanide (Amenable to Chlorination)	USEPA CLP	Exhibit D, Part E, Method 335.2	2 g	4 oz CWM	1		12 Days
Cyanizine by HPLC							
Dazomet by GC							
1,2-Dibromomethane & 1,2-Dibromo 3-Chloropropane by Microextraction/GC	EPA SW846	8011	5 g	4 oz CWM	1		14 days
1,2-Dibromomethane & 1,2-Dibromo 3-Chloropropane by Microextraction/GC							
2,4-Dichlorophenoxy-acetic Acid by Immunoassay	EPA SW846	4015	Requires test kit for screening				N.S
Dioxins & Furans by HRGC/HRMS	EPA SW846	8290	10 g	4 oz CWM	1	30 Days	45 Days
Dioxins & Furans by HRGC/LRMS	EPA SW846	8280A	30 g	8 oz CWM	1	30 Days	45 Days
DDT by Immunoassay	EPA SW846	4042	Requires test kit for screening				N.S
Diphenylamine by GC							
Fecal Streptococci							
Fluoride							

Guidelines for Environmental Samples

METHOD REFERENCE	METHOD	MINIMUM SAMPLE AMOUNT FOR ANALYSIS	CONTAINER TYPE	PRESER-VATION	EXTRACTION HOLDING TIME	ANALYSIS HOLDING TIME	PARAMETERS
			AQUEOUS SAMPLES Drinking, Surface, Ground, Saline, Domestic and Industrial Waste Water				
EPA SW846	9131, 9132	100 mL	125 mL HDPE	8, 1		6 Hours	Coliform, Fecal & Total
SM 19th Edition	9221, 9222	100 mL	125 mL HDPE	1		N.S.	Coliform, Fecal & Total
EPA WW	110.1, 110.2, 110.3	50 mL	100 mL HDPE or Glass	1		48 Hours	Color
SM 19th Edition	2120	500 mL	500 mL HDPE or Glass	1		48 Hours	Color
SM 19th Edition	2510 A-B	500 mL	500 mL HDPE or Glass	1		28 Days	Conductivity
EPA SW846	9050A	500 mL	1 L	1		28 Days	Conductance, Specific
EPA WW	120.1	100 mL	250 mL HDPE or Glass	1		28 Days	Conductance
							Corrosivity
EPA WW	335.2, 335.3, 335.4	500 mL	250 mL HDPE or Glass	4, 1, 16		14 Days	Cyanide (Total)
SM 19th Edition	4500-CN G-H	500 mL	1 L HDPE	4, 1, Dark		14 Days	Cyanide (Total)
EPA SW846	9010B, 9012A, 9014	500 mL	1 L HDPE or Glass	16 or 17, 4, 1	14 Days	28 Days	Cyanide (Amenable to Chlorination)
SM 19th Edition	4500-CN G-H	500 mL	1 L HDPE	Add 100mg $Na_2S_2O_3$/ L		14 Days	Cyanide (Amenable to Cholorinated)
							Cyanide for Soil and Oil
USEPA CLP	Exhibit D, Part E, Method 335.2	100 mL	250 mL HDPE or Glass	8, 4, 1		12 Days	Cyanide (Amenable to Chlorination)
40 CFR Pt. 136 App.A	629	1 L	1 L Amber Glass	1	7 Days	40 Days	Cyanizine by HPLC
40 CFR Pt. 136 App.A	1659	1 L	1 L Amber Glass	1,8, pH > 9	7 Days	40 Days	Dazomet by GC
EPA SW846	8011	5 mL	2 x 40 mL VOA Vial	8, 1		14 days	1,2-Dibromomethane & 1,2-Dibromo 3-Chloropropane by Microextraction/GC
SM 19th Edition	6231 A-C	5 mL	2 x 40 mL VOA Vial	1		28 Days	1,2-Dibromomethane & 1,2-Dibromo 3-Chloropropane by Microextraction/GC
EPA SW846	4015	Requires test kit for screening				N.S	2,4-Dichlorophenoxy-acetic Acid by Immunoassay
EPA SW846	8290	1 L	4 x 1 L Amber Glass	8, 1	30 Days	45 Days	Dioxins & Furans by HRGC/HRMS
EPA SW846	8280A	1 L	4 x 1 L Amber Glass	8, 1, pH 7-9	30 Days	45 Days	Dioxins & Furans by HRGC/LRMS
							DDT by Immunoassay
40 CFR Pt. 136 App.A	620	1 L	1 L Amber Glass	1, pH 6-8	N.S	N.S	Diphenylamine by GC
SM 19th Edition	9230	100 mL	125 mL HDPE	8, 1		6 Hours	Fecal Streptococci
SM 19th Edition	4500-F-A-F	300 mL	500 mL HDPE	None Required		6 Month	Fluoride

Sampling

PARAMETERS	SOLID SAMPLES Soil, Sediment, Sludge, Solid, Concentrated Liquids						
	METHOD REFERENCE	METHOD	MINIMUM SAMPLE AMOUNT FOR ANALYSIS	CONTAINER TYPE	PRESER-VATION	EXTRACTION HOLDING TIME	ANALYSIS HOLDING TIME
Fluoride							
Fluoride by Ion Selective Electrode	EPA SW846	9214	50 g	HDPE	1		28 Days
Halides in Soil (EOX) Extractable	EPA SW846	9023	5 g	250 mL Amber Glass	1		28 Days
Haloacetic Acids and Trichlorophenol							
Haloethers by GC	EPA SW846	8111	30 g	8 oz CWM	1	14 Days	40 Days
Haloethers by GC							
Halogenated & Aromatic Volatiles by PID/ELCD GC	EPA SW846	8021B	30 g	8 oz. CWM	1		14 Days
Hardness							
Hardness							
Hazardous Waste Ignitability							
Hazardous Waste Reactivity-Cyanide/ Sulfide	EPA SW846	Chapter 7 7.3, 3.2/ 7.3.4.2	10 g each	4 oz CWM	1, dark, Zero Headspace		
Hexahydro-1,3,5-Trinitro-1,3,5- Triazine (RDX) by Immunoassay	EPA SW846	4051	Requires test kit for screening				N.S.
Hydrogen Ion (pH)							
Iodide							
Iodide							
Ignitability	EPA SW846	1010, 1020A	N.S.	N.S.			N.S.
Ignitability for Solids	EPA SW846	1030	N.S.	N.S.	No Headspace		A.S.A.P.
Inorganic Anions by Ion Chromatography							
Liquid Release Test (LRT) Procedure	EPA SW846	9096		N.S.		1, Dark	3 Days
MBAS (Methylene Blue Active Substance)							
MBTS & TCMTB by LC							
Mercaptobenzothiazone by HPLC							
Mercury	EPA SW846	7470A, 7471A, 7472	2 g	4 oz CWM			28 Days
Mercury							

Sampling

Guidelines for Environmental Samples

METHOD REFERENCE	METHOD	MINIMUM SAMPLE AMOUNT FOR ANALYSIS	CONTAINER TYPE	PRESER-VATION	EXTRACTION HOLDING TIME	ANALYSIS HOLDING TIME	PARAMETERS
	**AQUEOUS SAMPLES** Drinking, Surface, Ground, Saline, Domestic and Industrial Waste Water						
EPA WW	340.1, 340.2, 340.3	300 mL	500 mL HDPE or Glass	None Required		28 Days	Fluoride
EPA SW846	9214	100 mL	250 mL HDPE or Glass	1		28 Days	Fluoride by Ion Selective Electrode
							Halides in Soil (EOX) Extractable
SM 19th Edition	6251 A-B	30 mL	4 x 40 mL VOA Vial	11, 1	9 Days	21 Days (at -11°C)	Haloacetic Acids and Trichlorophenol
EPA SW846	8111	1 L	1 L Amber Glass	8, 1	7 Days	40 Days	Haloethers by GC
40 CFR Pt. 136 App.A	611	1 L	1 L Amber Glass	8, 1	7 Days	40 Days	Haloethers by GC
EPA SW846	8021B	10 mL	2x 40 mL VOA Vial	8, 1, 10		14 Days	Halogenated & Aromatic Volatiles by PID/ELCD GC
EPA WW	130.1, 130.2	100 mL	250 mL HDPE or Glass	3, 1		6 Month	Hardness
SM 19th Edition	2340 A-C	100 mL	250 mL HDPE or Glass	3		6 Month	Hardness
EPA SW846	1010, 1020A	100 mL	500 mL B.R.	None Required		7 Days	Hazardous Waste Ignitability
EPA SW846	Chapter 7 7.3, 3.2/7.3.4.2	250 mL each	500 mL HDPE	1, Dark, Zero headspace		A.S.A.P.	Hazardous Waste Reactivity-Cyanide/ Sulfide
							Hexahydro1,3,5-Trinitro-1,3,5- Triazine (RDX) by Immunoassay
EPA WW	150.1, 150.2	25 mL	60 mL HDPE	None Required		Analyze Immediately	Hydrogen Ion (pH)
EPA WW	345.1	100 mL	250 mL HDPE or Glass	1		Analyze Immediately	Iodide
SM 19th Edition	4500-I- A-C	500 mL	500 mL HDPE or Glass	None Required		Analyze Immediately	Iodide
EPA SW846	1010, 1020A	2 mL	N.S.	N.S.		N.S.	Ignitability
							Ignitability for Solids
EPA SW846	9056	2-3 mL	N.S.	N.S.		N.S.	Inorganic Anions by Ion Chromatography
							Liquid Release Test (LRT) Procedure
EPA WW	425.1	250 mL	500 mL HDPE or Glass	1		48 Hours	MBAS (Methylene Blue Active Substance)
40 CFR Pt. 136 App. A	637	1 L	1 L Amber Glass	1, pH 6-8	7 Days	40 Days	MBTS & TCMTB by LC
40 CFR Pt. 136 App. A	640	1 L	1 L Amber Glass	1, pH 6-8	7 Days	40 Days	Mercaptobenzothiazole by LC
EPA SW846	7470A, 7471A, 7472	500 mL	1 L HDPE or Glass	3, 1		28 Days	Mercury
EPA WW	245.1, 245.2, 245.5	100 mL	250 mL HDPE or Glass	3		28 Days	Mercury

Sampling

Method Reference and Sampling

PARAMETERS	SOLID SAMPLES Soil, Sediment, Sludge, Solid, Concentrated Liquids						
	METHOD REFERENCE	METHOD	MINIMUM SAMPLE AMOUNT FOR ANALYSIS	CONTAINER TYPE	PRESER-VATION	EXTRACTION HOLDING TIME	ANALYSIS HOLDING TIME
Mercury by Cold Vapor Atomic Absorption Spectrometry	USEPA CLP	Exhibit D, Part D	2 g	4 oz CWM	1		26 Days
Metals (excl.Chromium VI & Mercury)	EPA SW846	7000A Series/6010B	2 g	4 oz CWM			6 Month
Metals by Flame Atomic Absorption Spectrometry/Furnace Technique							
Metals by Flame Atomic Absorption Spectrometry							
Metals by Cold Vapor Atomic Absorption Spectrometry							
Metals by Electrothermal Atomic Absorption Spectrometry							
Metals by Plasma Emission Spectroscopy							
Metals by Inductively Coupled Plasma							
Metals by Inductively Coupled Plasma	USEPA CLP	Exhibit D, Part A, ICP-AES	2 g	4 oz CWM	1		6 Month
Metals by Flame Atomic Absorption Methods, Furnace Technique	USEPA CLP	Exhibit D, Parts B-C	2 g	4 oz CWM	1		6 Month
Methane							
Metals by ICP/MS	EPA SW846	6020	2 g	4 oz CWM			6 Month
Metals by ICP/MS							
Nitrate by Ion Selective Electrode	EPA SW846	9211	50 g	4 oz CWM	1		N.S.
Nitrate							
Nitrogen (Organic							
Nitrogen, Nitrite							
Nitrogen, Nitrate							
Nitrogen, Nitrate - Nitrite							
Nitrogen, Nitrite							
Nitroaromatics & Cyclic Ketones by GC	EPA SW846	8091	30 g	8 oz CWM	1	14 Days	40 Days
Nitroaromatics & Isophorone							
Nitroaromatics & Nitroamine by HPLC	EPA SW846	8330	30 g	8 oz CWM	1	14 Days	40 Days
Nitrogen, Ammonia							

Guidelines for Environmental Samples

METHOD REFERENCE	METHOD	MINIMUM SAMPLE AMOUNT FOR ANALYSIS	CONTAINER TYPE	PRESER-VATION	EXTRACTION HOLDING TIME	ANALYSIS HOLDING TIME	PARAMETERS
AQUEOUS SAMPLES Drinking, Surface, Ground, Saline, Domestic and Industrial Waste Water							
USEPA CLP	Exhibit D, Part D	100 mL	250 mL HDPE or Glass	3		26 Days	Mercury by Cold Vapor Atomic Absorption Spectrometry
EPA SW846	7000A Series/ 6010B	100 mL	400 mL HDPE or Glass	3		6 Month	Metals (excl. Chromium VI & Mercury)
EPA WW	200 Series	200 mL	250 mL HDPE or Glass	3		6 Month	Metals by Flame Atomic Absorption Spectrometry/Furnace Technique
SM 19th Edition	3111A-E/ 3500	500 mL	1 L HDPE or Glass	3		6 Month	Metals by Flame Atomic Absorption Spectrometry
SM 19th Edition	3112A-B/ 3500	500 mL	1 L HDPE or Glass	3, 1		6 Month	Metals by Cold Vapor Atomic Absorption Spectrometry
SM 19th Edition	3113A-B/ 3500	500 mL	1 L HDPE or Glass	3		6 Month	Metals by Electrothermal Atomic Absorption Spectrometry
SM 19th Edition	6211 A-B	500 mL	1 L HDPE or Glass	3		6 Month	Metals by Plasma Emission Spectroscopy
EPA WW	200.7	100 mL	250 mL HDPE	3		6 Month	Metals by Inductively Coupled Plasma
USEPA CLP	Exhibit D, Part A, ICP-AES	100 mL	250 mL HDPE or Glass	3		6 Month	Metals by Inductively Coupled Plasma
USEPA CLP	Exhibit D, Parts B-C	100 mL	250 mL HDPE or Glass	3		6 Month	Metals by Flame Atomic Absorption Methods, Furnace Technique
SM 19th Edition	6211 A-B	5 mL	2 x 40 mL VOA Vial	1		14 Days	Methane
EPA SW846	6020	100 mL	250 mL HDPE or Glass			6 Month	Metals by ICP/MS
EPA WW	200.8	100 mL	250 mL HDPE or Glass			6 Month	Metals by ICP/MS
EPA SW846	9210	100 mL	250 mL HDPE or Glass	1,19		48 Hours	Nitrate by Ion Selective Electrode
SM 19th Edition	4500-NO3-A-H	100 mL	250 mL HDPE or Glass	Analyze Immedi-ately , or 1		48 Hours	Nitrate
SM 19th Edition	4500-Norg-A-D	500 mL	250 mL HDPE or Glass	1, 2		28 Days	Nitrogen (Organic)
SM 19th Edition	4500-NO2-A-C	100 mL	250 mL HDPE or Glass	1, 2		28 Days	Nitrogen, Nitrite
EPA WW	352.1	100 mL	250 mL HDPE or Glass	1, 2		48 Hours	Nitrogen, Nitrate
EPA WW	353.1, 353.2, 353.3	100 mL	250 mL HDPE or Glass	1, 2		28 Days	Nitrogen, Nitrate - Nitrite
EPA WW	354.1	50 mL	100 mL HDPE or Glass	1		48 Hours	Nitrogen, Nitrite
EPA SW846	8091	1 L	1 L Amber Glass	8, 1	7 Days	40 Days	Nitroaromatics & Cyclic Ketones by GC
40 CFR Pt. 136 App. A	609	1 L	1 L Amber Glass	1	7 Days	40 Days	Nitroaromatics & Isophorone
EPA SW846	8330	1 L	1 L Amber Glass	8, 1	7 Days	40 Days	Nitroaromatics & Nitroamine by HPLC
EPA WW	350.1, 350.2, 350.3	400 mL	500 mL HDPE or Glass	2,1		28 Days	Nitrogen, Ammonia

Sampling

PARAMETERS	SOLID SAMPLES Soil, Sediment, Sludge, Solid, Concentrated Liquids						
	METHOD REFERENCE	METHOD	MINIMUM SAMPLE AMOUNT FOR ANALYSIS	CONTAINER TYPE	PRESER-VATION	EXTRACTION HOLDING TIME	ANALYSIS HOLDING TIME
Nitrogen, Ammonia							
Nitrogen Organic & Kjeldahl							
Nitrogen, Kjeldahl, Total							
Nitroglycerin by HPLC	EPA SW846	8332	30 g	8 oz CWM		14 Days	40 Days
Nitrosamines by GC	EPA SW846	8070A	30 g	8 oz CWM	1	14 Days	40 Days
Nitrosamines by GC							
N-Methylcarbamates by HPLC	EPA SW846	8318	30 g	8 oz CWM	1	14 Days	40 Days
Nonhalogenated Volatile Organics by GC/FID	EPA SW846	8015B	5 g	4 oz CWM	1		14 Days
Non-Volatile Compounds by HPLC/TS/MS or UV (Solvent extractables)	EPA SW846	8321A	30 g	8 oz CWM	1	14 Days	40 Days
Non-Volatile Compounds by HPLC/PB/MS (Solvent extractables)							
NTA							
Odor							
Odor							
Oil & Grease	EPA SW846	9070, 9071A	20 g	4 oz CWM	1		28 Days
Oil & Grease							
Oil & Grease							
Oil & Grease							
Organophosphorus Compounds by GC	EPA SW846	8141A	30 g	8 oz CWM	1	7 Days	40 Days
Oryzaline by HPLC							
Oxygen, Dissolved							
Oxygen, Dissolved (Probe)							
Oxygen, Dissolved (Winkler)							
Ozone							
Paint Filter Liquid Test	EPA SW846	9095A	100 g	4 oz. CWM			N.S
Pesticides, Amine & Lethane by GC							
Pesticides, Carbamate & Urea by HPLC							

Guidelines for Environmental Samples

METHOD REFERENCE	METHOD	MINIMUM SAMPLE AMOUNT FOR ANALYSIS	CONTAINER TYPE	PRESER-VATION	EXTRACTION HOLDING TIME	ANALYSIS HOLDING TIME	PARAMETERS
AQUEOUS SAMPLES Drinking, Surface, Ground, Saline, Domestic and Industrial Waste Water							
SM 19th Edition	4500-NH₃ A-G	500 mL	500 mL HDPE or Glass	2, 1		28 Days	Nitrogen, Ammonia
SM 19th Edition	4500-N$_{org}$ A-C	500 mL	500 mL HDPE or Glass	2, 1		28 Days	Nitrogen Organic & Kjeldahl
EPA WW	351.1-351.4	500 mL	500 mL HDPE or Glass	2, 1		28 Days	Nitrogen, Kjeldahl, Total
EPA SW846	8332	40 mL	2 x 40 mL VOA Vial	1	7 Days	40 Days	Nitroglycerin by HPLC
EPA SW846	8070A	1 L	2.5 L A.J.	8, pH 7-10, 1, Dark	7 Days	40 Days	Nitrosamines by GC
40 CFR Pt. 136 App. A	607	1 L	1 L Amber Glass	8, pH 7-10, 1, Dark	7 Days	40 Days	Nitrosamines by GC
EPA SW846	8318	1 L	1 L Amber Glass	18, 1, pH 4-5	7 Days	40 Days	N-Methylcarbamates by HPLC
EPA SW846	8015B	10 mL	2 x 40 mL, VOA Vial	8, 1, pH<2, 10		14 Days	Nonhalogenated Volatile Organics by GC/FID
EPA SW846	8321A	1 L	1 L Amber Glass	8, 1	7 Days	40 Days	Non-Volatile Compounds by HPLC/TS/MS or UV (Solvent extractables)
EPA SW846	8325	1 L	1 L Amber Glass	8, 1	7 Days	40 Days	Non-Volatile Compounds by HPLC/PB/MS (Solvent extractables)
EPA WW	430.1, 430.2	50 mL	100 mL HDPE or Glass	1		24 Hours	NTA
EPA WW	140.1	200 mL	250 mL Glass	1		24 Hours	Odor
SM 19th Edition	2150 A-B	500 mL	500 mL Glass	1		Analyze ASAP	Odor
EPA SW846	9070, 9071A	1 L	1 L Glass	1, 10		28 Days	Oil & Grease
SM 19th Edition	5520 A-F	1 L	1 L CWM	1, 10		28 Days	Oil & Grease
EPA WW	413.1, 413.2	1 L	1 L Glass	1, 10		28 Days	Oil & Grease
EPA821-R-95-036	1664	1L	1L Glass	1 and 2 or 10		28 Days	Oil & Grease
EPA SW846	8141A	1 L	1 L Amber Glass	8, 1, pH~ 5-8	7 Days	40 Days	Organophosphorus Compounds by GC
40 CFR Pt. 136 App. A	638	1 L	1 L Amber Glass	1, pH 6-8	7 Days	40 Days	Oryzalin by HPLC
SM 19th Edition	4500-O A-G	300 mL	300 mL BOD Btl.	(1) None (2) No Iodine Demand 9, 20, 21 (3) Iodine Demand 22, 23, then 9, 20, 21		(1)Analyze Immediately (2) 8 Hours or less (3) 8 Hours or less	Oxygen, Dissolved
EPA WW	360.1	300 mL	300 mL BOD Btl.	None Required		Analyze Immediately	Oxygen, Dissolved (Probe)
EPA WW	360.2	300 mL	300 mL BOD Btl.	14, 9/Dark		8 Hours	Oxygen, Dissolved (Winkler)
SM 19th Edition	4500-O₃ A-B	1 L	1 L Glass	None Required		Analyze Immediately	Ozone
EPA SW846	9095A	100 mL	500 mL Glass			N.S.	Paint Filter Liquid Test
40 CFR Pt. 136 App. A	645	1 L	1 L Amber Glass	1, pH 6-8	7 Days	40 Days	Pesticides, Amine & Lethane by GC
40 CFR Pt. 136 App. A	632	1 L	1 L Amber Glass	1	7 Days	40 Days	Pesticides, Carbamate & Urea by HPLC

Sampling

PARAMETERS	SOLID SAMPLES Soil, Sediment, Sludge, Solid, Concentrated Liquids						
	METHOD REFERENCE	METHOD	MINIMUM SAMPLE AMOUNT FOR ANALYSIS	CONTAINER TYPE	PRESER-VATION	EXTRACTION HOLDING TIME	ANALYSIS HOLDING TIME
Pesticides, Carbamate & Amide by HPLC							
Pesticides containing Carbon, Hydrogen & Oxygen by GC							
Pesticides, Chlorinated by GC	EPA SW846	8081A	30 g	8 oz CWM	1	14 Days	40 Days
Pesticides, Chlorinated							
Pesticides, Chlorinated							
Pesticides, Chlorinated	USEPA-CLP	Exhibit D, PEST	30 g	8 oz CWM	1	10 Days	40 Days
Pesticides, Dinitroaniline by GC							
Pesticides, Dinitro Aromatic by GC							
Pesticides, Dithiocarbamate by UV							
Pesticides, Dithiocarbamate by GC							
Pesticides, Neutral Nitrogen containing by GC							
Pesticide Organophosporus by GC							
Pesticide Organophosporus by GC	40 CFR Pt. 136 App . A	1657	30 g	8 OZ cwm	1	7 Days	40 Days
Pesticides, Organohalide by GC	40 CFR Pt. 136 App. A	1656	30 g	8 oz CWM	1	7 Days	40 Days
Pesticide, Organohalide by GC							
Pesticides, Organonitrogen by							
Pesticides, Volatile by GC							
Pesticides, Thiocarbate by GC							
Pesticides, Thiophosphate by GC							
Pesticide, Triazine by GC							
Petroleum Hydrocarbons (TPH)							
Petroleum Hydrocarbon Total, Recoverable by Infrared Spectrophotometer	EPA SW846	8440	10 g	4 oz CWM	1	ASAP	ASAP
Petroleum Hydrocarbons by Immunoassay	EPA SW846	4030	Requires test kit for screening				N.S.
pH Electrometric measurement							
pH Electrometric measurement							

Guidelines for Environmental Samples

METHOD REFERENCE	METHOD	MINIMUM SAMPLE AMOUNT FOR ANALYSIS	CONTAINER TYPE	PRESER-VATION	EXTRACTION HOLDING TIME	ANALYSIS HOLDING TIME	PARAMETERS
AQUEOUS SAMPLES Drinking, Surface, Ground, Saline, Domestic and Industrial Waste Water							
40 CFR Pt. 136 App. A	632.1	1 L	1 L Amber Glass	1, 8, pH 2-4	7 Days	30 Days	Pesticides, Carbamate & Amide by HPLC
40 CFR Pt. 136 App. A	616	1 L	1 L Amber Glass	1, pH 2-4	7 Days	30 Days	Pesticides containing Carbon, Hydrogen & Oxygen by GC
EPA SW846	8081A	1 L	1 L Amber Glass	8, 1	7 Days	40 Days	Pesticides, Chlorinated by GC
40 CFR Pt. 136 App. A	608, 608.1, 608.2	1 L	1 L Amber Glass	8, pH 5-9, 1	7 Days	40 Days	Pesticides, Chlorinated
SM 19th Edition	6630 A-D	1 L	1 L Amber Glass	8, 1	7 Days	40 Days	Pesticides, Chlorinated
USEPA-CLP	Exhibit D, PEST	1 L	1 L Amber Glass	1	5 Days	40 Days	Pesticides, Chlorinated
40 CFR Pt. 136 App. A	627	1 L	1 L Amber Glass	1	7 Days	40 Days	Pesticides, Dinitroaniline by GC
40 CFR Pt. 136 App. A	646	1 L	1 L Amber Glass	1	30 Days	N.S.	Pesticides, Dinitro Aromatic by GC
40 CFR Pt. 136 App. A	630,	1 L	1 L Amber Glass	1		7 Days	Pesticides, Dithiocarbamate by UV
40 CFR Pt. 136 App. A	630.1	1 L	1 L Amber Glass	1, pH 12-13		N.S.	Pesticides, Dithiocarbamate by GC
40 CFR Pt. 136 App. A	633.1	1 L	1 L Amber Glass	1, pH 6-8	7 Days	40 Days	Pesticides, Neutral Nitrogen containing by GC
40 CFR Pt. 136 App. A	614, 614.1, 622	1 L	1 L Amber Glass	1, pH 6-8	7 Days	40 Days	Pesticides, Organophosphorus by GC
40 CFR Pt. 136 App. A	1657	1 L	1 L Amber Glass	1, pH 5-9	7 Days	40 Days	Pesticides, Organophosphorus by GC
40 CFR Pt. 136 App. A	1656	1 L	1 L Amber Glass	1, pH 5-9, 8	7 Days	40 Days	Pesticides, Organohalide by GC
40 CFR Pt. 136 App. A	617	1 L	1 L Amber Glass	1, pH 6-8	7 Days	40 Days	Pesticides, Organohalide by GC
40 CFR Pt. 136 App. A	633	1 L	1 L Amber Glass	1	7 Days	40 Days	Pesticides, Organonitrogen by GC
40 CFR Pt. 136 App. A	618	20 mL	2 x 40 mL VOA Vial	1		14 Days	Pesticides, Volatile by GC
40 CFR Pt. 136 App. A	634	1 L	1 L Amber Glass	1, pH 6-8	7 Days	40 Days	Pesticides, Thiocarbate by GC
40 CFR Pt. 136 App. A	622.1	1 L	1 L Amber Glass	1	7 Days	40 Days	Pesticides, Thiophosphate by GC
40 CFR Pt. 136 App. A	619	1 L	1 L Amber Glass	1	7 Days	40 Days	Pesticides, Triazine by GC
EPA WW	418.1	1 L	1 L Glass	1, 10		28 Days	Petroleum Hydrocarbons (TPH)
							Petroleum Hydrocarbon Total, Recoverable by Infrared Spectrophotometer
							Petroleum Hydrocarbons by Immunoassay
EPA SW846	9040B	25 mL	60 mL HDPE or Glass	None Required		Analyze Immediately	pH Electrometric measurement
EPA WW	150.1, 150.2	25 mL	60 mL HDPE or Glass	None Required		Analyze Immediately	pH Electrometric measurement

Sampling

Method Reference and Sampling

PARAMETERS	SOLID SAMPLES Soil, Sediment, Sludge, Solid, Concentrated Liquids						
	METHOD REFERENCE	METHOD	MINIMUM SAMPLE AMOUNT FOR ANALYSIS	CONTAINER TYPE	PRESER-VATION	EXTRACTION HOLDING TIME	ANALYSIS HOLDING TIME
pH Electrometric measurement							
pH Paper Method							
pH Soil and Waste	EPA SW846	9045C	20 g	4 oz. CWM			Analyze Immediately
Phenolics							
Phenolics							
Phenols by GC	EPA SW846	8041	30 g	8 oz CWM	1	14 Days	40 Days
Phenols by GC							
Phenols							
Phenols							
Pentachlorophenol Screening by Immunoassay	EPA SW846	4010A	Requires test kit for screening				N.S.
Phosphorus, Total							
Phosphorus, All Forms							
Phosphorus							
Phthalate Esters by GC/ECD	EPA SW846	8061A	30 g	8 oz CWM	1	14 Days	40 Days
Phthalate Esters by GC							
Picloram by HPLC							
Polychlorinated Biphenyls (PCB) by GC	EPA SW846	8082	30 g	8 oz CWM	1	14 Days	40 Days
Polychlorinated Biphenyls (PCB) by GC							
Polychlorinated Biphenyls (PCB) Organohalide by GC							
Polychlorinated Biphenyls (PCB)							
Polychlorinated Biphenyls (PCB)	USEPA-CLP	Exhibit D, PEST	30 g	8 oz CWM	1	10 Days	40 Days
Polychlorinated Biphenyls (PCB) by Immunoassay	EPA SW846	4020	Requires test kit for screening				N.S
Polychlorinated Biphenyls (PCB) in Soil by Screening Test Method	EPA SW846	9078	Requires test kit for screening				N.S.
Polychlorinated Biphenyls (PCB) in Oil by Screening Test Method	EPA SW846	9079	Requires test kit for screening				

Sampling

Guidelines for Environmental Samples

		AQUEOUS SAMPLES					
		Drinking, Surface, Ground, Saline, Domestic and Industrial Waste Water					PARAMETERS
METHOD REFERENCE	METHOD	MINIMUM SAMPLE AMOUNT FOR ANALYSIS	CONTAINER TYPE	PRESER-VATION	EXTRACTION HOLDING TIME	ANALYSIS HOLDING TIME	
SM 19th Edition	4500-H+ A-B	50 mL	100 mL HDPE or Glass	None Required		Analyze Immediately	pH Electrometric measurement
EPA SW846	9041A	25 mL	60 mL HDPE	None Required		N.S.	pH Paper Method
							pH Soil and Waste
EPA SW846	9065, 9066, 9067	500 mL	1 L B.R.	1, 2		28 Days	Phenolics
EPA WW	420.1, 420.2, 420.3, 420.4	500 mL	1 L Glass	1, 12, 13		24 Hours/ 28 Days	Phenolics
EPA SW846	8041	1 L	1 L Amber Glass	8, 1	7 Days	40 Days	Phenols by GC
40 CFR Pt. 136 App. A	604, 604.1	1 L	1 L Amber Glass	8, 1	7 Days	40 Days	Phenols by GC
SM 19th Edition	5530 A-D	500 mL	1 L HDPE or Glass	1, 2		28 Days	Phenols
SM 19th Edition	6420 A-C	1 L	1 L Amber Glass	8, 1	7 Days	40 Days	Phenols
EPA SW846	4010A	Requires test kit for screening				N.S	Pentachlorophenol Screening by Immunoassay
EPA WW	365.1, 365.4	50 mL	100 mL HDPE or Glass	1, 2		28 Days	Phosphorus, Total
EPA WW	365.1, 365.2, 365.3	50 mL	100 mL HDPE or Glass	Filter, 1, 2		48 Hours	Phosphorus, All Forms
SM 19th Edition	4500-P A-F	100 mL	125 mL Glass	Filter Imme-diately/1		48 Days	Phosphorus
EPA SW846	8061A	1 L	1 L Amber Glass	8, 1	7 Days	40 Days	Phthalate Esters by GC/ECD
40 CFR Pt. 136 App. A	606	1 L	1 L Amber Glass	1	7 Days	40 Days	Phthalate Esters by GC
40 CFR Pt. 136 App. A	644	1 L	1 L Amber Glass	1, pH 1-3	7 Days	30 Days	Picloram by HPLC
EPA SW846	8082	1 L	1 L Amber Glass	8, 1	7 Days	40 Days	Polychlorinated Biphenyls (PCB) by GC
40 CFR Pt. 136 App. A	608	1 L	1 L Amber Glass	8, pH 5-9, 1	7 Days	40 Days	Polychlorinated Biphenyls (PCB) by GC
40 CFR Pt. 136 App. A	617	1 L	1 L Amber Glass	8, pH 5-9, 1	7 Days	40 Days	Polychlorinated Biphenyls (PCB) Organohalide by GC
SM 19th Edition	6431 A-C 6630 A-D	1 L	1 L Amber Glass	8, 1	7 Days	40 Days	Polychlorinated Biphenyls (PCB)
USEPA-CLP	Exhibit D, PEST	1 L	1 L Amber Glass	1	5 Days	40 Days	Polychlorinated Biphenyls (PCB)
							Polychlorinated Biphenyls (PCB) by Immunoassay
							Polychlorinated Biphenyls (PCB) in Soil by Screening Test Method
							Polychlorinated Biphenyls (PCB) in Oil by Screening Test Method

Sampling

PARAMETERS	SOLID SAMPLES Soil, Sediment, Sludge, Solid, Concentrated Liquids						
	METHOD REFERENCE	METHOD	MINIMUM SAMPLE AMOUNT FOR ANALYSIS	CONTAINER TYPE	PRESER-VATION	EXTRACTION HOLDING TIME	ANALYSIS HOLDING TIME
Polynuclear Aromatic Hydrocarbons (PAH/PNA) by GC	EPA SW846	8100	30 g	8 oz CWM	1	14 Days	40 Days
Polynuclear Aromatic Hydrocarbons (PAH/PNA) by HPLC	EPA SW846	8310	30 g	8 oz CWM	1	14 Days	40 Days
Polynuclear Aromatic Hydrocarbons PAH by GC or HPLC							
Polynuclear Aromatic Hydrocarbons (PAH/PNA)							
Polynuclear Aromatic Hydrocarbons by Immunoassay	EPA SW846	4035	Requires test kit for screening				N.S
Purgeable Aromatic Hydrocarbons							
Purgeable Aromatic Hydrocarbons by GC							
Purgeable Halocarbons							
Purgeable Halocarbons by GC							
Purgeable Halides (PO$_X$)							
Purgeable Organics by GC/MS							
Pyrethines & Pyrethroids by HPLC							
Radioactive Cesium							
Radioactive Iodine							
Radioactive Radium							
Radioactive Strontium							
Radioactive Tritium							
Radioactive Uranium							
Radiological Test, Gross Alpha & Beta							
Radiological Test, Gross Alpha & Beta; (Total, Suspended & Dissolved)							
Radium, Total							
Residue, Filterable (TDS)							
Residue, Non-Filterable (TSS)							
Residue, Settleable Matter							
Residue, Total							

Sampling

Guidelines for Environmental Samples

METHOD REFERENCE	METHOD	MINIMUM SAMPLE AMOUNT FOR ANALYSIS	CONTAINER TYPE	PRESER-VATION	EXTRACTION HOLDING TIME	ANALYSIS HOLDING TIME	PARAMETERS
AQUEOUS SAMPLES Drinking, Surface, Ground, Saline, Domestic and Industrial Waste Water							
EPA SW846	8100	1 L	1 L Amber Glass	8, 1	7 Days	40 Days	Polynuclear Aromatic Hydrocarbons (PAH/PNA) by GC
EPA SW846	8310	1 L	1 L Amber Glass	8, 1	7 Days	40 Days	Polynuclear Aromatic Hydrocarbons (PAH/PNA) by HPLC
40 CFR Pt. 136 App. A	610	1 L	1 L Amber Glass	8, 1, Dark	7 Days	40 Days	Polynuclear Aromatic Hydrocarbons PAH by GC or HPLC
SM 19th Edition	6440 A-C	1 L	1 L Amber Glass	8, 1	7 Days	40 Days	Polynuclear Aromatic Hydrocarbons (PAH/PNA)
							Polynuclear Aromatic Hydrocarbons by Immunoassay
SM 19th Edition	6220 A-E	5 mL	2 x 40 mL VOA Vial	8, 1		14 Days	Purgeable Aromatic Hydrocarbons
40 CFR Pt. 136 App. A	602	5 mL	2 x 40 mL VOA Vial	8, 10, 1		14 Days	Purgeable Aromatic Hydrocarbons by GC
SM 19th Edition	6230 A-E	5 mL	2 x 40 mL VOA Vial	8, 1		14 Days	Purgeable Halocarbons
40 CFR Pt. 136 App. A	601	5 mL	2 x 40 mL VOA Vial	8, 1		14 Days	Purgeable Halocarbons by GC
EPA SW846	9021	5 mL	2 x 40 mL VOA Vial	1		14 Days	Purgeable Halides (PO$_X$)
40 CFR Pt. 136 App. A	624	5 mL	2 x 40 mL VOA Vial	8, 1, 10		14 Days	Purgeable Organics by GC/MS
40 CFR Pt. 136 App.A	1660	1 L	1 L Amber Glass	1,8, pH 5-7	7 Days	40 Days	Pyrethines & Pyrethroids by HPLC
SM 19th Edition	7500-Cs A-B	1 L	1 L HDPE or Glass	10		1 Year	Radioactive Cesium
SM 19th Edition	7500-I A-D	1 L - 2 L	2 x 1 L HDPE or Glass	None		14 Days	Radioactive Iodine
SM 19th Edition	7500-Ra A-D	1 L	1 L HDPE or Glass	3 or 10		1 Year	Radioactive Radium
SM 19th Edition	7500-Sr A-B	1 L	1 L HDPE or Glass	3 or 10		1 Year	Radioactive Strontium
SM 19th Edition	7500-^{3}H A-B	100 mL	500 mL Glass	None		1 Year	Radioactive Tritium
SM 19th Edition	7500-U A-B	1 L	1 L HDPE or Glass	3 or 10		1 Year	Radioactive Uranium
EPA SW846	9310	1 L	2 L HDPE or Glass	3		6 Month	Radiological Test, Gross Alpha & Beta
SM 19th Edition	7110 A-B	1 L	1 L HDPE or Glass	3		1 Year	Radiological Test, Gross Alpha &Beta, (Total, Suspended & Dissolved)
EPA SW846	9315, 9320	1 L	2 L HDPE or Glass	3		6 Month	Radium, Total
EPA WW	160.1	100 mL	250 mL HDPE or Glass	1		7 Days	Residue, Filterable (TDS)
EPA WW	160.2	100 mL	250 mL HDPE or Glass	1		7 Days	Residue, Non-Filterable (TSS)
EPA WW	160.5	1 L	1 L HDPE or Glass	1		48 Hours	Residue, Settleable Matter
EPA WW	160.3	100 mL	250 mL HDPE or Glass	1		7 Days	Residue, Total

Sampling

Method Reference and Sampling

PARAMETERS	SOLID SAMPLES Soil, Sediment, Sludge, Solid, Concentrated Liquids						
	METHOD REFERENCE	METHOD	MINIMUM SAMPLE AMOUNT FOR ANALYSIS	CONTAINER TYPE	PRESER-VATION	EXTRACTION HOLDING TIME	ANALYSIS HOLDING TIME
Residue, Volatile							
Rotenone by HPLC							
Salinity							
Semi-Volatile Organics by GC/MS	USEPA-CLP	Exhibit D, SVOA	30 g	8 oz CWM	1	10 Days	40 Days
Semi-Volatile Organics (BNA) by GC							
Semi-Volatile Organics (BNA) Extractable Base/Neutrals and Acids							
Semi-Volatile Organics Compounds by Isotope Dilution GC/MS							
Semi-Volatile Organics by (TE/GC/MS)	EPA SW846	8275A	30 g	8 oz CWM	1	14 Days	40 Days
Semi-Volatile Organics by (GC/FT-IR)	EPA SW846	8410	30 g	8 oz CWM	1	14 Days	40 Days
Semivolatile Organic Compounds by GC/MS	EPA SW846	8270C	30 g	8 oz CWM	1	14 Days	40 Days
Silica							
Silica							
Solids							
Sulfate							
Sulfate							
Sulfate							
Sulfides, Acid Soluble & Insoluble (Distillation)	EPA SW846	9030B, 9034	25 g	4 oz CWM	1,6, Zero Headspace		7 Days
Sulfides, Extractable	EPA SW846	9031	25 g	4 oz CWM	1,6, Zero Headspace		7 Days
Sulfide by Ion Selective Electrode							
Sulfide							
Sulfide							
Sulfite							
Sulfite							
Surfactants (Linear Alkyl Sulfonates, MBAS)							
Surfactants (Linear Alkyl Sulfonates, MBAS)							
Taste							

Sampling

Guidelines for Environmental Samples

METHOD REFERENCE	METHOD	MINIMUM SAMPLE AMOUNT FOR ANALYSIS	CONTAINER TYPE	PRESER-VATION	EXTRACTION HOLDING TIME	ANALYSIS HOLDING TIME	PARAMETERS
	**AQUEOUS SAMPLES** **Drinking, Surface, Ground, Saline, Domestic and Industrial Waste Water**	
EPA WW	160.4	100 mL	250 mL HDPE or Glass	1		7 Days	Residue, Volatile
40 CFR Pt. 136 App.A	635	1 L	1 L Amber Glass	1	7 Days	40 Days	Rotenone by HPLC
SM 19th Edition	2520 A - D	240 mL	300 mL	Wax Seal		Analyze Immediately	Salinity
USEPA-CLP	Exhibit D, SVOA	1 L	1 L Amber Glass	1	5 Days	40 Days	Semi-Volatile Organics by GC/MS
40 CFR Pt. 136 App.A	625	1 L	1 L Amber Glass	8, 1	7 Days	40 Days	Semi-Volatile Organics (BNA) by GC
SM 19th Edition	6410 A-B	1 L	1 L Amber Glass	8, 1	7 Days	40 Days	Semi-Volatile Organics (BNA) Extractable Base/Neutrals and Acids
40 CFR Pt. 136 App.A	1625	1 L	1 L Amber Glass	8, 1	7 Days	40 Days	Semi-Volatile Organics Compounds by Isotope Dilution GC/MS
EPA SW846	8275A	1 L	1 L Amber Glass	8, 1	7 Days	40 Days	Semi-Volatile Organics by (TE/GC/MS)
EPA SW846	8410	1 L	1 L Amber Glass	8, 1	7 Days	40 Days	Semi-Volatile Organics by (GC/FT-IR)
EPA SW846	8270C	1 L	1 L Amber Glass	8, 1	7 Days	40 Days	Semivolatile Organic Compounds by GC/MS
EPA WW	370.1	50 mL	100 mL HDPE	1		28 Days	Silica
SM 19th Edition	4500-Si A-G	200 mL	225 mL HDPE	1		28 Days	Silica
SM 19th Edition	2540 A - G	200 mL	250 mL HDPE or Glass	1		7 Days	Solids
EPA SW846	9035, 9036, 9038	100 mL	250 mL HDPE or Glass	1		28 Days	Sulfate
SM 19th Edition	4500-SO$_4$ A-F	100 mL	125 mL HDPE or Glass	1		28 Days	Sulfate
EPA WW	375.1, 375.2, 375.3, 375.4	50 mL	100 mL HDPE or Glass	1		28 Days	Sulfate
EPA SW846	9030B, 9034	500 mL	1 L HDPE	4, 6, 1		7 Days	Sulfides, Acid Soluble & Insoluble (Distillation)
EPA SW846	9031	500 mL	1 L HDPE	4, 6, 1		7 Days	Sulfides, Extractble
EPA SW846	9215	500 mL	1 L HDPE	4, 6, 1		7 Days	Sulfide by Ion Selective Electrode
SM 19th Edition	4500-S A-H	100 mL	125 mL	4, 6, 1		7 Days	Sulfide
EPA WW	376.1, 376.2	500 mL	1 L HDPE or Glass	4, 6, 1		7 Days	Sulfide
EPA WW	377.1	50 mL	125 mL HDPE or Glass	None Required		Analyze Immediately	Sulfite
SM 19th Edition	4500-SO$_3$ A-C	100 mL	125 mL HDPE or Glass	1 mL EDTA/100 mL sample		N/S	Sulfite
EPA WW	425.1	250 mL	500 mL HDPE	1		48 Hours	Surfactants (Linear Alkyl Sulfonates, MBAS)
SM 19th Edition	5540 A-D	250 mL	500 mL HDPE or Glass	1		48 Hours	Surfactants (Linear Alkyl Sulfonates, MBAS)
SM 19th Edition	2160A	500 mL	500 mL CWM	None Required		24 Hours	Taste

Sampling

PARAMETERS	SOLID SAMPLES Soil, Sediment, Sludge, Solid, Concentrated Liquids						
	METHOD REFERENCE	METHOD	MINIMUM SAMPLE AMOUNT FOR ANALYSIS	CONTAINER TYPE	PRESER- VATION	EXTRACTION HOLDING TIME	ANALYSIS HOLDING TIME
TCLP Extraction (Hazardous Waste Toxicity)	EPA SW846	1311	300 g	16 oz CWM		Organic 14 Days/Hg 28 Days/ Metals 180 Days	
Temperature							
Temperature							
2,3,7,8- Tetrachlorodibenzo-p-dioxin by GC/MS							
Tetrazene by HPLC	EPA SW846	8331	2 g	4 oz CWM	1	14 Days	40 Days
Thiabendazole by LC							
TNT Explosives by Immunoassay	EPA SW846	4050	Requires test kit for screening				N.S
Trinitrotoluene (TNT) by Colorimetric Screening Method	EPA SW846	8515	Requires test kit for screening				N.S.
Total Organic Halogens (TOX)							
Toxaphene by Immunoassay	EPA SW846	4040	Requires test kit for screening				N.S
Trihalomethanes and Chlorinated Organic Solvents (THMs)							
Turbidity							
Turbidity							
Volatile Organics							
Volatile Organic Compounds by GC/MS	EPA SW846	8260B	30 g	4 oz CWM	1		14 Days
Volatile Organic Compounds by GC/MS	USEPA CLP	Exhibit D, VOA	10 g	4 oz CWM	1		10 Days
Volatile Organic Compounds by Isotope Dilution GC/MS							

Sampling

Guidelines for Environmental Samples

METHOD REFERENCE	METHOD	MINIMUM SAMPLE AMOUNT FOR ANALYSIS	CONTAINER TYPE	PRESER-VATION	EXTRACTION HOLDING TIME	ANALYSIS HOLDING TIME	PARAMETERS
AQUEOUS SAMPLES Drinking, Surface, Ground, Saline, Domestic and Industrial Waste Water							
EPA SW846	1311	4 L	4 x 1 L Amber Glass	None Required	Organic 14 Days/Hg 28 Days/ Metals 180 Days		TCLP Extraction (Hazardous Waste Toxicity)
EPA WW	170.1	1 L	1 L HDPE or Glass	None Required		Analyze Immediately	Temperature
SM 19th Edition	2550 A-B		HDPE or Glass	Analyze Imme-diately		Analyze Immediately	Temperature
40 CFR Pt. 136 App. A	613	1 L	1 L Amber Glass	8, 1	7 Days	40 Days	2,3,7,8-Tetrachlorodibenzo-p-dioxin by GC/MS
EPA SW846	8331	10 mL	250 mL Amber Glass	1	7 Days	40 Days	Tetrazene by HPLC
40 CFR Pt. 136 App. A	641	1 L	1 L Amber Glass	1, 8, pH 1-3	7 Days	40 Days	Thiabendazole by LC
							TNT Explosives by Immunoassay
							Trinitrotoluene (TNT) by Colorimetric Screening Method
EPA SW846	9020B, 9022	250 mL	1 L B.R.	2, 1		28 Days	Total Organic Halides (TOX)
							Toxaphene by Immunoassay
SM 19th Edition	6232 A-D	5 mL	2 x 40 mL VOA Vial	8, 1		14 Days	Trihalomethanes and Chlorinated Organic Solvents (THMs)
EPA WW	180.1	100 mL	250 mL HDPE or Glass	1		48 Hours	Turbidity
SM 19th Edition	2130 A-B	100 mL	250 mL HDPE or Glass	1, Dark		48 Hours	Turbidity
SM 19th Edition	6210 A-D	5 mL	2 x 40 mL VOA Vial	24		14 Days	Volatile Organics
EPA SW846	8260B	10 mL	2 x 40 mL VOA Vial	8, 1, pH<2		14 Days	Volatile Organic Compounds by GC/MS
USEPA CLP	Exhibit D, VOA	5 mL	2 x 40 mL VOA Vial	1 pH<2		10 Days	Volatile Organic Compounds by GC/MS
40 CFR Pt. 136 App. A	1624	5 mL	2 x 40 mL VOA Vial	1, (8,10)		14 Days	Volatile Organic Compounds by Isotope Dilution GC/MS

Sampling

Method Reference and Sampling Guidelines for Environmental Samples

PARAMETERS	DRINKING WATER						
	METHOD REFERENCE	METHOD	MINIMUM SAMPLE AMOUNT FOR ANALYSIS	CONTAINER TYPE	PRESER-VATION	EXTRACTION HOLDING TIME	ANALYSIS HOLDING TIME
Asbestos (DW)	EPA DW	600/4-83-43	2 L	2 x 1 L Amber	1		48 Hours
Benzidines & Nitrogen Containing Pesticides	EPA DW	553	1 L	1 L Amber Glass	8 or 5, 1	7 Days	23 Days
Carbonyl Compounds (Aldehydes)	EPA DW	554	100 mL	250 mL Amber Glass	8, 1	3 Days	
Chlorinated Acids (Herbicides)	EPA DW	515.1	1 L	1 L Amber Glass	15, 1, 8	14 Days	28 Days
Chlorinated Acids (Herbicides)	EPA DW	515.2	250 mL	1 L Amber Glass	8 ,10, 1	14 Days	14 Days
Chlorinated Acids (Herbicides)	EPA DW	555	100 mL	150 mL Amber Glass	5, 10, 1, Dark		14 Days
Chlorinated Pesticides	EPA DW	508	1 L	1 L Amber Glass	15, 1, 8	7 Days	14 Days
Chlorinated Pesticides, Herbicides, Organohalides	EPA DW	508.1	1 L	1 L Amber Glass	10,1, 5	14 Days	30 Days
Chlorination Disinfection Byproducts and Chlorinated Solvents	EPA DW	551	35 mL	2 x 40 mL VOA Vial	See Method for Preservation		14 Days
Chlorination Disinfection Byproducts and Chlorinated Solvents	EPA DW	551.1	35 mL	2 x 40 mL VOA Vial	See Method for Preservation	14 Days	14 Days
Diquat & Paraquat	EPA DW	549.1	250 mL	1 L Silanized Amber Glass	8, 2, 1	7 Days	21 Days
EDB, DBCP & 123TCP	EPA DW	504.1	40 mL	40 mL VOA Vial	1, 8		14 Days
Endothall	EPA DW	548.1	100 mL	1 L Amber Glass	8, 1, 10	7 Days	14 Days
Ethylene Thiourea	EPA DW	509	50 mL	60 mL Glass	1, Dark		ASAP
Glyphosate	EPA DW	547	50 mL	60 mL Glass Vial	8, 1,		14 Days
Haloacetic Acids & Dalapon	EPA DW	552	100 mL	250 mL Amber Glass	11, 1, Dark	28 Days	48 Hours
Haloacetic Acids & Dalapon	EPA DW	552.1	100 mL	250 mL Amber Glass	11, 1, Dark	28 Days	48 Hours
Haloacetic Acids & Dalapon	EPA DW	552.2	100 mL	250 mL Amber Glass	11, 1, Dark	14 Days	48 Hours
N & P Containing Pesticides	EPA DW	507	1 L	1 L Amber Glass	15, 1, 8	14 Days	14 Days
N-Methylcarbamoyl oximes and N-Methylcarbamates	EPA DW	531.1	50 mL	60 mL Glass Vial	Acid Buffer pH=3, 8, -10°C		28 Days
Organic Compounds by Liquid/Solid Extraction,	EPA DW	524.2	25 mL	2 x 40 mL VOA Vial	16 or 8, 10, 1		14 Days
Organic Compounds by Liquid/Solid Extraction, GC/MS	EPA DW	525.2	1 L	1 L Amber Glass	1, 5, 10	14 Days or Immediately for Sensitive Analytes	30 Days from Sample Collection
Organohalide Pesti-cides/Polychlorinated Biphenyls	EPA DW	505	40 mL	40 mL VOA Vial	8, 1		7 Days/ 14 Days
Phthalate and Adipate Esters by Liquid/Liquid or Solid/Liquid	EPA DW	506	1 L	1 L Amber Glass	8, 1	14 Days	14 Days
Polychlorinated Biphenyls Screening Method	EPA DW	508A	1 L	1 L Amber Glass	1	14 Days	30 Days
Polycyclic Aromatic Hydrocarbons	EPA DW	550	1 L	1 L Amber Glass	8, 10, 1, Dark	7 Days	30 Days
Polycyclic Aromatic Hydrocarbons	EPA DW	550.1	1 L	1 L Amber Glass	8, 10, 1, Dark	7 Days	40 Days

Sampling

Method Reference and Sampling Guidelines for Environmental Samples (Continued)

PARAMETERS	DRINKING WATER						
	METHOD REFERENCE	METHOD	MINIMUM SAMPLE AMOUNT FOR ANALYSIS	CONTAINER TYPE	PRESER-VATION	EXTRACTION HOLDING TIME	ANALYSIS HOLDING TIME
Volatile Organic Compounds by Purge & Trap	EPA DW	502.2	5 mL	2 x 40 mL VOA Vial	8 or 5 pH<2, 1		14 Days

Key to Codes

Footnotes:

Preservation	
(1) 4 °C	(13) H_3PO_4 <pH 4-5
(2) H_2SO_4 <pH 2	(14) 2 mL Iodide Azide
(3) HNO_3 < pH 2	(15) 1 mL $HgCl_2$ 10 mL/L
(4) NaOH > pH 12	(16) Ascorbic Acid
(5) 1.0mL Na_2SO_3	(17) 0.1 N $NaAsO_2$/ L
(6) 4 drops 2N Zinc Acetate/ Liter	(18) 1 N $ClCH_2COOH$
(7) 2 Drops 1:1 HCL	(19) 1M Boric Acid
(8) 0.008% $Na_2S_2O_3$/Liter if residual chlorine present	(20) 3.0 mL Alkali-Iodide Solution
(9) 2.0 mL $MnSO_4$	(21) 2.0 mL conc. H_2SO_4
(10) HCL pH<2	(22) 1.0 mL Sodium Azide (NaN_3)
(11) 65 mg NH_4Cl/ Vial	(23) 0.7 mL conc. H_2SO_4
(12) $CuSO_4$ 1 g/1 L	(24) HCl: 4 drops 6N (no residual chlorine). If residual chlorine: ascorbic acid (25 mg/40 mL), or sodium thiosulfate (3 mg/40mL), or sodium sulfate (3 mg/40mL)

Abbreviations:

HDPE	High Density Polyethylene Bottle
CWM	Clear Wide Mouth Glass Jar
A.J.	Amber Glass Jug
B.R.	Boston Round Bottle
L	Liter(s)
mL	Milliliter(s)
oz	Ounce(s)
g	Gram(s)
SM	Standard Methods for the Examination of Water and Waste Water, 19th Edition
EPA WW	EPA Methods for the Chemical Analysis of Water and Wastes
EPA DW	EPA Methods for the Determination of Organic Compounds in Drinking Water
USEPA CLP	USEPA Contract Laboratory Program
40 CFR	Code of Federal Regulations, Title 40, Protection of Environment
EPA SW846	Test Methods for Evaluating Solid Waste Physical/Chemical Methods (SW-846) Third Edition
N.S	Not Specified
ASAP	As Soon As Possible

Sampling

Sample Extraction

EPA Method Name	EPA Method No.	SW-846 SOLIDS Extraction Methods (Soil, Sediment, Sludge, Solid, Sample Train Component)						
		3540C Soxhlet	3541 Automated Soxhlet	3542 Semi-VOST	3545 Pressurized Fluid Extraction	3550B Ultrasonic Extraction	3560 Super Critical Fluid for TPH	3561 Super Critical Fluid for PAH
1,2-Dibromomethane & 1,2-Dibromo-3-chloropropane by Micro-extraction and GC	8011	This test method does not specify solid matrices.						
Nonhalogenated Volatile Organics by GC/FID	8015B	•	•		•	•	•	
Aromatic and Halogenated Volatiles by GC (PID/ELCD)	8021B							
Acrylonitrile by GC	8031		(1)					
Acrylamide by GC	8032A		(2)					
Acetonitrile by GC/NPD	8033	(3)		•				
Phenols by GC	8041	•				•		
Phthalate Esters by GC/ECD	8061A	•	•		•	•		
Nitrosamines by GC	8070A	•	•		•	•		
Organochlorine Pesticides by GC	8081A	•	•		•	•		
Polychlorinated Biphenyls as Aroclors and Congeners by GC	8082	•	•		•			
Nitroaromatics and Cyclic Ketones by GC	8091	•	•		•	•		
Polynuclear Aromatic Hydrocarbons	8100	•	•		•	•		•
Haloethers by GC	8111	•	•		•	•		
Chlorinated Hydrocarbons by GC Capillary Column	8121	•	•			•		
Aniline and Selected Derivitives	8131	•	•		•	•		
Organophosphorus Pesticides by GC Capillary Column	8141A	•	•		•			
Chlorinated Herbicides by GC using Methylation or Pentafluoro-benzylation Derivitization	8151A		(4)					
Volatile Organic Compounds by GC/MS	8260B							
Semivolatile Organic Compounds by GC/MS	8270C	•	•		•	•	•	•
Semivolatile Organic Compounds (PAHs and PCBs) by TE/GC/MS	8275A	(5)						

Footnotes:

(1) This method is designed for aqueous samples but may be applied to other matrices.

(2) This method is designed for aqueous samples but may be applied to other matrices.

(3) Direct aqueous Injection is used as the sample introduction technique.

(4) Method 8151A provides the extraction procedures for this test. An ultrasonic or wrist shaker extraction is performed prior to derivitization.

(5) Method 8275A provides the extraction procedures for this test. A thermal extraction technique employed.

Sampling

Sampling Procedures and Preparation

Methods

SW-846						EPA Method No.
SOLIDS **Extraction Methods** (Soil, Sediment, Sludge, Solid, Sample Train Component)						
5021 Head space	5030B[6] Purge and Trap	5031 Azeotropic Distillation	5032 Vacuum Distillation	5035 Closed System Purge and Trap	5041 VOST	
						8011
•	•		•	•		8015B
•	•		•	•		8021B
						8031
						8032
					•	8033
						8041
						8061A
						8070A
						8081A
						8082
						8091
						8100
						8111
						8121
						8131
						8141A
						8151A
•	•	•	•	•	•	8260B
						8270C
						8275A

(6) Method 5030 may be used to analyze extracts from high level soil/solid samples. A measured sample amount is extracted with methanol or PEG as described in method 5035 then diluted into organic free reagent water. The dilute extract is introduced into the purge and trap system as an aqueous sample.

Sampling

Sample Extraction

EPA Method Name	EPA Method No.	SW-846 SOLIDS Extraction Methods (Soil, Sediment, Sludge, Solid, Sample Train Component)						
		3540C Soxhlet	3541 Automated Soxhlet	3542 Semi-VOST	3545 Pressurized Fluid Extraction	3550B Ultrasonic Extraction	3560 Super Critical Fluid for TPH	3561 Super Critical Fluid for PAH
Polychlorinated Dibenzo-p-Dioxins & Polychlorinated Dibenzofurans by HRGC/LRMS	8280A		(7)					
Polychlorinated Dibenzodioxins & Polychlorinated Dibenzofurans by HRGC/HRMS	8290	(8)						
Polynuclear Aromatic Hydrocarbons	8310	•	•	•	•	•		•
Carbonyl Compounds by HPLC	8315A		(9)					
Acylamide, Acrylonitrile & Acrolein by HPLC	8316	(10)						
N-Methylcarbamates by HPLC	8318	(11)						
Solvent Extractable Non-Volatile Compounds by HPLC/TSP/MS or Ultraviolet (UV) Detection	8321A (12)	•				•		
Solvent Extractable Non-volatile Compounds HPLC/PB/MS	8325	•	•					
Nitroaromatic & Nitroamines by HPLC	8330	(13)						
Tetrazene by Reverse Phase HPLC	8331	(14)						
Nitroglycerine by HPLC	8332	(15)						
Semivolatile Organics by GC/FT-IR Capillary Column	8410	•	•	•		•		
Bis(2-Chloroethyl) Ether and Hydrolysis products by GC /FTIR	8430	(10)						
Total Recoverable Petroleum Hydrocarbons by IR	8440						•	
Colorimetric Screening Method for TNT in Soil	8515	(16)						
Continuous Measurement of Formaldehyde in Ambient Air	8520	(10)						

Footnotes:

(7) Method 8280 provides matrix-specific extraction techniques. A Soxhlet/Dean-Stark water separator apparatus is employed.

(8) Method 8290 provides matrix-specific extraction techniques. A Soxhlet or Soxhlet/Dean-Stark water separator apparatus is employed.

(9) Method 8315 provides matrix -specific extraction techniques. Method 1311 (TCLP Extraction) is employed prior to derivitization using dinitrophenylhydrazine (DNPH).

(10) Method does not specify solid matrices.

(11) Method 8318: N-methylcarbamates are extracted from aqueous samples with methylene chloride, and from soils, oily solid wastes and oils with acetonitrile.

(12) Method 8321A encompasses several groups of analytes. Guidance for choice of extraction method based upon target analyte(s) is provided in the method.

(13) Method 8330 provides sample extraction procedures. Soil and sediment samples are extracted using acetonitrile in an ultrasonic bath, filtered then filtered and analyzed.

(14) Method 8331 provides sample extraction procedures. 2 g of soil are extracted with 55:45 v/v methanol-water and 1-decanesulfonic acid on a platform shaker, filtered, and eluted on a C-18 column using ion pairing reverse phase HPLC, and quantitated at 280 nm.

(15) Method 8332 is designed for aqueoues samples but may be applied to other matrices.

(16) Method 8515 utilizes a commercial test kit for sample preparation. Refer to the method and the test kit instructions for guidance.

Sampling

Methods

SW-846						EPA Method No.
SOLIDS Extraction Methods (Soil, Sediment, Sludge, Solid, Sample Train Component)						
5021 Head space	5030B Purge and Trap	5031 Azeotropic Distillation	5032 Vacuum Distillation	5035 Closed System Purge and Trap	5041 VOST	
						8280
						8290
						8310
						8315 A
						8316
						8318
						8321 A
						8325
						8330
						8331
						8332
						8410
						8430
						8440
						8515
						8520

Sampling

Sample Extraction Methods

EPA Method Name	EPA Method No.	SW-846 LIQUIDS Extraction Methods (Water, Wastewater, Oils)							
		3510C Separatory Funnel Liquid/ Liquid	3520C Continuous Liquid/ Liquid	3535 Solid Phase Extraction	3580A Waste Dilution	3585 Waste Dilution for Volatile Organics	5030B Purge and Trap	5031 Volatile Organics by Azeotropic Distillation	5032 Volatile Organics by Vacuum Distillation
1,2-Dibromomethane & 1,2-Dibromo-3-chloropropane by Micro-extraction and GC	8011		(1)						•
Nonhalogenated Volatile Organics by GC/FID	8015B	•	•				•		
Aromatic and Halogenated Volatiles by GC (PID/ELCD)	8021B					•	•		•
Acrylonitrile by GC	8031		(2)						
Acrylamide by GC	8032		(3)						
Acetonitrile by GC/NPD	8033	(4)							
Phenols by GC	8041	•	•		•				
Phthalate Esters by GC/ECD	8061A	•	•	•	•				
Nitrosamines by GC	8070A	•	•		•				
Organochlorine Pesticides by GC	8081A	•	•	•					
Polychlorinated Biphenyls as Aroclors and Congeners by GC	8082	•	•	•					
Nitroaromatics and Cyclic Ketones by GC	8091	•	•		•				
Polynuclear Aromatic Hydrocarbons	8100	•	•		•				
Haloethers by GC	8111	•	•						
Chlorinated Hydrocarbons by GC Capillary Column	8121								
Aniline and Selected Derivitives	8131	•	•		•			•(7)	
Organophosphorus Pesticides by GC Capillary Column	8141A	•	•		•				
Chlorinated Herbicides by GC using Methylation or Pentafluoro-benzylation Derivitization	8151A		(5)		•				
Volatile Organic Compounds by GC/MS	8260B					•	•	•	•
Semivolatile Organic Compounds by GC/MS	8270C	•	•		•				
Semivolatile Organic Compounds (PAHs and PCBs) by TE/GC/MS	8275A	(6)							

Footnotes:
(1) Method 8011 provides extraction procedures for this test. Thirty five mL of sample is extracted with 2 mL of hexane.
(2) Method 8031 provides extraction procedures for this test. A measured sample volume is microextracted with methyl tert-butyl ether.
(3) Method 8032 is based on bromination of the acrylamide double bond. The reaction product (2,3-dibromopropionamide) is extracted from the reaction mixture with ethyl acetate, after salting out with sodium sulfate.
(4) Sample is introduced by direct aqueous injection.
(5) Method 8151A provides extraction procedures for this test. A separatory funnel extraction method is used for aqueous samples prior to derivitization.
(6) Test method does not specify liquid matrices.
(7) Method 5031 is applicable to Aniline only.

Sampling

Sample Extraction Methods (Continued)

EPA Method Name	EPA Method No.	SW-846 LIQUIDS Extraction Methods (Water, Wastewater, Oils)							
		3510C Separatory Funnel Liquid /Liquid	3520C Contin-uous Liquid/ Liquid	3535 Solid Phase Extraction	3580A Waste Dilution	3585 Waste Dilution for Volatile Organics	5030B Purge and Trap	5031 Volatile Organics by Azeotropic Distillation	5032 Volatile Organics by Vacuum Distillation
Polychlorinated Dibenzo-p-Dioxins & Polychlorinated Dibenzofurans by HRGC/LRMS	8280		(1)						
Polychlorinated Dibenzodioxins & Polychlorinated Dibenzofurans by HRGC/HRMS	8290	(2)							
Polynuclear Aromatic Hydrocarbons	8310	•	•						
Carbonyl Compounds by HPLC	8315 A		(3)						
Acylamide, Acrylonitrile & Acrolein by HPLC	8316	(4)							
N-Methylcarbamates by HPLC	8318	(5)							
Solvent Extractable Non-Volatile Compounds by HPLC/TSP/MS or Ultraviolet (UV) Detection	8321 A	•(6)	•(6)						
Solvent Extractable Non-volatile Compounds HPLC/PB/MS	8325	•	•	•					
Nitroaromatic & Nitroamines by HPLC	8330	(7)							
Tetrazene by Reverse Phase HPLC	8331	(8)							
Nitroglycerine by HPLC	8332	(4)							
Semivolatile Organics by GC/FT-IR Capillary Column	8410	•	•	•	•				
Bis(2-Chloroethyl) Ether and Hydrolysis products by GC /FTIR	8430	(4)							
Total Recoverable Petroleum Hydrocarbons by IR	8440	(9)							
Colorimetric Screening Method for TNT in Soil	8515	(9)							
Continuous Measurement of Formaldehyde in Ambient Air	8520	(9)							

Footnotes:

(1) Method 8280 provides specific extraction techniques using separatory funnel or continuous liquid/liquid extraction.

(2) Method 8290 provides specific extraction techniques using continuous liquid/liquid extraction.

(3) Method 8315 provides specific extraction techniques. Liquid/Liquid or solid phase techniques are employed along with a derivitization procedure using dinitrophenylhydrazine (DNPH).

(4) Sample introduced by direct aspiration.

(5) Method 8318: N-methylcarbamates are extracted from aqueous samples with methylene chloride, and from oils with acetonitrile.

(6) Method 8321A encompasses several groups of analytes. Guidance for choice of extraction method based upon target analyte(s) is provided in the method.

(7) Method 8330 Low level Salting-out Method with no evaporation: Aqueous samples of low concentration are extracted by a salting-out extraction procedure with acetonitrile and sodium chloride.

(8) Method 8331 provides sample extraction procedures. A 10 mL sample is filtered, eluted on a C-18 column using ion pairing reverse phase HPLC, and quantitated at 280 nm.

(9) Method does not specify liquid matrices.

Sampling

Clean-up Methods

EPA Method Name	EPA Method No.	SW-846 CLEAN-UP Methods							
		3610B Alumina Cleanup	3611B Alumina Cleanup & Separation	3620B Florisil	3630C Silica Gel	3640A Gel Permeation	3650B Acid/Base Partition	3660B Sulfur	3665A Sulfuric Acid/ Permanganate
1,2-Dibromomethane & 1,2-Dibromo-3-chloropropane by Micro-extraction and GC	8011								
Nonhalogenated Volatile Organics by GC/FID	8015B								
Aromatic and Halogenated Volatiles by GC (PID/ELCD)	8021B								
Acrylonitrile by GC	8031								
Acrylamide by GC	8032A (1)								
Acetonitrile by GC/NPD	8033								
Phenols by GC	8041				•(2)	•	•		
Phthalate Esters by GC/ECD	8061A	•	•	•		•		•	
Nitrosamines by GC	8070A	•		•		•			
Organochlorine Pesticides by GC	8081A	•		•	•	•		•	
Polychlorinated Biphenyls as Aroclors and Congeners by GC	8082	•		•	•	•		•	•
Nitroaromatics and Cyclic Ketones by GC	8091			•		•			
Polynuclear Aromatic Hydrocarbons	8100		•		•	•			
Haloethers by GC	8111			•		•		•	
Chlorinated Hydrocarbons by GC Capillary Column	8121			•		•			
Aniline and Selected Derivitives	8131			•		•			
Organophosphorus Pesticides by GC Capillary Column	8141A			•					
Chlorinated Herbicides by GC using Methylation or Pentafluoro-benzylation Derivitization	8151A (3)			•					

Footnotes:

(1) Method 8032 provides a specific florisil column cleanup technique.
(2) Method 3630C is applicable to derivitized phenols.
(3) Method 8151A provides specific cleanup procedure along with the derivitization technique, including solvent wash procedures.

Sampling

Clean-up Methods (Continued)

EPA Method Name	EPA Method No.	SW-846 CLEAN-UP Methods							
		3610B Alumina Cleanup	3611B Alumina Cleanup & Separation	3620B Florisil	3630C Silica Gel	3640A Gel Permeation	3650B Acid/Base Partition	3660B Sulfur	3665A Sulfuric Acid/ Permanganate
Volatile Organic Compounds by GC/MS	8260B								
Semivolatile Organic Compounds by GC/MS	8270B	•[4]	•[4]	•[4]	•[4]	•[4]	•[4]	•[4]	•[4]
Semivolatile Organic Compounds (PAHs and PCBs) by TE/GC/MS	8275								
Polychlorinated Dibenzo-p-Dioxins & Polychlorinated Dibenzofurans by HRGC/LRMS	8280A [5]								
Polychlorinated Dibenzodioxins & Polychlorinated Dibenzofurans by HRGC/HRMS	8290 [6]								
Polynuclear Aromatic Hydrocarbons	8310		•		•	•			
Carbonyl Compounds by HPLC	8315A [7]								
Acylamide, Acrylonitrile & Acrolein by HPLC	8316								
N-Methylcarbamates by HPLC	8318 [8]								
Solvent Extractable Non-Volatile Compounds by HPLC/TSP/MS or Ultraviolet (UV) Detection	8321A			•[9]					
Solvent Extractable Non-volatile Compounds HPLC/PB/MS	8325								
Nitroaromatic & Nitroamines by HPLC	8330								
Tetrazene by Reverse Phase HPLC	8331								
Nitroglycerine by HPLC	8332								
Semivolatile Organics by GC/FT-IR Capillary Column	8410					•			
Bis(2-Chloroethyl) Ether and Hydrolysis products by GC /FTIR	8430								
Total Recoverable Petroleum Hydrocarbons by IR	8440 [10]								
Colorimetric Screening Method for TNT in Soil	8515								
Continuous Measurement of Formaldehyde in Ambient Air	8520								

Footnotes: *continued on next page*

Sampling

Clean-up Methods (Continued)

Footnotes *continued*

(4) Extract cleanup – Extracts may be cleaned up by any of the following methods prior to GC/MS analysis.

Compounds	Methods
Aniline & analine derivitives	3620B
Phenols	3630C, 3640A, 8041[a]
Pthalate esters	3610B, 3620B, 3640A
Nitrosamines	3610B, 3620B, 3640A
Organochlorine pesticides & PCBs	3610B, 3620B, 3630C, 3660B, 3665A
Nitroaromatics and cyclic ketones	3620B, 3640A
Polynuclear aromatic hydrocarbons	3611B,
Haloethers	3620B, 3640A
Chlorinated hydrocarbons	3620B, 3640A
Organophosphorus pesticides	3620B
Petroluem waste	3611B, 3650B
All base, neutral and acid priority pollutants	3640A

[a] Method 8041 includes a derivatization technique followed by GC/ECD analysis, if interference is encountered on GC/FID.

(5) Method 8280A provides several cleanup steps including acid/base/salt wash, silica gel column, and florisil column.

(6) Method 8290 provides several cleanup steps including acid/base/salt wash, silica gel/alumina column and carbon column.

(7) Method 8315A: If chromatographic interference is observed cleanup procedures may be necessary (not-specified).

(8) Method 8318 provides extract cleanup procedures using reverse phase C-18 cartridge.

(9) Method 8321: Florisil column cleanup is not recommended for all compounds, consult method for guidance.

(10) Method 8440 includes a silica gel cleanup step which is provided in the determinative method.

Sampling

GENIUM PUBLISHING CORPORATION

Sampling

Sample Extraction and Clean-up

Standard Method Name	EPA Method Number/Section	STANDARD METHODS 19th Edition Extraction Methods GROUND WATER, SURFACE WATER, WASTE WATER, RAW SOURCE WATER						
		Purge-and-Trap	Combustible Gas Indicator	Hexane Liq/Liq Extraction	Pentane Liq/Liq Extraction	Liq/Liq Separatory Funnel	Soxhlet Extraction	Distillation
Oil and Grease	5520 B,C					•		
Oil and Grease	5520 D,E						•	
Phenols	5530 B							•
Volatile Organic Compounds	6210 B,C,D	•						
Methane	6211 B		•					
Volatile Aromatic Organic Compounds	6220 B,C,D	•						
Volatile Halocarbons	6230 B,C,D	•						
1,2-Dibromoethane (EDB) and 1,2-Dibromo-3-chloropropane (DBCP)	6231 B			•				
1,2-Dibromoethane (EDB) and 1,2-Dibromo-3-chloropropane (DBCP)	6231 C	•						
Trihalomethanes and Chlorinated Solvents	6232 B				•			
Trihalomethanes and Chlorinated Solvents	6232 C,D	•						
Haloacetic Acids and Trichlorophenol	6251 B	(1)						
Extractable Base/Neutrals and Acids	6410 B					•		
Phenols	6420 B,C				•			
Polychlorinated Biphenyls	6431 B,C							
Polychlorinated Biphenyls	6630 B,C					•		
Polynuclear Aromatic Hydrocarbons	6440 B,C					•		
Organochlorine Pesticides	6630 B					•		
Organochlorine Pesticides	6630 C,D					•		
Acidic Herbicide Compounds	6640 B	(2)						
Glyphosate Herbicides	6651 B	No Extraction or Cleanup						

Footnotes:

(1) The sample is extracted with methyl tertiary-butyl ether (MtBE) at an acidic pH to extract the nondissociated acidic compounds to be determined. A salting agent is added to increase extraction efficiency. The extracted compounds are methylated with diazomethane solution to produce methyl ether or ether derivatives that can be separated chromatographically.

(2) A 30 mL sample portion is extracted with 3 mL methyl tertiary-butyl ether (MtBE), at an acidic pH to extract the nondissociated acid, with a salting agent to increase extraction efficiency. Extracted compounds are esterified with diazomethane solution to produce methyl ester derivatives that can be chromatographed.

Sampling

Methods for Standard Methods

Cleanup Methods			Method Number/ Section	Standard Method Name
Silica Gel Column	Magnesia Silica Gel Column	Mercury Shake		
			5520 B,C	Oil and Grease
			5520 D,E	Oil and Grease
			5530 B	Phenols
			6210 B,C,D	Volatile Organic Compounds
			6211 B	Methane
			6220 B,C,D	Volatile Aromatic Organic Compounds
			6230 B,C,D	Volatile Halocarbons
			6231 B	1,2-Dibromoethane (EDB) and 1,2-Dibromo-3-chloropropane (DBCP)
			6231 C	1,2-Dibromoethane (EDB) and 1,2-Dibromo-3-chloropropane (DBCP)
			6232 B	Trihalomethanes and Chlorinated Solvents
			6232 C,D	Trihalomethanes and Chlorinated Solvents
			6251 B	Haloacetic Acids and Trichlorophenol
			6410 B	Extractable Base/Neutrals and Acids
•			6420 B,C	Phenols
			6431 B,C	Polychlorinated Biphenyls
	•	•	6630 B,C	Polychlorinated Biphenyls
•			6440 B,C	Polynuclear Aromatic Hydrocarbons
	•		6630 B	Organochlorine Pesticides
	•	•	6630 C,D	Organochlorine Pesticides
			6640 B	Acidic Herbicide Compounds
			6651 B	Glyphosate Herbicides

Sampling

Sample Extraction and Clean-up

EPA Method Name	EPA Method Number	EPA DRINKING WATER 500 SERIES						
		Preparation and Extraction Methods						
		Purge-and-Trap	Liq/Liq Extraction	Liq/Solid Extraction	Esterification	Micro-extraction	Perchlorination	Extrelut Column
Trihalomethanes	501.3	•						
Volatile Organic Compounds	502.2	•						
1,2-Dibromoethane & 1,2-Dibromo-3-chloropropane	504					•		
1,2-Dibromoethane & 1,2-Dibromo-3-chloropropane and 1,2,3-Trichloropropane	504.1					•		
Organohalide Pesticides/ Polychlorinated Biphenyls	505					•		
Phthalate and Adipated Esters by Liquid/Liquid or Solid/Liquid	506		•	•				
N & P Containing Pesticides	507		•					
Chlorinated Pesticides	508		•					
Polychlorinated Biphenyls Screening	508A		•				•	
Chlorinated Pesticides, Herbicides and Organohalides	508.1			•				
Ethylene Thiourea	509							•
Tetrachlorodibenzo-p-dioxin	513		•	•				
Chlorinated Acids (Herbicides)	515.1		•	•	•			
Chlorinated Acids (Herbicides)	515.2		•	•	•			
Purgeable Organic Compounds	524.2	•						
Organic Compounds (Extractable)	525.1			•				
Organic Compounds (Extractable)	525.2			•				
N-Methylcarbamoyloximes and N-Methylcarbamates	531.1							
Glyphosate	547							
Endothall	548			•				
Endothall	548.1		•	•				
Diquat and Paraquat	549			•				
Diquat and Paraquat	549.1			•				
Polycyclic Aromatic Hydrocarbons	550		•					
Polycyclic Aromatic Hydrocarbons	550.1			•				
Chlorination Disinfection By Products and Chlorinated Solvents	551		•					
Chlorination Disinfection By Products, Chlorinated Solvents, and Halogenated Pesticides/Herbicides	551.1		•					
Haloacetic Acids & Dalapon	552		•		•			
Haloacetic Acids & Dalapon	552.1			•				
Haloacetic Acids & Dalapon	552.2					•		
Benzidines & Nitrogen Containing Pesticides	553		•	•				
Carbonyl Compounds (Aldehydes)	554			•				
Chlorinated Acids (Herbicides)	555			•				

Methods for Drinking Water

Cleanup Methods				EPA Method Number	EPA Method Name
Florisil Column Cleanup	Alumina Column Cleanup	Silica Gel Column	Filtration		
				501.3	Trihalomethanes
				502.2	Volatile Organic Compounds
				504	1,2-Dibromoethane & 1,2-Dibromo-3-chloropropane
				504.1	1,2-Dibromoethane & 1,2-Dibromo-3-chloropropane and 1,2,3-Trichloropropane
				505	Organohalide Pesticides/ Polychlorinated Biphenyls
•	•			506	Phthalate and Adipated Esters by Liquid/Liquid or Solid/Liquid
				507	N & P Containing Pesticides
				508	Chlorinated Pesticides
				508A	Polychlorinated Biphenyls Screening
				508.1	Chlorinated Pesticides, Herbicides and Organohalides
				509	Ethylene Thiourea
	•	•		513	Tetrachlorodibenzo-p-dioxin
				515.1	Chlorinated Acids (Herbicides)
				515.2	Chlorinated Acids (Herbicides)
				524.2	Purgeable Organic Compounds
				525.1	Organic Compounds (Extractable)
				525.2	Organic Compounds (Extractable)
				531.1	N-Methylcarbamoyloximes and N-Methylcarbamates
			•	547	Glyphosate
				548	Endothall
				548.1	Endothall
				549	Diquat and Paraquat
				549.1	Diquat and Paraquat
				550	Polycyclic Aromatic Hydrocarbons
				550.1	Polycyclic Aromatic Hydrocarbons
				551	Chlorination Disinfection By Products and Chlorinated Solvents
				551.1	Chlorination Disinfection By Products, Chlorinated Solvents, and Halogenated Pesticides/Herbicides
				552	Haloacetic Acids & Dalapon
				552.1	Haloacetic Acids & Dalapon
				552.2	Haloacetic Acids & Dalapon
				553	Benzidines & Nitrogen Containing Pesticides
				554	Carbonyl Compounds (Aldehydes)
				555	Chlorinated Acids (Herbicides)

Sampling

Sample Extraction and Clean-up Methods for Municipal and Industrial Discharge

EPA Method Name	Method #	Preparation and Extraction				Cleanup Methods			
		Purge-and-Trap	Liquid/Liquid Extraction	Base Wash	Derivatization	Florisil Column Cleanup	Alumina Column Cleanup	Silica Gel Column Cleanup	Sulfur Cleanup (Hg)
Purgeable Halocarbons	601	•							
Purgeable Aromatics	602	•							
Acrolein and Acrylonitrile	603	•							
Phenols	604		•		•			•	
Hexachlorophene and Dichlorophen	604.1*		•	•					
Benzidines	605		•						
Phthalate Esters	606		•			•	•		
Nitrosamines	607		•			•	•		
Organochlorine Pesticides and PCBs	608		•			•			•
Organochlorine Pesticides	608.1		•			•			
Organochlorine Pesticides	608.2		•			•		•	
Nitroaromatics and Isophorone	609		•			•			
Polynuclear Aromatic Hydrocarbons	610		•					•	
Haloethers	611		•			•			
Chlorinated Hydrocarbons	612		•			•			
2,3,7,8-Tetrachlorodibenzo-p-dioxin	613		•	•			•	•	
Organophosphorus Pesticides	614*		•			•			•
Organophosphorus Pesticides	614.1*		•					•	
Chlorinated Herbicides	615*		•		•				
Carbon, Hydrogen, and Oxygen containing pesticides	616*		•					•	
Organohalide pesticides and PCBs	617*		•			•			•
Volatile Pesticides	618*		•						
Triazine Pesticides	619*		•			•			
Diphenylamine	620*		•					•	
Organophosphorus Pesticides	622*		•						
Purgeables	624	•							
Base Neutrals and Acids	625		•						
Dinitroaniline Pesticides	627*		•						
Cyanazine	629*		•						
Dithiocarbamate Pesticides	630*[1]								
Dithiocarbamate Pesticides	630.1*[1]								

Sampling

Sample Extraction and Clean-up Methods for Municipal and Industrial Discharge (Cont.)

EPA Method Name	Method #	Preparation and Extraction				Cleanup Methods			
		Purge-and-Trap	Liquid/Liquid Extraction	Base Wash	Derivati-zation	Florisil Column Cleanup	Alumina Column Cleanup	Silica Gel Column Cleanup	Sulfur Cleanup (Hg)
Benomyl and Carbendazim	631*		•						
Carbamate and Urea Pesticides	632*		•			•			
Carbamate and Urea Pesticides	632.1*		•						
Organonitrogen Pesticides	633*		•						
Neutral Nitrogen Containing Pesticides	633.1*		•			•			
Thiocarbate Pesticides	634*		•					•	
Rotenone	635*		•					•	
Bensulide	636*		•			•			
MBTS and TCMTB	637*		•					•	
Oryzalin	638*		•			•			
Bendiocarb	639*		•			•			
Mercaptobenzothiazole	640*		•					•	
Thiabendazole	641*		•						
Biphenyl and Ortho-Phenylphenol	642*		•						
Bentazon	643*		•	•					
Picloram	644*		•						
Amine Pesticides and Lethane	645*		•			•			
Dinitro Aromatic Pesticides	646*		•			•			
Volatile Organic Compounds	1624	•							
Semivolatile Organic Compounds	1625		•						
Organohalide Pesticides[2]	1656*		•			•	•		•
Organophosphorus Pesticides[3]	1657*		•						
Phenoxy-Acid Pesticides[3]	1658*		•		•	•			
Dazomet	1659*		•						
Pyrethrins & Pyrethoids	1660*		•						
Bromoxynil	1661*		•						
Oil & Grease and TPH[2]	1664		•						

Footnotes:
(1) Sample digested to yield carbon disulfide using a hydrolysis method. H_2S is measured via UV-Vis spectroscopy.
(2) Methods 1656 and 1664 provide option for Solid Phase Extraction (SPE)
(3) Methods 1657 and 1658 provide option for Solid Phase Extraction (SPE) and Gel Permeation Chromatography (GPC) cleanup
   * **Proposed Method**: Method is proposed for promulgation in 40CFR136 at the time of publication.

Sampling

Contract Lab Program (CLP) Sample Extraction and Clean-up Methods

Contract Laboratory Protocol	CLP Method	Extraction Methods				Cleanup Methods		
		Purge-and Trap	Continuous Liquid/ Liquid	Liquid/ Liquid Sep Funnel	Sonication	GPC Cleanup	Florisil Cartridge	Sulfur Cleanup with Hg or Cu
Pesticides/PCBs	USEPA-CLP/PEST		•	•	•	•	•	•
Semivolatile Compounds	USEPA-CLP/SVOA		•		•	•		
Volatile Organic Compounds	USEPA-CLP/VOA	•						

GENIUM PUBLISHING CORPORATION

Sample Preparation and Digestion Methods for Elemental Analysis

Method	Acid(s) used in digestion	A	G	I	M	SW846 Method 3000 Series	EPA Method 200 Series	EPA Method 200.7/ 200.8	Std. Methods Method 3000	CLP Inorganic Method (3)
Filtration and analysis for dissolved metals (water)	HNO_3,HCl	•		•	•	3005A				
	HNO_3	•	•	•	•		4.1.1			
	HNO_3	•	•	•					3030 B	
Filtration and digestion for suspended metals (water)	HNO_3,HCl	•	•	•	•		4.1.2			
Filtration for dissolved metals (water)	none			•	•			Sec 8.2		
Filtrate prep. for dissolved metals (water)	HNO_3			•	•			Sec 11.1		
Filtration for suspended metals (water)	none	•	•	•					3030 B	
Hot plate treatment for acid-extractable metals (water)	HNO_3,HCl	•	•	•					3030 C	
Sample preparation for acid-soluble metals (1) water)	HNO_3,HCl			•			200.1			
	HNO_3	•	•				200.1			
Hot plate treatment for total recoverable metals (water)	HNO_3,HCl	•		•	•	3005A				
	HNO_3,HCl	•	•	•	•		200.2			
	HNO_3,HCl			•	•			Sec. 11.2		
	HNO_3,HCl	•	•	•					3030 F (5)	
Hot plate treatment for total metals (water)	HNO_3,HCl	•		•	•	3010A				Part A
	HNO_3		•			3020A				Part A
	HNO_3,HCl	•	•	•	•		200.2			
	HNO_3,HCl			•	•			Sec. 11.3		
	HNO_3,HCl	•	•	•					3030 F	
	HNO_3	•	•	•					3030 E	
Hot plate treatment for total metals (soil)	HNO_3,HCl	•		•		3050 B				Part B
	HNO_3,HCl	•	•	•	•		200.2			
	HNO_3,HCl			•	•			Sec. 11.3		
	HNO_3,HCl	•	•	•					3030 F	
	HNO_3		•		•	3050 B				
	HNO_3		•							Part B
	HNO_3	•	•	•					3030 E	
Microwave digestion for total metals (water)	HNO_3	•	•	•	•	3015				Part C
	HNO_3	•	•	•					3030 K	
Microwave digestion for total metals (soil)	HNO_3	•	•	•	•	3051 (4)				Part C
Microwave digestion for total metals (siliceous and organically based matrices)	HNO_3,HF, HCl	•	•	•	•	3052 (7)				
Hotplate digestion for metals (oil)	HNO_3,HCl, H_2SO_4	•	•	•		3031 (6)				
Dissolution procedure (2)	-		•		•	3040 A				

Notes:

Legend: A Flame atomic absorption spectroscopy
       G Graphite furnace atomic absorption spectroscopy
       I Inductively coupled argon plasma spectroscopy
      M Inductively coupled argon plasma mass-spectrometry

(1) Applicable to the analysis of arsenic, cadmium, chromium, copper and lead in ambient water and aqueous wastes.

Sampling

Notes *continued*

(2) Method is used for the preparation of samples containing oil, greases or waxes, for analysis of dissolved metals.
(3) ILM 4.0 Exhibit D section III.
(4) Method is used for the digestion of sediments, sludges, soils, and oils.
(5) Method 3030F section b is a less rigorous digestion.
(6) Method is used for the preparation of oils, oil sludges, tars, waxes, paints, sludges, and other viscous petroleum products.
(7) Method is used for the preparation of ashes, biological tissues, oils, oil contaminated soil, sediments, sludges, and soils.

Sampling

APPENDIX A

This is a compilation of some of the tables, exhibits, etc., found in the methods. If the method comparison refers to a specific table, that table will be found in this appendix. Only tables and exhibits referred to in the method comparison are found in this appendix.

Listing of Tables and Exhibits from the Methods included in this Appendix

Tables Extracted from Original Documentation

Pesticides – Organic Statement of Work – *Exhibit C*

Appendix A

Target Compound List (TCL) and Contract Required Quantitation Limits (CRQL)

Pesticides/Aroclors	CAS Number	Quantitation Limits [1]		
		Water μg/L	Soil μg/Kg	On Column (pg)
α-BHC	319-84-6	0.05	1.7	5
β-BHC	319-85-7	0.05	1.7	5
δ-BHC	319-86-8	0.05	1.7	5
γ-BHC (Lindane)	58-89-9	0.05	1.7	5
Heptachlor	76-44-8	0.05	1.7	5
Aldrin	309-00-2	0.05	1.7	5
Heptachlor epoxide	1024-57-3	0.05	1.7	5
Endosulfan I	959-98-8	0.05	1.7	5
Dieldrin	60-57-1	0.10	3.3	10
4,4'-DDE	72-55-9	0.10	3.3	10
Endrin	72-20-8	0.10	3.3	10
Endosulfan II	33213-65-9	0.10	3.3	10
4,4'-DDD	72-54-8	0.10	3.3	10
Endosulfan sulfate	1031-07-8	0.10	3.3	10
4,4'-DDT	50-29-3	0.10	3.3	10
Methoxychlor	72-43-5	0.50	17.0	50
Endrin ketone	53494-70-5	0.10	3.3	10
Endrin aldehyde	7421-36-3	0.10	3.3	10
α-Chlordane	5103-71-9	0.05	1.7	5
γ-Chlordane	5103-74-2	0.05	1.7	5
Toxaphene	8001-35-2	5.0	170.0	500
Aroclor-1016	12674-11-2	1.0	33.0	100
Aroclor-1221	11104-28-2	2.0	67.0	200
Aroclor-1232	11141-16-5	1.0	33.0	100
Aroclor-1242	53469-21-9	1.0	33.0	100
Aroclor-1248	12672-29-6	1.0	33.0	100
Aroclor-1254	11097-69-1	1.0	33.0	100
Aroclor-1260	11096-82-5	1.0	33.0	100

(1) Quantitation limits listed for soil/sediment are based on wet weight. The quantitation limits calculated by the laboratory for soil/sediment, calculated on dry weight basis as required by the contract, will be higher.

There is no differentiation between the preparation of low and medium soil samples in this method for the analysis of Pesticides/Aroclors.

Appendix A

Matrix Spike Recovery and Relative Percent Difference Limits

Compound	%Recovery Water	RPD Water	%Recovery Soil	RPD Soil
γ-BHC (Lindane)	56-123	15	46-127	50
Heptachlor	40-131	20	35-130	31
Aldrin	40-120	22	34-132	43
Dieldrin	52-126	18	31-134	38
Endrin	56-121	21	42-139	45
4,4'-DDT	38-127	27	23-134	50

Semi-volatile Organic Analysis – Organic Statement of Work – *Exhibit C*

Target Compound List (TCL) and Contract Required Quantitation Limits (CRQL)

Semivolatiles	CAS Number	Quantitation Limits[1]			
		Water µg/L	Low Soil µg/Kg	Med. Soil µg/Kg	On Column (ng)
Phenol	108-95-2	10	330	10000	(20)
bis(2-Chloroethyl) ether	111-44-4	10	330	10000	(20)
2-Chlorophenol	95-57-8	10	330	10000	(20)
1,3-Dichlorobenzene	541-73-1	10	330	10000	(20)
1,4-Dichlorobenzene	106-46-7	10	330	10000	(20)
1,2-Dichlorobenzene	95-50-1	10	330	10000	(20)
2-Methylphenol	95-48-7	10	330	10000	(20)
2,2′-oxybis (1-Chloropropane)[2]	108-60-1	10	330	10000	(20)
4-Methylphenol	106-44-5	10	330	10000	(20)
N-Nitroso-di-n-propylamine	621-64-7	10	330	10000	(20)
Hexachloroethane	67-72-1	10	330	10000	(20)
Nitrobenzene	98-95-3	10	330	10000	(20)
Isophorone	78-59-1	10	330	10000	(20)
2-Nitrophenol	88-75-5	10	330	10000	(20)
2,4-Dimethylphenol	105-67-9	10	330	10000	(20)
bis(2-Chloroethoxy) methane	111-91-1	10	330	10000	(20)
2,4-Dichlorophenol	120-83-2	10	330	10000	(20)
1,2,4-Trichlorobenzene	120-82-1	10	330	10000	(20)
Naphthalene	91-20-3	10	330	10000	(20)
4-Chloroaniline	106-47-8	10	330	10000	(20)
Hexachlorobutadiene	87-68-3	10	330	10000	(20)
4-Chloro-3-methylphenol	59-50-7	10	330	10000	(20)
2-Methylnaphthalene	91-57-6	10	330	10000	(20)
Hexachlorocyclopentadiene	77-47-4	10	330	10000	(20)
2,4,6-Trichlorophenol	88-06-2	10	330	10000	(20)
2,4,5-Trichlorophenol	95-95-4	25	800	25000	(50)
2-Chloronaphthalene	91-58-7	10	330	10000	(20)
2-Nitroaniline	88-74-4	25	800	25000	(50)
Dimethyl phthalate	131-11-3	10	330	10000	(20)
Acenaphthylene	208-96-8	10	330	10000	(20)
2,6-Dinitrotoluene	606-20-2	10	330	10000	(20)
3-Nitroaniline	99-09-2	25	800	25000	(50)
Acenaphthene	83-32-9	10	330	10000	(20)

NOTES:

(1) Quantitation limits listed for soil/sediment are based on wet weight. The quantitation limits calculated by the laboratory for soil/sediment, calculated on dry weight basis as required by the contract, will be higher.

(2) Previously known by the name bis(2-Chloroisopropyl) ether

Appendix A

Target Compound List (TCL) and Contract Required Quantitation Limits (CRQL) (Continued)

Semivolatiles	CAS Number	Quantitation Limits[1]			
		Water µg/L	Low Soil µg/Kg	Med. Soil µg/Kg	On Column (ng)
2,4-Dinitrophenol	51-28-5	25	800	25000	(50)
4-Nitrophenol	100-02-7	25	800	25000	(50)
Dibenzofuran	132-64-9	10	330	10000	(20)
2,4-Dinitrotoluene	121-14-2	10	330	10000	(20)
Diethylphthalate	84-66-2	10	330	10000	(20)
4-Chlorophenyl-phenyl ether	7005-72-3	10	330	10000	(20)
Fluorene	86-73-7	10	330	10000	(20)
4-Nitroaniline	100-01-6	25	800	25000	(50)
4,6-Dinitro-2-methylphenol	534-52-1	25	800	25000	(50)
N-nitrosodiphenylamine	86-30-6	10	330	10000	(20)
4-Bromophenyl-phenylether	101-55-3	10	330	10000	(20)
Hexachlorobenzene	118-74-1	10	330	10000	(20)
Pentachlorophenol	87-86-5	25	800	25000	(50)
Phenanthrene	85-01-8	10	330	10000	(20)
Anthracene	120-12-7	10	330	10000	(20)
Carbazole	86-74-8	10	330	10000	(20)
Di-n-butylphthalate	84-74-2	10	330	10000	(20)
Fluoranthene	206-44-0	10	330	10000	(20)
Pyrene	129-00-0	10	330	10000	(20)
Butylbenzylphthalate	85-68-7	10	330	10000	(20)
3,3'-Dichlorobenzidine	91-94-1	10	330	10000	(20)
Benzo(a)anthracene	56-55-3	10	330	10000	(20)
Chrysene	218-01-9	10	330	10000	(20)
bis(2-Ethylhexyl)phthalate	117-81-7	10	330	10000	(20)
Di-n-octylphthalate	117-84-0	10	330	10000	(20)
Benzo(b)fluoranthene	205-99-2	10	330	10000	(20)
Benzo(k)fluoranthene	207-08-9	10	330	10000	(20)
Benzo(a)pyrene	50-32-8	10	330	10000	(20)
Indeno(1,2,3-cd)pyrene	193-39-5	10	330	10000	(20)
Dibenz(a,h)anthracene	53-70-3	10	330	10000	(20)
Benzo(g,h,i)perylene	191-24-2	10	330	10000	(20)

NOTES:

(1) Quantitation limits listed for soil/sediment are based on wet weight. The quantitation limits calculated by the laboratory for soil/sediment, calculated on dry weight basis as required by the contract, will be higher.

DFTPP Key Ions and Ion Abundance Criteria for Quadrapole Mass Spectrometers

Mass	Ion Abundance Criteria
51	30.0-80.0 percent of mass 198
68	Less than 2.0 percent of mass 69
69	Present
70	Less than 2.0 percent of mass 69
127	25.0-75.0 percent of mass 198
197	Less than 1.0 percent of mass 198
198	Base Peak, 100 percent relative abundance (see note)
199	5.0-9.0 percent of mass 198
275	10.0-30.0 percent of mass 198
365	Greater than 0.75 percent of mass 198
441	Present but less than mass 443
442	40.0-110.0 percent of mass 198
443	15.0-24.0 percent of mass 442

NOTE: All ion abundances MUST be normalized to m/z 198, the nominal base peak, even though the ion abundances of m/z 442 may be up to 110 percent that of m/z 198.

Relative Response Factor Criteria for Initial and Continuing Calibration of Semivolatile Target Compounds

Semivolatile Compounds	Minimum RRF	Maximum %RSD	Maximum %Diff
Phenol	0.800	20.5	25.0
bis(2-Chloroethyl)ether	0.700	20.5	25.0
2-Chlorophenol	0.800	20.5	25.0
1,3-Dichlorobenzene	0.600	20.5	25.0
1,4-Dichlorobenzene	0.500	20.5	25.0
1,2-Dichlorobenzene	0.400	20.5	25.0
2-Methylphenol	0.700	20.5	25.0
2,2′-oxybis(1-Chloropropane)	0.010	none	none
4-Methylphenol	0.600	20.5	25.0
N-Nitroso-di-n-propylamine	0.500	20.5	25.0
Hexachloroethane	0.300	20.5	25.0
Nitrobenzene	0.200	20.5	25.0
Isophorone	0.400	20.5	25.0
2-Nitrophenol	0.100	20.5	25.0
2,4-Dimethylphenol	0.200	20.5	25.0
bis(2-Chloroethoxy)methane	0.300	20.5	25.0
2,4-Dichlorophenol	0.200	20.5	25.0
1,2,4-Trichlorobenzene	0.200	20.5	25.0
Naphthalene	0.700	20.5	25.0
4-Chloroaniline	0.010	none	none
Hexachlorobutadiene	0.010	none	none
4-Chloro-3-methylphenol	0.200	20.5	25.0
2-Methylnaphthalene	0.400	20.5	25.0
Hexachlorocyclopentadiene	0.010	none	none
2,4,6-Trichlorophenol	0.200	20.5	25.0
2,4,5-Trichlorophenol	0.200	20.5	25.0
2-Chloronaphthalene	0.800	20.5	25.0
2-Nitroaniline	0.010	none	none
Dimethylphthalate	0.010	none	none
Acenaphthylene	0.900	20.5	25.0
3-Nitroaniline	0.010	none	none
2,6-Dinitrotoluene	0.200	20.5	25.0
Acenaphthene	0.900	20.5	25.0
2,4-Dinitrophenol	0.010	none	none
4-Nitrophenol	0.010	none	none
Dibenzofuran	0.800	20.5	25.0
2,4-Dinitrotoluene	0.200	20.5	25.0
Diethylphthalate	0.010	none	none
4-Chlorophenyl-phenylether	0.400	20.5	25.0

**Relative Response Factor Criteria for Initial and Continuing Calibration
of Semivolatile Target Compounds (Continued)**

Semivolatile Compounds	Minimum RRF	Maximum %RSD	Maximum %Diff
Fluorene	0.900	20.5	25.0
4-Nitroaniline	0.010	none	none
4,6-Dinotro-2-methylphenol	0.010	none	none
N-Nitrosodiphenylamine	0.010	none	none
4-Bromophenyl-phenylether	0.100	20.5	25.0
Hexachlorobenzene	0.100	20.5	25.0
Pentachlorophenol	0.050	20.5	25.0
Phenanthrene	0.700	20.5	25.0
Anthracene	0.700	20.5	25.0
Carbazole	0.010	none	none
Di-n-butylphthalate	0.010	none	none
Fluoranthene	0.600	20.5	25.0
Pyrene	0.600	20.5	25.0
Butylbenzylphthalate	0.010	none	none
3,3'-Dichlorobenzidine	0.010	none	none
Benzo(a)anthracene	0.800	20.5	25.0
bis(2-Ethylhexyl)phthalate	0.010	none	none
Chrysene	0.700	20.5	25.0
Di-n-octylphthalate	0.010	none	none
Benzo(b)fluoranthene	0.700	20.5	25.0
Benzo(k)fluoranthene	0.700	20.5	25.0
Benzo(a)pyrene	0.700	20.5	25.0
Indeno(1,2,3-cd)pyrene	0.500	20.5	25.0
Dibenzo(a,h)anthracene	0.400	20.5	25.0
Benzo(g,h,i)perylene	0.500	20.5	25.0
SURROGATES			
Nitrobenzene-d_5	0.200	20.5	25.0
2-Fluorobiphenyl	0.700	20.5	25.0
Terphenyl-d_{14}	0.500	20.5	25.0
Phenol-d_5	0.800	20.5	25.0
2-Fluorophenol	0.600	20.5	25.0
2,4,6-Tribromophenol	0.010	none	none
2-Chlorophenol-d_4	0.800	20.5	25.0
1,2-Dichlorobenzene-d_4	0.400	20.5	25.0

Semi-volatile Organic Analysis – Organic Statement of Work – *Table 6*

Matrix Spike Recovery and Relative Percent Difference Limits

Compound	%Recovery Water	RPD Water	%Recovery Soil	RPD Soil
Phenol	12-110	42	26-90	35
2-Chlorophenol	27-123	40	25-102	50
1,4-Dichlorobenzene	36-97	28	28-104	27
N-Nitroso-di-n-propylamine	41-116	38	41-126	38
1,2,4-Trichlorobenzene	39-98	28	38-107	23
4-Chloro-3-methylphenol	23-97	42	26-103	33
Acenaphthene	46-118	31	31-137	19
4-Nitrophenol	10-80	50	11-114	50
2,4-Dinitrotoluene	24-96	38	28-89	47
Pentachlorophenol	9-103	50	17-109	47
Pyrene	26-127	31	35-142	36

Semi-volatile Organic Analysis – Organic Statement of Work – *Table 7*

Surrogate Recovery Limits

Compound	%Recovery Water	%Recovery Soil
Nitrobenzene-d$_5$	35-114	23-120
2-Fluorobiphenyl	43-116	30-115
Terphenyl-d$_{14}$	33-141	18-137
Phenol-d$_5$	10-110	24-113
2-Fluorophenol	21-110	25-121
2,4,6-Tribromophenol	10-123	19-122
2-Chlorophenol-d$_4$	33-110	20-130 (advisory)
1,2-Dichlorobenzene-d$_4$	16-110	20-130 (advisory)

Volatile Organic Analysis – Organic Statement of Work – *Exhibit C*

Appendix A

Target Compound List (TCL) and Contract Required Quantitation Limits (CRQL)

Volatiles	CAS Number	Quantitation Limits[1]			
		Water µg/L	Low Soil µg/Kg	Med. Soil µg/Kg	On Column (ng)
Chloromethane	74-87-3	10	10	1200	(50)
Bromomethane	74-83-9	10	10	1200	(50)
Vinyl chloride	75-01-4	10	10	1200	(50)
Chloroethane	75-00-3	10	10	1200	(50)
Methylene chloride	75-09-2	10	10	1200	(50)
Acetone	67-64-1	10	10	1200	(50)
Carbon disulfide	75-15-0	10	10	1200	(50)
1,1-Dichloroethene	75-35-4	10	10	1200	(50)
1,1-Dichloroethane	75-34-3	10	10	1200	(50)
1,2-Dichloroethene (total)	540-59-0	10	10	1200	(50)
Chloroform	67-66-3	10	10	1200	(50)
1,2-Dichloroethane	107-06-2	10	10	1200	(50)
2-Butanone	78-93-3	10	10	1200	(50)
1,1,1-Trichloroethane	71-55-6	10	10	1200	(50)
Carbon tetrachloride	56-23-5	10	10	1200	(50)
Bromodichloromethane	75-27-4	10	10	1200	(50)
1,2-Dichloropropane	78-87-5	10	10	1200	(50)
cis-1,3-Dichloropropene	10061-01-5	10	10	1200	(50)
Trichloroethene	79-01-6	10	10	1200	(50)
Dibromochloromethane	124-48-1	10	10	1200	(50)
1,1,2-Trichloroethane	79-00-5	10	10	1200	(50)
Benzene	71-43-2	10	10	1200	(50)
trans-1,3-Dichloropropene	10061-02-6	10	10	1200	(50)
Bromoform	75-25-2	10	10	1200	(50)
4-Methyl-2-pentanone	108-10-1	10	10	1200	(50)
2-Hexanone	591-78-6	10	10	1200	(50)
Tetrachloroethene	127-18-4	10	10	1200	(50)
Toluene	108-88-3	10	10	1200	(50)
1,1,2,2-Tetrachloroethane	79-34-5	10	10	1200	(50)
Chlorobenzene	108-90-7	10	10	1200	(50)
Ethyl benzene	100-41-4	10	10	1200	(50)
Styrene	100-42-5	10	10	1200	(50)
Xylenes (total)	1330-20-7	10	10	1200	(50)

NOTE:

(1) Quantitation limits listed for soil/sediment are based on wet weight. The quantitation limits calculated by the laboratory for soil/sediment, calculated on dry weight basis as required by the contract, will be higher.

Appendix A

BFB Key Ions and Ion Abundance Criteria

Mass	Ion Abundance Criteria
50	8.0-40.0 percent of mass 95
75	30.0-66.0 percent of mass 95
95	Base Peak, 100 percent relative abundance
96	5.0-9.0 percent of mass 95 (see note)
173	less than 2.0 percent of mass 174
174	50.0-120.0 percent of mass 95
175	4.0-9.0 percent of mass 174
176	93.0-101.0 percent of mass 174
177	5.0-9.0 percent of mass 176

Note: All ion abundances must be normalized to m/z 95, the nominal base peak, even though the ion abundance of m/z 174 may be up to 120 percent that of m/z 95.

Volatile Organic Analysis – Organic Statement of Work – *Table 2*

Characteristic Ions for Volatile Target Compounds

Analyte	Primary Quantitation Ion	Secondary Ion(s)
Chloromethane	50	52
Bromomethane	94	96
Vinyl chloride	62	64
Chloroethane	64	66
Methylene chloride	84	49, 51, 86
Acetone	43	58
Carbon disulfide	76	78
1,1-Dichloroethene	96	61, 98
1,1-Dichloroethane	63	65, 83, 85, 98, 100
1,2-Dichloroethene (total)	96	61, 98
Chloroform	83	85
1,2-Dichloroethane	62	64, 100, 98
2-Butanone	43*	57
1,1,1-Trichloroethane	97	99, 117, 119
Carbon tetrachloride	117	119, 121
Bromodichloromethane	83	85
1,1,2,2-Tetrachloroethane	83	85, 131, 133, 166
1,2-Dichloropropane	63	65, 114
trans-1,3-Dichloropropene	75	77
Trichloroethene	130	95, 97, 132
Dibromochloromethane	129	208, 206
1,1,2-Trichloroethane	97	83, 85, 99, 132, 134
Benzene	78	–
cis-1,3-Dichloropropene	75	77
Bromoform	173	171, 175, 250, 252, 254, 256
2-Hexanone	43	58, 57, 100
4-Methyl-2-pentanone	43	58, 100
Tetrachloroethene	164	129, 131, 166
Toluene	91	92
Chlorobenzene	112	114
Ethylbenzene	106	91
Styrene	104	78, 103
Total Xylenes	106	91

*m/z 43 is used for quantitation of 2-Butanone, but m/z 72 must be present for positive identification.

Volatile Internal Standards with Corresponding Target Compounds and System Monitoring Compounds Assigned for Quantitation

Bromochloromethane	1,4-Difluorobenzene	Chlorobenzene-d5
Chloromethane	1,1,1-Trichloroethane	2-Hexanone
Bromomethane	Carbon tetrachloride	4-Methyl-2-pentanone
Vinyl chloride	Bromodichloromethane	Tetrachloroethene
Chloroethane	1,2-Dichloropropane	1,1,2,2-Tetrachloroethane
Methylene chloride	*trans*-1,3-Dichloropropene	Toluene
Acetone	Trichloroethene	Chlorobenzene
Carbon disulfide	Dibromochloromethane	Ethylbenzene
1,1-Dichloroethene	1,1,2-Trichloroethane	Styrene
1,1-Dichloroethane	Benzene	Xylenes (total)
1,2-Dichloroethene (total)	*cis*-1,3-Dichloropropene	4-Bromofluorobenzene (SMC)
Chloroform	Bromoform	Toluene-d8 (SMC)
1,2-Dichloroethane		
2-Butanone		
1,2-Dichloroethane-d4 (SMC)		

(SMC) = system monitoring compound

Volatile Organic Analysis – Organic Statement of Work – *Table 4*

Characteristic Ions for System Monitoring Compounds and Internal Standards for Volatile Organic Compounds with CAS Numbers

Compound	Primary Quantitation Ion	Secondary Ion(s)	CAS Number
System Monitoring Compounds			
4-Bromofluorobenzene	95	174, 176	460-00-4
1,2-Dichloroethane-d4	65	102	17060-07-0
Toluene-d8	98	70, 100	2037-26-5
Internal Standards			
Bromochloromethane	128	49, 130, 51	74-97-5
1,4-Difluorobenzene	114	63, 88	540-36-3
Chlorobenzene-d5	117	82, 119	3114-55-4

Appendix A

Relative Response Factor Criteria for Initial and Continuing Calibration of Volatile Organic Compounds

Volatile Compound	Minimum RRF	Maximum %RSD	Maximum %Diff
Chloromethane	0.010	none	none
Bromomethane	0.100	20.5	25.0
Vinyl chloride	0.100	20.5	25.0
Chloroethane	0.010	none	none
Methylene chloride	0.010	none	none
Acetone	0.010	none	none
Carbon disulfide	0.010	none	none
1,1-Dichloroethene	0.100	20.5	25.0
1,1-Dichloroethane	0.200	20.5	25.0
1,2-Dichloroethene (total)	0.010	none	none
Chloroform	0.200	20.5	25.0
1,2-Dichloroethane	0.100	20.5	25.0
2-Butanone	0.010	none	none
1,1,1-Trichloroethane	0.100	20.5	25.0
Carbon tetrachloride	0.100	20.5	25.0
Bromodichloromethane	0.200	20.5	25.0
1,2-Dichloropropane	0.010	none	none
cis-1,3-Dichloropropene	0.200	20.5	25.0
Trichloroethene	0.300	20.5	25.0
Dibromochloromethane	0.100	20.5	25.0
1,1,2-Trichloroethane	0.100	20.5	25.0
Benzene	0.500	20.5	25.0
trans-1,3-Dichloropropene	0.100	20.5	25.0
Bromoform	0.100	20.5	25.0
4-Methyl-2-pentanone	0.010	none	none
2-Hexanone	0.010	none	none
Tetrachloroethene	0.200	20.5	25.0
1,1,2,2-Tetrachloroethane	0.300	20.5	25.0
Toluene	0.400	20.5	25.0
Chlorobenzene	0.500	20.5	25.0
Ethylbenzene	0.100	20.5	25.0
Styrene	0.300	20.5	25.0
Xylenes (total)	0.300	20.5	25.0
System Monitoring Compounds			
Bromofluorobenzene	0.200	20.5	25.0
Toluene-d_8	0.0.10	none	none
1,2-Dichloroethane-d_4	0.0.10	none	none

Volatile Organic Analysis – Organic Statement of Work – *Table 7*

System Monitoring Compound Recovery Limits

Compound	%Recovery Water	%Recovery Soil
Toluene-d$_8$	88-110	84-138
Bromofluorobenzene	86-115	59-113
1,2-Dichloroethane-d$_4$	76-114	70-121

Volatile Organic Analysis – Organic Statement of Work – *Table 8*

Matrix Spike Recovery and Relative Percent Difference Limits

Compound	%Recovery Water	RPD Water	%Recovery Soil	RPD Soil
1,1-Dichloroethane	61-145	14	59-172	22
Trichloroethene	71-120	14	62-137	24
Benzene	76-127	11	66-142	21
Toluene	76-125	13	59-139	21
Chlorobenzene	75-130	13	60-133	21

Inorganic Statement of Work – *Exhibit C*

Appendix A

Inorganic Target Analyte List (TAL)

Analyte	Contract Required Detection Limit (µg/L) (1,2)
Aluminum	200
Antimony	60
Arsenic	10
Barium	200
Beryllium	5
Cadmium	5
Calcium	5000
Chromium	10
Cobalt	50
Copper	25
Iron	100
Lead	3
Magnesium	5000
Manganese	15
Mercury	0.2
Nickel	40
Potassium	5000
Selenium	5
Silver	10
Sodium	5000
Thallium	10
Vanadium	50
Zinc	20
Cyanide	10

(1) Subject to the restrictions specified in the first page of Part G, Section IV of Exhibit D (Alternate Methods - Catastrophic Failure) any analytical method specified in SOW Exhibit D may be utilized as long as the documented instrument or method detection limits meet the Contract Required Detection Limit (CRDL) requirements. Higher detection limits may only be used in the following circumstance:

> If the sample concentration exceeds five times the detection limit of the instrument or method in use, the value may be reported even though the instrument or method detection limit may not equal the Contract Required Detection Limit. This is illustrated in the example below:

> For lead:

> Method in use = ICP Instrument Detection Limit (IDL) = 40 Sample concentration = 220 Contract Required Detection Limit (CRDL) = 3

> The value of 220 may be reported even though the instrument detection limit is greater than CRDL. The instrument or method detection limit must be documented as described in Exhibits B and E.

(2) The CRDL's are the instrument detection limits obtained in pure water that must be met using the procedure in Exhibit E. The detection limits for samples may be considerably higher depending on the sample matrix.

Atomic Absorption Concentration Ranges[1]

Metal	Direct Aspiration			Furnace Procedure[4,5]	
	Detection Limit mg/L	Sensitivity mg/L	Optimum Concentration Range mg/L	Detection Limit µg/L	Optimum Concentration Range µg/L
Aluminum	0.1	1	5-50	3	20-200
Antimony	0.2	0.5	1-40	3	20-300
Arsenic[2]	0.002		0.002-0.02	1	5-100
Barium(p)	0.1	0.4	1-20	2	10-200
Beryllium	0.005	0.025	0.05-2	0.2	1-30
Cadmium	0.005	0.025	0.05-2	0.1	0.5-10
Calcium	0.01	0.08	0.2-7		
Chromium	0.05	0.25	0.5-10	1	5-100
Cobalt	0.05	0.2	0.5-5	1	5-100
Copper	0.02	0.1	0.2-5	1	5-100
Gold	0.1	0.25	0.5-20	1	5-100
Iridium(p)	3	8	20-500	30	100-1500
Iron	0.03	0.12	0.3-5	1	5-100
Lead	0.1	0.5	1-20	1	5-100
Magnesium	0.001	0.007	0.02-0.5		
Manganese	0.01	0.05	0.1-3	0.2	1-30
Mercury[3]	0.0002		0.0002-0.01		
Molybdenum(p)	0.1	0.4	1-40	1	3-60
Nickel(p)	0.04	0.15	0.3-5	1	5-50
Osmium	0.3	1	2-100	20	50-500
Palladium(p)	0.1	0.25	0.5-15	5	20-400
Platinum(p)	0.2	2	5-75	20	100-2000
Potassium	0.01	0.04	0.1-2		
Rhenium(p)	5	15	50-1000	200	500-5000
Rhodium(p)	0.05	0.3	1-30	5	20-400
Ruthenium	0.2	0.5	1-50	20	100-2000
Selenium[2]	0.002		0.002-0.02	2	5-100
Silver	0.01	0.06	0.1-4	0.2	1-25
Sodium	0.002	0.015	0.03-1		
Thallium	0.1	0.5	1-20	1	5-100
Tin	0.8	4	10-300	5	20-300
Titanium(p)	0.4	2	5-100	10	50-500
Vanadium(p)	0.2	0.8	2-100	4	10-200
Zinc	0.005	0.02	0.05-1	0.05	0.2-4

(1) The concentrations shown are not contrived values and should be obtainable with any satisfactory atomic absorption spectrophotometer.
(2) Gaseous hydride method.
(3) Cold vapor technique.
(4) For furnace sensitivity values consult instrument operating manual.
(5) The listed furnace values are those expected when using a 20 µL injection and normal gas flow except in the case of arsenic and selenium where gas interrupt is used.
(p) Indicates the use of pyrolytic graphite with the furnace procedure.

Method 200.7 – *Table 1*

Recommended Wavelengths[1] and Estimated Instrumental Detection Limits

Element	Wavelength, nm[1]	Estimated Detection Limit (μg/L)[2]	Calibrate[3] to (mg/L)
Aluminum	308.215	45	10
Antimony	206.833	32	5
Arsenic	193.759	53	10
Barium	493.409	2.3	1
Beryllium	313.042	0.27	1
Boron	249.678	5.7	1
Cadmium	226.502	3.4	2
Calcium	315.887	30	10
Cerium	413.765	48	2
Chromium	205.552	6.1	5
Cobalt	228.616	7.0	2
Copper	324.754	5.4	2
Iron	259.940	6.2	10
Lead	220.353	42	10
Lithium	670.784	3.7[4]	5
Magnesium	279.079	30	10
Manganese	257.610	1.4	2
Mercury	194.227	2.5	2
Molybdenum	203.844	12	10
Nickel	231.604	76	2
Phosphorus	214.914	76	10
Potassium	766.491	700[5]	20
Selenium	196.090	75	5
Silica (SiO_2)	251.611	26[4] (SiO_2)	10
Silver	328.068	7.0	0.5
Sodium	588.995	29	10
Strontium	421.552	0.77	1
Thallium	190.864	40	5
Tin	189.980	25	4
Titanium	334.941	3.8	10
Vanadium	292.402	7.5	2
Zinc	213.856	1.8	5

(1) The wavelengths listed are recommended because of their sensitivity and overall acceptance. Other wavelengths may be substituted if they can provide the needed sensitivity and are treated with the same corrective techniques for spectral interference.
(2) These estimated 3-sigma instrumental detection limits are provided only as a guide to instrumental limits. The method detection limits are sample dependent and may vary as the sample matrix varies. Detection limits for solids can be estimated by dividing these values by the grams extracted per liter, which depends upon the extraction procedure. Divide solution detection limits by 10 for 1 g extracted to 100 mL for solid detection limits.
(3) Suggested concentration for instrument calibration. Other calibration limits in the linear ranges may be used.
(4) Calculated from 2-sigma data.
(5) Highly dependent on operating conditions and plasma position.

Method 200.8 – *Table 1*

Estimated Instrument Detection Limits

Appendix A

Element	Recommended Analytical Mass	Estimated IDLS(μg/L)	
		Scanning Mode[1]	**Selective Ion Monitoring Mode[2]**
Aluminum	27	0.05	0.02
Antimony	123	0.08	0.008
Arsenic[3]	75	0.9	0.02
Barium	137	0.5	0.03
Beryllium	9	0.1	0.02
Cadmium	111	0.1	0.02
Chromium	52	0.07	0.04
Cobalt	59	0.03	0.002
Copper	63	0.03	0.004
Lead	206, 207, 208	0.08	0.015
Manganese	55	0.1	0.007
Mercury	202	n.a.	0.2
Molybdenum	98	0.1	0.005
Nickel	60	0.2	0.07
Selenium[3]	82	5	1.3
Silver	107	0.05	0.004
Thallium	205	0.09	0.014
Thorium	232	0.03	0.005
Uranium	238	0.02	0.005
Vanadium	51	0.02	0.006
Zinc	66	0.2	0.07

Instrument detection limits (3σ) estimated from seven replicate integrations of the blank (1% v/v nitric acid) following calibration of the instrument with three replicate integrations of a multi-element standard.

[1] Instrument operating conditions and data acquisition mode are given in Table 6.

[2] IDLs determined using state-of-the-art instrumentation (1994). Data for ^{75}As, ^{77}Se, and ^{82}Se were acquired using a dwell time of 4.096 sec with 1500 area count per sec ^{83}Kr present in argon supply. All other data were acquired using a dwell time of 1.024 sec per AMU monitored.

Method 200.8 – *Table 2*

Common Molecular Ion Interferences in ICP-MS

Background Molecular Ions		
Molecular Ion	**Mass**	**Element Interference(1)**
NH^+	15	
OH^+	17	
OH_2^+	18	
C_2^+	24	
CN^+	26	
CO^+	28	
N_2^+	28	
N_2H^+	29	
NO^+	30	
NOH^+	31	
O_2^+	32	
O_2H^+	33	
$^{36}ArH^+$	37	
$^{38}ArH^+$	39	
$^{40}ArH^+$	41	
CO_2^+	44	
CO_2H^+	45	Sc
ArC^+, ArO^+	52	Cr
ArN^+	54	Cr
$ArNH^+$	55	Mn
ArO^+	56	
$ArOH^+$	57	
$^{40}Ar^{36}Ar^+$	76	Se
$^{40}Ar^{38}Ar^+$	78	Se
$^{40}Ar_2^+$	80	Se

Matrix Molecular Ions

BROMIDE[12]

Molecular Ion	Mass	Element Interference
$^{81}BrH^+$	82	Se
$^{79}BrO^+$	95	Mo
$^{81}BrO^+$	97	Mo
$^{81}BrOH^+$	98	Mo
$Ar^{81}Br^+$	121	Sb

CHLORIDE

Molecular Ion	Mass	Element Interference
$^{35}ClO^+$	51	V
$^{35}ClOH^+$	52	Cr
$^{37}ClO^+$	53	Cr
$^{37}ClOH^+$	54	Cr
$Ar^{35}Cl^+$	75	As
$Ar^{37}Cl^+$	77	Se

SULPHATE

Molecular Ion	Mass	Element Interference
$^{32}SO^+$	48	
$^{32}SOH^+$	49	
$^{34}SO^+$	50	V, Cr
$^{34}SOH^+$	51	V
SO_2^+, S_2^+	64	Zn
$Ar^{32}S^+$	72	
$Ar^{34}S^+$	74	

PHOSPHATE

Molecular Ion	Mass	Element Interference
PO^+	47	
POH^+	48	
PO_2^+	63	Cu
ArP^+	71	

GROUP I, II METALS

Molecular Ion	Mass	Element Interference
$ArNa^+$	63	Cu
ArK^+	79	
$ArCa^+$	80	

MATRIX OXIDES(2)

Molecular Ion	Mass	Element Interference
TiO	62-66	Ni, Cu, Zn
ZrO	106-112	Ag, Cd
MoO	108-116	Cd

(1) method elements or internal standards affected by the molecular ions.
(2) Oxide interferences will normally be very small and will only impact the method elements when present at relatively high concentrations. Some examples of matrix oxides are listed of which the analyst should be aware. It is recommended that Ti and Zr isotopes are monitored in solid waste samples, which are likely to contain high levels of these elements. Mo is monitored as a method analyte.

Method 200.8 – *Table 3*

Internal Standards and Limitations of Use

Internal Standards	Mass	Possible Limitation
^{6}Lithium	6	(a)
Scandium	45	polyatomic ion interference
Yttrium	89	(a), (b)
Rhodium	103	
Indium	115	isobaric interference by Sn
Terbium	159	
Holmium	165	
Lutetium	175	
Bismuth	209	(a)

(a)   May be present in environmental samples.

(b)   In some instruments Yttrium may form measurable amounts of YO$^+$ (105 amu) and YOH$^+$ (106 amu). If this is the case, care should be taken in the use of the cadmium elemental correction equation.

Internal standards recommended for use with this method are shown in bold face. Preparation procedures for these are included in Section 7.3.

Method 200.8 – *Table 4*

Recommended Analytical Isotopes and Additional Masses that Must Be Monitored

Isotope	Element of Interest
27	Aluminum
121, 123	Antimony
75	Arsenic
135, 137	Barium
9	Beryllium
106, 108, 111, 114	Cadmium
52, 53	Chromium
59	Cobalt
63, 65	Copper
206, 207, 208	Lead
55	Manganese
95, 97, 98	Molybdenum
60, 62	Nickel
77, 82	Selenium
107, 109	Silver
203, 205	Thallium
232	Thorium
238	Uranium
51	Vanadium
66, 67, 68	Zinc
83	Krypton
99	Ruthenium
105	Palladium
118	Tin

NOTE: Isotopes recommended for analytical determinations are underlined.

Method 200.8 – *Table 5*

Recommended Elemental Equations for Data Calculations

Element	Elemental Equation	Note
Al	$(1.000)(^{27}C)$	
Sb	$(1.000)(^{123}C)$	
As	$(1.000)(^{75}C)-(3.127)[(^{77}C)-(0.815)(^{82}C)]$	(1)
Ba	$(1.000)(^{137}C)$	
Be	$(1.000)(^{9}C)$	
Cd	$(1.000)(^{111}C)-(1.073)[(^{108}C)-(0.712)(^{106}C)]$	(2)
Cr	$(1.000)(^{52}C)$	(3)
Co	$(1.000)(^{59}C)$	
Cu	$(1.000)(^{63}C)$	
Pb	$(1.000)(^{206}C)+(1.000)(^{207}C)+(1.000)(^{208}C)$	(4)
Mn	$(1.000)(^{55}C)$	
Mo	$(1.000)(^{98}C)-(0.146)(^{9}C)$	(5)
Ni	$(1.000)(^{60}C)$	
Se	$(1.000)(^{82}C)$	(6)
Ag	$(1.000)(^{107}C)$	
Tl	$(1.000)(^{205}C)$	
Th	$(1.000)(^{232}C)$	
U	$(1.000)(^{238}C)$	
V	$(1.000)(^{51}C)-(3.127)[(^{53}C)-(0.113)(^{52}C)]$	(7)
Zn	$(1.000)(^{66}C)$	

INTERNAL STANDARDS

Element	Element Equation	Note
Bi	$(1.000)(^{209}C)$	
In	$(1.000)(^{115}C)-(0.016)(^{118}C)$	(8)
Sc	$(1.000)(^{45}C)$	
Tb	$(1.000)(^{159}C)$	
Y	$(1.000)(^{89}C)$	

C   Calibration blank subtracted counts at specified mass.

(1)   Correction for chloride interference with adjustment for ^{77}Se. ArCl 75/77 ratio may be determined from the reagent blank. Isobaric mass 82 must be from Se only and not BrH^{+}.

(2)   Correction for MoO interference. Isobaric mass 106 must be from Cd only not ZrO^{+}. An additional isobaric elemental correction should be made if palladium is present.

(3)   In 0.4% v/v HCL, the background from ClOH will normally be small. However the contribution may be estimated from the reagent blank. Isobaric mass must be from Cr only not ArC^{+}.

(4)   Allowance for isotopic variability of lead isotopes.

(5)   Isobaric elemental correction for ruthenium.

(6)   Some argon supplies contain krypton as an impurity. Selenium is corrected for ^{85}Kr by background subtraction.

(7)   Correction for chloride interference with adjustment for ^{53}Cr. ClO 51/53 ratio may be determined from the reagent blank. Isobaric mass 52 must be from Cr only not ArC^{+}.

(8)   Isobaric elemental correction for tin.

Method 200.8 – *Table 6*

Appendix A

Instrument Operating Conditions for Precision and Recovery Data[1]

Instrument	VG PlasmaQuad Type I
Plasma forward power	1.35 kW
Coolant flow rate	13.5 L/min
Auxiliary flow rate	0.6 L/min
Nebulizer flow rate	0.78 L/min
Solution uptake rate	0.6 mL/min
Spray Chamber temperature	15 °C
Data Acquisition	
Detector mode	Pulse counting
Replicate integrations	3
Mass range	8 - 240 amu
Dwell time	320 µs
Number of MCA channels	2048
Number of scan sweeps	85
Total acquisition time	3 minutes per sample

(1) The described instrument and operating conditions were used to determine the scanning mode MDL data listed in Table 7 and the precision and recovery data given in Tables 9 and 10.

Method 200.8 – *Table 7*

Method Detection Limits

AMU Element	Scanning Mode[1] Total Recoverable		Selective Ion Monitoring Mode[2]	
	Aqueous µg/L	Solids mg/kg	Total Recoverable Aqueous µg/L	Direct Analysis[3] Aqueous µg/L
27 Al	1.0	0.4	1.7	0.04
123 Sb	0.4	0.2	0.04	0.02
75 As	1.4	0.6	0.4	0.1
137 Ba	0.8	0.4	0.04	0.04
9 Be	0.3	0.1	0.02	0.03
111 Cd	0.5	0.2	0.03	0.03
52 Cr	0.9	0.4	0.08	0.08
59 Co	0.09	0.04	0.004	0.003
63 Cu	0.5	0.2	0.02	0.01
206, 207, 208 Pb	0.6	0.3	0.05	0.02
55 Mn	0.1	0.05	0.02	0.04
202 Hg	n.a.	n.a.	n.a.	0.2
98 Mo	0.3	0.1	0.01	0.01
60 Ni	0.5	0.2	0.06	0.03
82 Se	7.9	3.2	2.1	0.5
107 Ag	0.1	0.05	0.005	0.005
205 Tl	0.3	0.1	0.02	0.01
232 Th	0.1	0.05	0.02	0.01
238 U	0.1	0.05	0.01	0.01
51 V	2.5	1.0	0.9	0.05
66 Zn	1.8	0.7	0.1	0.2

(1) Data acquisition mode given in Table 6. Total recoverable MDL concentrations are computed for original matrix with allowance for sample dilution during preparation. Listed MDLs for solids calculated from determined aqueous MDLs.

(2) MDLs determined using state-of-the-art instrumentation (1994). Data for ^{75}As, ^{77}Se, and ^{82}Se were acquired using a dwell time of 4.096 sec with 1500 area count per sec ^{83}Kr present in argon supply. All other data were acquired using a dwell time of 1.024 sec per AMU monitored.

(3) MDLs were determined from analysis of 7 undigested aqueous sample aliquots.

n.a. Not applicable. Total recoverable digestion not suitable for organo-mercury compounds

Method 200.8 – *Table 8*

Appendix A

Acceptance Limits for QC Check Sample

Element	Method Performance (μg/L)[1]			
	QC Check Sample Conc.	Average Recovery	Standard Deviation[2] (S_r)	Acceptance Limits[3] μg/L
Aluminum	100	100.4	5.49	84-117
Antimony	100	99.9	2.40	93-107
Arsenic	100	101.6	3.66	91-113
Barium	100	99.7	2.64	92-108
Beryllium	100	105.9	4.13	88-112[4]
Cadmium	100	100.8	2.32	94-108
Chromium	100	102.3	3.91	91-114
Cobalt	100	97.7	2.66	90-106
Copper	100	100.3	2.11	94-107
Lead	100	104.0	3.42	94-114
Manganese	100	98.3	2.71	90-106
Molybdenum	100	101.0	2.21	94-108
Nickel	100	100.1	2.10	94-106
Selenium	100	103.5	5.67	86-121
Silver	100	101.1	3.29	91-111[5]
Thallium	100	98.5	2.79	90-107
Thorium	100	101.4	2.60	94-109
Uranium	100	102.6	2.82	94-111
Vanadium	100	100.3	3.26	90-110
Zinc	100	105.1	4.57	91-119

(1) Method performance characteristics calculated using regression equations from collaborative study, reference 11.

(2) Single-analyst standard deviation, S_r.

(3) Acceptance limits calculated as average recovery ±3 standard deviations.

(4) Acceptance limits centered at 100% recovery.

(5) Statistics estimated from summary statistics at 48 and 64 μg/L.

Method 502.2 – *Table 2*

Single Laboratory Accuracy, Precision and Method Detection Limits for Volatile Organic Compounds in Reagent Water for Column 1[a][b]

Analyte	Photoionization Detector			Electrolytic Conductivity Detector		
	Average Recovery, %	Rel. Std. Deviation	MDL (µg/L)	Average Recovery, %	Rel. Std. Deviation	MDL (µg/L)
Benzene	99	1.2	0.01	(c)	(c)	(c)
Bromobenzene	99	1.7	0.01	97	2.7	0.03
Bromochloromethane	(c)	(c)	(c)	96	3.0	0.01
Bromodichloromethane	(c)	(c)	(c)	97	2.9	0.02
Bromoform	(c)	(c)	(c)	106	5.2	1.6
Bromomethane	(c)	(c)	(c)	97	3.8	1.1
n-Butylbenzene	100	4.4	0.02	(c)	(c)	(c)
sec-Butylbenzene	97	2.7	0.02	(c)	(c)	(c)
tert-Butylbenzene	98	2.3	0.06	(c)	(c)	(c)
Carbon tetrachloride	(c)	(c)	(c)	92	3.6	0.01
Chlorobenzene	100	1.0	0.01	103	3.6	0.01
Chloroethane	(c)	(c)	(c)	96	3.9	0.1
Chloroform	(c)	(c)	(c)	98	2.5	0.02
Chloromethane	(c)	(c)	(c)	96	9.2	0.03
2-Chlorotoluene	ND	ND	ND	97	2.7	0.01
4-Chlorotoluene	101	1.0	0.02	97	3.2	0.01
1,2-Dibromo-3-chloropropane	(c)	(c)	(c)	86	11.3	3.0
Dibromochloromethane	(c)	(c)	(c)	102	3.3	0.3
1,2-Dibromoethane	(c)	(c)	(c)	97	2.8	0.8
Dibromomethane	(c)	(c)	(c)	109	6.7	2.2
1,2-Dichlorobenzene	102	2.1	0.05	100	1.5	0.02
1,3-Dichlorobenzene	104	1.6	0.02	106	4.0	0.02
1,4-Dichlorobenzene	103	2.1	0.01	98	2.3	0.01
Dichlorodifluoromethane	(c)	(c)	(c)	89	6.6	0.05
1,1-Dichloroethane	(c)	(c)	(c)	100	5.7	0.07
1,2-Dichloroethane	(c)	(c)	(c)	100	3.8	0.03
1,1-Dichloroethene	100	2.4	ND	103	2.8	0.07
cis-1,2-Dichloroethene	ND	ND	0.02	105	3.3	0.01
trans-1,2-Dichloroethene	93	4.0	0.05	99	3.7	0.06
1,2-Dichloropropane	(c)	(c)	(c)	103	3.7	0.01
1,3-Dichloropropane	(c)	(c)	(c)	100	3.4	0.03

(a) Recoveries and relative standard deviations were determined from seven samples fortified at 10 µg/L of each analyte. Recoveries were determined by internal standard method. Internal standards were: Fluorobenzene for PID, 2-Bromo-1-chloropropane for ELCD.

(b) Column 1: 60m long x 0.75mm ID VOCOL (Supelco, Inc.) wide-bore capillary column with 1.5 µm film thickness.

(c) Detector does not respond.

ND Not determined.

Method 502.2 – *Table 2*

Single Laboratory Accuracy, Precision and Method Detection Limits for Volatile Organic Compounds in Reagent Water for Column 1[a][b] (Continued)

Analyte	Photoionization Detector			Electrolytic Conductivity Detector		
	Average Recovery, %	Rel. Std. Deviation	MDL (μg/L)	Average Recovery, %	Rel. Std. Deviation	MDL (μg/L)
2,2-Dichloropropane	(c)	(c)	(c)	105	3.4	0.05
1,1-Dichloropropene	103	3.5	0.02	103	3.3	0.02
Ethyl benzene	101	1.4	0.01	(c)	(c)	(c)
Hexachlorobutadiene	99	9.5	0.06	98	8.3	0.02
Isopropylbenzene	98	0.9	0.05	(c)	(c)	(c)
Methylene chloride	(c)	(c)	(c)	97	2.9	0.02
Naphthalene	102	6.2	0.06	(c)	(c)	(c)
n-Propylbenzene	103	2.0	0.01	(c)	(c)	(c)
Styrene	104	1.3	0.01	(c)	(c)	(c)
1,1,1,2-Tetrachloroethane	(c)	(c)	(c)	99	2.3	0.01
1,1,2,2-Tetrachloroethane	(c)	(c)	(c)	99	6.8	0.01
Tetrachloroethene	101	1.8	0.05	97	2.5	0.04
Toluene	99	0.8	0.01	(c)	(c)	(c)
1,2,3-Trichlorobenzene	106	1.8	ND	98	3.1	0.03
1,2,4-Trichlorobenzene	104	2.2	0.02	102	2.1	0.03
1,1,1-Trichloroethane	(c)	(c)	(c)	104	3.3	0.03
1,1,2-Trichloroethane	(c)	(c)	(c)	109	5.6	ND
Trichloroethene	100	0.78	0.02	96	3.6	0.01
Trichlorofluoromethane	(c)	(c)	(c)	96	3.5	0.03
1,2,3-Trichloropropane	(c)	(c)	(c)	99	2.3	0.04
1,2,4-Trimethylbenzene	99	1.2	0.05	(c)	(c)	(c)
1,3,5-Trimethylbenzene	101	1.4	0.01	(c)	(c)	(c)
Vinyl chloride	109	5.0	0.02	95	5.9	0.04
o-Xylene	99	0.8	0.02	(c)	(c)	(c)
m-Xylene	100	1.4	0.01	(c)	(c)	(c)
p-Xylene	99	0.9	0.01	(c)	(c)	(c)

(a) Recoveries and relative standard deviations were determined from seven samples fortified at 10 μg/L of each analyte. Recoveries were determined by internal standard method. Internal standards were: Fluorobenzene for PID, 2-Bromo-1-chloropropane for ELCD.

(b) Column 1: 60m long x 0.75mm ID VOCOL (Supelco, Inc.) wide-bore capillary column with 1.5 μm film thickness.

(c) Detector does not respond.

ND Not determined.

Method 502.2 – *Table 4*

Single Laboratory Accuracy, Precision and Method Detection Limits for Volatile Organic Compounds in Reagent Water for Column 2[a][b]

Analyte	Photoionization Detector			Electrolytic Conductivity Detector		
	Average Recovery, %	Rel. Std. Deviation	MDL (µg/L)	Average Recovery, %	Rel. Std. Deviation	MDL (µg/L)
Bromobenzene	98	1.1	0.04	96	3.2	0.14
Bromochloromethane	(c)	(c)	(c)	95	2.5	0.01
Bromodichloromethane	(c)	(c)	(c)	96	2.6	0.10
Bromoform	(c)	(c)	(c)	98	4.0	0.09
Bromomethane	(c)	(c)	(c)	97	2.4	0.19
n-Butylbenzene	95	2.4	0.03	(c)	(c)	(c)
sec-Butylbenzene	96	2.1	0.03	(c)	(c)	(c)
tert-Butylbenzene	98	2.1	0.06	(c)	(c)	(c)
Carbon tetrachloride	(c)	(c)	(c)	97	2.4	0.02
Chlorobenzene	98	1.5	0.02	98	2.2	ND
Chloroethane	(c)	(c)	(c)	97	3.2	0.13
Chloroform	(c)	(c)	(c)	92	4.2	0.01
Chloromethane	(c)	(c)	(c)	98	2.3	0.10
2-Chlorotoluene	94	3.1	0.03	99	2.3	0.04
4-Chlorotoluene	97	1.6	0.02	97	2.3	0.07
1,2-Dibromo-3-chloropropane	(c)	(c)	(c)	97	2.2	0.05
Dibromochloromethane	(c)	(c)	(c)	99.	2.0	0.05
1,2-Dibromoethane	(c)	(c)	(c)	99	2.8	0.17
Dibromomethane	(c)	(c)	(c)	98	3.5	0.10
1,2-Dichlorobenzene	97	1.4	0.03	98	2.0	0.04
1,3-Dichlorobenzene	97	1.6	0.02	97	2.2	0.07
1,4-Dichlorobenzene	97	1.5	0.03	97	2.2	0.04
Dichlorodifluoromethane	(c)	(c)	(c)	96	3.2	0.29
1,1-Dichloroethane	(c)	(c)	(c)	97	2.3	0.03
1,2-Dichloroethane	(c)	(c)	(c)	98	1.8	0.03
1,1-Dichloroethene	96	2.2	0.01	97	2.3	0.04
cis-1,2-Dichloroethene	97	1.7	0.03	96	3.3	0.05
trans-1,2-Dichloroethene	97	1.8	0.03	98	1.5	0.05
1,2-Dichloropropane	(c)	(c)	(c)	98	1.8	0.03
1,3-Dichloropropane	(c)	(c)	(c)	100	1.3	0.02
2,2-Dichloropropane	(c)	(c)	(c)	95	14.2	ND
1,1-Dichloropropene	96	2.1	0.05	97	2.6	0.02

(a) Recoveries and relative standard deviations were determined from seven samples fortified at 10 µg/L of each analyte. Recoveries were determined by external standard method.

(b) Column 2: 105m long x 0.53mm ID RTX-502.2 (RESTEK Corporation) mega-bore capillary column with 3.0 µm film thickness.

(c) Detector does not respond.

ND = Not determined.

Method 502.2 – *Table 4*

Single Laboratory Accuracy, Precision and Method Detection Limits for Volatile Organic Compounds in Reagent Water for Column 2[a][b] (Continued)

Analyte	Photoionization Detector			Electrolytic Conductivity Detector		
	Average Recovery, %	Rel. Std. Deviation	MDL (µg/L)	Average Recovery, %	Rel. Std. Deviation	MDL (µg/L)
cis-1,3-Dichloropropene	98	1.6	0.06	98	2.0	0.08
trans-1,3-Dichloropropene	99	1.7	0.06	97	1.4	0.10
Hexachlorobutadiene	95	2.6	0.09	97	2.3	0.05
Isopropylbenzene	97	1.4	0.02	(c)	(c)	(c)
4-Isopropyltoluene	96	2.0	0.02	(c)	(c)	(c)
Methylene chloride	(c)	(c)	(c)	100	3.1	0.01
Naphthalene	96	2.1	0.02	(c)	(c)	(c)
n-Propylbenzene	97	1.8	0.03	(c)	(c)	(c)
Styrene	96	1.9	0.10	(c)	(c)	(c)
1,1,1,2-Tetrachloroethane	(c)	(c)	(c)	98	2.2	ND
1,1,2,2-Tetrachloroethane	(c)	(c)	(c)	100	2.8	0.02
Tetrachloroethene	97	1.6	0.04	97	1.9	0.02
Toluene	98	1.3	0.02	(c)	(c)	(c)
1,2,3-Trichlorobenzene	95	2.3	0.05	98	2.8	0.06
1,2,4-Trichlorobenzene	94	3.0	0.06	96	2.5	0.08
1,1,1-Trichloroethane	(c)	(c)	(c)	96	2.6	0.01
1,1,2-Trichloroethane	(c)	(c)	(c)	99	1.6	0.04
Trichloroethene	97	1.7	0.03	98	1.2	0.06
Trichlorofluoromethane	(c)	(c)	(c)	97	6.0	0.34
1,2,3-Trichloropropane	(c)	(c)	(c)	100	2.0	0.02
1,2,4-Trimethylbenzene	96	2.0	0.02	(c)	(c)	(c)
1,3,5-Trimethylbenzene	98	1.6	0.03	(c)	(c)	(c)
Vinyl chloride	95	1.1	0.01	96	2.6	0.18
o-Xylene	98	1.1	0.02	(c)	(c)	(c)
m-Xylene	98	1.1	0.02	(c)	(c)	(c)
p-Xylene	98	0.9	0.02	(c)	(c)	(c)

(a) Recoveries and relative standard deviations were determined from seven samples fortified at 10 µg/L of each analyte. Recoveries were determined by external standard method.

(b) Column 2: 105m long x 0.53mm ID RTX-502.2 (RESTEK Corporation) mega-bore capillary column with 3.0 µm film thickness.

(c) Detector does not respond.

ND = Not determined.

Method 508 – *Table 2*

Single Laboratory Accuracy, Precision and Estimated Detection Limits (EDLS) for Analytes from Reagent Water and Synthetic Groundwaters[1]

Analyte	EDL[2] µg/L	Conc. µg/L	Reagent Water R[3]	Reagent Water S_R[4]	Synthetic Water 1[5] R	Synthetic Water 1[5] S_R	Synthetic Water 2[6] R	Synthetic Water 2[6] S_R
Aldrin	0.075	0.15	86	9.5	100	11.0	69	9.0
Chlordane-α	0.0015	0.15	99	11.9	96	12.5	99	7.9
Chlordane-γ	0.0015	0.15	99	11.9	96	12.5	99	6.9
Chlorneb	0.5	5	97	11.6	95	6.7	75	8.3
Chlorobenzilate	5	10	108	5.4	98	10.8	102	9.2
Chlorthalonil	0.025	0.25	91	8.2	103	10.3	71	9.2
DCPA	0.025	0.25	103	12.4	100	13.0	101	6.1
4,4'-DDD	0.0025	0.25	107	6.4	96	8.6	101	7.1
4,4'-DDE	0.01	0.1	99	11.9	96	12.5	99	6.9
4,4'-DDT	0.06	0.6	112	16.8	98	11.8	84	7.4
Dieldrin	0.02	0.2	87	8.7	103	9.3	82	8.4
Endosulfan I	0.015	0.15	87	8.7	102	8.2	84	8.4
Endosulfan sulfate	0.015	0.15	102	15.3	94	1.3	72	12.2
Endrin	0.015	0.15	88	8.8	98	9.8	104	9.4
Endrin aldehyde	0.025	0.25	88	7.9	103	11.3	84	9.2
Endosulfan II	0.024	0.15	92	10.1	98	10.8	76	6.8
Etridiazole	0.025	0.25	103	6.2	91	6.4	98	3.9
HCH-α	0.025	0.05	92	10.1	106	7.4	86	7.7
HCH-β	0.01	0.1	95	6.7	92	5.5	100	6.0
HCH-δ	0.01	0.1	102	11.2	99	11.9	103	6.2
HCH-γ	0.015	0.15	89	9.8	115	6.9	85	7.7
Heptachlor	0.01	0.1	98	11.8	85	11.1	85	7.7
Heptachlor epoxide	0.015	0.15	87	8.7	103	7.2	82	9.8
Hexachlorobenzene	0.0077	0.05	99	21.8	82	9.8	68	4.8
Methoxychlor	0.05	0.5	105	13.7	101	10.1	104	6.2
cis-Permethrin	0.5	5	91	9.1	96	11.5	86	9.5
trans-Permethrin	0.5	5	111	6.7	97	9.7	102	7.1
Propachlor	0.5	5	103	9.3	116	4.6	95	7.6
Trifluralin	0.025	0.25	103	5.2	86	10.3	87	9.6

(1) Data corrected for amount detected in blank and represent the mean of 7-8 samples.

(2) EDL = estimated detection limit; defined as either MDL (Appendix B to 40 CFR Part 136 -Definition and Procedure for the Determination of the Method Detection Limit - Revision 1.11) or a level of compound in a sample yielding a peak in the final extract with signal-to-noise ratio of approximately 5, whichever value is higher. The concentration level used in determining the EDL is not the same as the concentration level presented in this table.

(3) R = average percent recovery

(4) S_R = standard deviation of the percent recovery

(5) Corrected for amount found in blank; Absopure Nature Artesian Spring Water obtained from the Absopure Water Company in Plymouth, Michigan.

(6) Corrected for amount found in blank; reagent water fortified with fulvic acid at the 1 mg/L concentration level. A well-characterized fulvic acid, available from the International Humic Substances Society (associated with the United States Geological Survey in Denver, Colorado), was used.

Method 508 – *Table 3*

Single Laboratory Accuracy, Precision, Method Detection Limits (MDLS) and Estimated Detection limits (EDLs) for Analytes from Reagent Water

Analyte	Fortified Conc. μg/L	N[1]	Recovery %	RSD %	MDL[2] μg/L	EDL[3] μg/L
Aldrin	0.075	7	66	9	0.014	0.075
Chlordane-alpha	0.015	7	117	8	0.0041	0.0015
Chlordane-gamma	0.015	7	109	3	0.0016	0.0015
Chlorneb	0.50	7	47	34	0.25	0.5
Chlorobenzilate	5.0	8	99	5	2.2	5.0
Chlorothalonil	0.025	7	119	12	0.011	0.025
DCPA	0.025	7	112	4	0.0032	0.025
4,4'-DDD	0.025	7	115	5	0.0044	0.025
4,4'-DDE	0.010	7	127	6	0.0025	0.01
4,4'-DDT	0.60	7	87	23	0.039	0.06
Dieldrin	0.20	7	77	22	0.011	0.02
Endosulfan I	0.015	7	78	25	0.0092	0.015
Endosulfan Sulfate	0.015	7	129	4	0.0024	0.015
Endrin	0.015	7	72	18	0.0062	0.015
Endrin Aldehyde	0.025	7	95	15	0.011	0.025
Endosulfan II	0.015	7	148	35	0.024	0.024
Etridiazole	0.025	7	96	17	0.013	0.025
HCH-alpha	0.025	8	94	8	0.0053	0.025
HCH-beta	0.010	7	95	12	0.0036	0.01
HCH-delta	0.010	7	84	7	0.0020	0.01
HCH-gamma	0.015	7	80	16	0.0060	0.015
Heptachlor	0.010	7	67	7	0.0015	0.01
Heptachlor Epoxide	0.015	7	71	18	0.0059	0.015
Hexachlorobenzene	0.0050	7	115	43	0.0077	0.0077
Methoxychlor	0.050	7	120	11	0.022	0.05
cis-Permethrin	0.50	7	64	24	0.25	0.50
trans-Permethrin	0.50	7	122	9	0.18	0.50
Propachlor	0.50	7	90	18	0.25	0.50
Trifluralin	0.025	7	108	3	0.0026	0.025

(1)   Number of replicates.

(2)   With these data, the method detection limits (MDL) in the tables were calculated using the formula:

$$MDL = S\, t_{(n-1,\ 1-alpha=0.99)}$$

where:

$t_{(n-1,\ 1-alpha=0.99)}$ = Student's t value for the 99% confidence level with n-1 degrees of freedom.

(3)   EDL = estimated detection limit: defined as either MDL (Appendix B to 40 CFR Part 136 - Definition and Procedure for the Determination of the Method Detection Limit - Revision 1.11) or a level of compound in a sample yielding a peak in the final extract with signal-to-noise ratio of approximately 5, whichever is higher.

s   Standard deviation of replicate analytes.

Method 508 – *Table 4*

Laboratory Performance Check Solution

Test	Analyte	Conc, µg/L	Requirements
Sensitivity	Chlorpyrifos	0.0020	Detection of analyte; S/N > 3
Chromatographic Performance	DCPA	0.0500	PSF between 0.80 and 1.15 [a]
Column Performance	Chlorothalonil HCH-delta	0.0500 0.0400	Resolution > 0.50 [b]

(a) PGF - peak Gaussian factor. Calculated using the equation: $PGF = ((1.83) \times W(1/2))/(W(1/10))$. Where W(1/2) is the peak width at half height and W(1/10) is the peak width at tenth height.

(b) Resolution between the two peaks as defined by the equation: $R = t/W$. Where t is the difference in elution times between the two peaks and W is the average peak width, at the baseline, of the two peaks.

Method 515.1 – *Table 2*

Single Laboratory Accuracy, Precision and Estimated Detection Limits (EDLS) for Analytes from Reagent Water and Synthetic Groundwaters[a]

Analyte	EDL µg/L [b]	Concentration µg/L	Reagent Water R[c]	Reagent Water S$_R$[d]	Synthetic Water[e] R	Synthetic Water[e] S$_R$
Acifluorfen	0.096	0.2	121	15.7	103	20.6
Bentazon	0.2	1	120	16.8	82	37.7
Chloramben	0.093	0.4	111	14.4	112	10.1
2,4-D	0.2	1	131	27.5	110	5.5
Dalapon	1.3	10	100	20	128	30.7
2,4-DB	0.8	4	87	13.1	0	0
DCPA acid metabolites	0.02	0.2	74	9.7	81	21.9
Dicamba	0.081	0.4	135	32.4	92	17.5
3,5-Dichlorobenzoic acid	0.061	0.6	102	16.3	82	7.4
Dichlorprop	0.26	2	107	20.3	106	5.3
Dinoseb	0.19	0.4	42	14.3	89	13.4
5-Hydroxydicamba	0.04	0.2	103	16.5	88	5.3
4-Nitrophenol	0.13	1	131	23.6	127	34.3
Pentachlorophenol (PCP)	0.076	0.04	130	31.2	84	9.2
Picloram	0.14	0.6	91	15.5	97	23.3
2,4,5-T	0.08	0.4	117	16.4	96	3.8
2,4,5-TP	0.075	0.2	134	30.8	105	6.3

(a) Data corrected for amount detected in blank and represent the mean of 7-8 samples.

(b) EDL = estimated detection limit; defined as either MDL (Appendix B to 40 CFR Part 136 - Definition and Procedure for the Determination of the Method Detection Limit - Revision 1.11) or a level of compound in a sample yielding a peak in the final extract with signal-to-noise ratio of approximately 5, whichever value is higher. The concentration used in determining the EDL is not the same as the concentration presented in this table.

(c) R = average percent recovery

(d) SR = standard deviation of the percent recovery

(e) Corrected for amount found in blank; Absopure Nature Artesian Spring Water obtained from the Absopure Water Company in Plymouth, Michigan.

Method 515.1 – *Table 3*

Single Laboratory Accuracy, Precision, Method Detection Limits (MDLS) and Estimated Detection limits (EDLs) for Analytes from Reagent Water

Analyte	Fortified Conc. μg/L	N[1]	Recovery μg/L	RSD %	MDL[2] μg/L	EDL[3] μg/L
Acifluorfen	0.040	7	88	23	0.097	0.04
Bentazon	0.20	7	92	19	0.11	0.20
Chloramben	0.080	7	118	33	0.097	0.097
2,4-D	0.20	7	90	14	0.078	0.20
Dalapon	1.0	6	90	45	1.3	1.3
2,4-DB	4.0	6	87	15	1.7	4.0
DCPA acid metabolites	0.20	6	74	13	0.067	0.20
Dicamba	0.080	7	155	22	0.085	0.085
3,5-Dichlorobenzoic acid	0.060	6	69	38	0.065	0.065
Dichlorprop	0.40	7	110	16	0.22	0.40
Dinoseb	0.40	6	42	34	0.33	0.40
5-Hydroxydicamba	0.040	7	49	28	0.017	0.040
4-Nitrophenol	0.20	7	148	15	0.14	0.20
Pentachlorophenol (PCP)	0.10	6	82	9	0.032	0.10
Picloram	0.12	7	166	24	0.15	0.15
2,4,5-T	0.080	7	87	21	0.045	0.80
2,4,5-TP	0.20	6	134	23	0.21	0.21

(1) Number of replicates.

(2) With these data, the method detection limits (MDL) in the tables were calculated using the formula:

$$MDL = S\, t_{(n-1,\ 1-alpha=0.99)}$$

where:

$t_{(n-1,\ 1-alpha=0.99)}$ = Student's t value for the 99% confidence level with n-1 degrees of freedom.

(3) EDL = estimated detection limit: defined as either MDL (Appendix B to 40 CFR Part 136 - Definition and Procedure for the Determination of the Method Detection Limit - Revision 1.11) or a level of compound in a sample yielding a peak in the final extract with signal-to-noise ratio of approximately 5, whichever is higher.

S   Standard deviation of replicate analytes.

Method 515.1 – *Table 11*

Laboratory Performance Check Solution

Test	Analyte	Conc. μg/mL	Requirements
Sensitivity	Dinoseb	0.004	Detection of analyte; S/N > 3
Chromatographic performance	4-Nitrophenol	1.6	0.70 < PGF < 1.05 [b]
Column performance	3,5-Dichlorobenzoic acid	0.6	Resolution >0.40 [b]
	4-Nitrophenol	1.6	

(a) PGF = peak Gaussian factor. Calculated using the equation: $PGF = \dfrac{1.83 \times W(1/2)}{W(1/10)}$, where W(1/2) is the peak width at half height, and W(1/10) is the peak width at tenth height.

(b) Resolution between the two peaks as defined by the equation: R = t/w, where t is the difference in elution times between the two peaks, and W is the average peak width, at the baseline, of the two peaks.

Method 515.2 – *Table 2*

Single Laboratory Recovery, Precision Data and Method Detection Limit with Fortified Reagent Water - Level 1

Analyte	Fortified Conc. µg/L	Mean[a] Recovery %	Relative Std. Dev. %	MDL µg/L
Acifluorfen	0.50	70	21	0.25
Bentazon	2.50	70	11	0.63
2,4-D	0.25	96	38	0.28
2,4-DB	2.50	79	12	0.72
Dacthal[b]	0.25	96	16	0.13
Dicamba	0.75	109	11	0.28
3,5-Dichlorobenzoic Acid	1.25	126	24	1.23
Dichlorprop	0.25	106	15	0.13
Dinoseb	0.50	87	22	0.28
5-Hydroxydicamba	0.75	90	12	0.25
Pentachlorophenol	0.25	103	18	0.16
Picloram	0.75	95	15	0.35
2,4,5-T	0.25	116	18	0.16
2,4,5-TP	0.25	98	9	0.06

(a) Based on the analyses of seven replicates.
(b) Measurement includes the mono- and diacid- metabolites.

Method 515.2 – *Table 11*

Laboratory Performance Check Solution

Test	Analyte	Conc. µg/mL	Requirements
Sensitivity	Dinoseb	0.004	Detection of analyte; S/N > 3
Chromatographic performance	4-Nitrophenol	1.6	0.70 < PGF < 1.05 [a]
Column performance	3,5-Dichlorobenzoic acid	0.6	Resolution >0.40 [b]
	4-Nitrophenol	1.6	

(a) PGF = peak Gaussian factor. Calculated using the equation: $PGF = \dfrac{1.83 \times W(1/2)}{W(1/10)}$, where W(1/2) is the peak width at half height, and W(1/10) is the peak width at tenth height.

(b) Resolution between the two peaks as defined by the equation: R = t/w, where t is the difference in elution times between the two peaks, and W is the average peak width, at the baseline, of the two peaks. Source: U.S. ENVIRONMENTAL PROTECTION AGENCY. 1989. Determination of Chlorinated Acids by Gas Chromatography with an Electron Capture Detector. Method 515.1, rev. 4.0. Environmental Monitoring Systems Lab., Off. Research & Development, Cincinnati, Ohio

Method 524.2 – *Table 1*

Molecular Weights and Quantitation Ions for Method Analytes

Compound	MW[a]	Primary Quantitation Ion	Secondary Quantitation Ions
Internal Standard			
Fluorobenzene	96	96	77
Surrogates			
4-Bromofluorobenzene	174	95	174, 176
1,2-Dichlorobenzene-d$_4$	150	152	115, 150
Target Analytes			
Acetone	58	43	58
Acrylonitrile	53	52	53
Allyl Chloride	76	76	49
Benzene	78	78	77
Bromobenzene	156	156	77, 158
Bromochloromethane	128	128	49, 130
Bromodichloromethane	162	83	85, 127
Bromoform	250	173	173, 252
Bromomethane	94	94	96
2-Butanone	72	43	57, 72
n-Butylbenzene	134	91	134
sec-Butylbenzene	134	105	134
tert-Butylbenzene	134	119	91
Carbon disulfide	76	76	—
Carbon tetrachloride	152	117	119
Chloroacetonitrile	75	48	75
Chlorobenzene	112	112	77, 114
1-Chlorobutane	92	56	49
Chloroethane	64	64	66
Chloroform	118	83	85
Chloromethane	50	50	52
2-Chlorotoluene	126	91	126
4-Chlorotoluene	126	91	126
Dibromochloromethane	206	129	127
1,2-Dibromo-3-chloropropane	234	75	155, 157
1,2-Dibromoethane	186	107	109, 188
Dibromomethane	172	93	95, 174
1,2-Dichlorobenzene	146	146	111, 148
1,3-Dichlorobenzene	146	146	111, 148
1,4-Dichlorobenzene	146	146	111, 148
trans-1,4-Dichloro-2-butene	124	53	88, 75
Dichlorodifluoromethane	120	85	87
1,1-Dichloroethane	98	63	65, 83
1,2-Dichloroethane	98	62	98
1,1-Dichloroethene	96	96	61, 63

(a) Monoisotopic molecular weight calculated from the atomic masses of the iostopes with the smallest masses.

Method 524.2 – *Table 1*

Molecular Weights and Quantitation Ions for Method Analytes (Continued)

Compound	MW(a)	Primary Quantitation Ion	Secondary Quantitation Ions
cis-1,2-Dichloroethene	96	96	61, 98
trans-1,2-Dichloroethene	96	96	61, 98
1,2-Dichloropropane	112	63	112
1,3-Dichloropropane	112	76	78
2,2-Dichloropropane	112	77	97
1,1-Dichloropropene	110	75	110, 77
1,2-Dichloropropanone	126	43	83
cis-1,3-Dichloropropene	110	75	110
trans-1,3-Dichloropropene	110	75	110
Diethyl ether	74	59	45, 73
Ethylbenzene	106	91	106
Ethyl methacrylate	114	69	99
Hexachlorobutadiene	258	225	260
Hexachloroethane	234	117	119, 201
2-Hexanone	100	43	58
Isopropylbenzene	120	105	120
4-Isopropyltoluene	134	119	134, 91
Methacrylonitrile	67	67	52
Methyl acrylate	86	55	85
Methylene Chloride	84	84	86, 49
Methyl Iodide	142	142	127
Methylmethacrylate	100	69	99
4-Methyl-2-pentanone	100	43	58, 85
Methyl-t-butyl ether	88	73	57
Naphthalene	128	128	—
Nitrobenzene	123	51	77
2-Nitropropane	89	46	—
Pentachloroethane	200	117	119, 167
Propionitrile	55	54	—
n-Propylbenzene	120	91	120
Styrene	104	104	78
1,1,1,2-Tetrachloroethane	166	131	133, 119
1,1,2,2-Tetrachloroethane	166	83	131, 85
Tetrachloroethene	164	166	168, 129
Tetrahydrofuran	72	71	72, 42
Toluene	92	92	91
1,2,3-Trichlorobenzene	180	180	182
1,2,4-Trichlorobenzene	180	180	182
1,1,1-Trichloroethane	132	97	99, 61
1,1,2-Trichloroethane	132	83	97, 85

(a) Monoisotopic molecular weight calculated from the atomic masses of the iostopes with the smallest masses.

Method 524.2 – *Table 1*

Molecular Weights and Quantitation Ions for Method Analytes (Continued)

Compound	MW[a]	Primary Quantitation Ion	Secondary Quantitation Ions
Trichloroethene	130	95	130, 132
Trichlorofluoromethane	136	101	103
1,2,3-Trichloropropane	146	75	77
1,2,4-Trimethylbenzene	120	105	120
1,3,5-Trimethylbenzene	120	105	120
Vinyl Chloride	62	62	64
o-Xylene	106	106	91
m-Xylene	106	106	91
p-Xylene	106	106	91

(a) Monoisotopic molecular weight calculated from the atomic masses of the iostopes with the smallest masses.

Method 524.2 – *Table 3*

Ion Abundance Criteria for 4-Bromofluorobenzene (BFB)

Mass (M/z)	Relative Abundance Criteria
50	15-40% of mass 95
75	30-80% of mass 95
95	Base Peak, 100% Relative Abundance
96	5-9% of mass 95
173	<2% of mass 174
174	>50% of mass 95
175	5-9% of mass 174
176	> 95% but < 101% of mass 174
177	5-9% of mass 176

Appendix A

Accuracy and Precision Data From 16-31
Determinations of the Method Analytes in Reagent Water
Using Wide Bore Capillary Column 1[a]

Compound	True Conc. Range, (μg/L)	Mean Accuracy (% of true value)	Rel. Std. Dev. (%)	Method Det. Limit (μg/L)[b]
Benzene	0.1 - 10	97	5.7	0.04
Bromobenzene	0.1 - 10	100	5.5	0.03
Bromochloromethane	0.5 - 10	90	6.4	0.04
Bromodichloromethane	0.1 - 10	95	6.1	0.08
Bromoform	0.5 - 10	101	6.3	0.12
Bromomethane	0.5 - 10	95	8.2	0.11
n-Butylbenzene	0.5 - 10	100	7.6	0.11
sec-Butylbenzene	0.5 - 10	100	7.6	0.13
tert-Butylbenzene	0.5 - 10	102	7.3	0.14
Carbon tetrachloride	0.5 - 10	84	8.8	0.21
Chlorobenzene	0.1 - 10	98	5.9	0.04
Chloroethane	0.5 - 10	89	9.0	0.10
Chloroform	0.5 - 10	90	6.1	0.03
Chloromethane	0.5 - 10	93	8.9	0.13
2-Chlorotoluene	0.1 - 10	90	6.2	0.04
4-Chlorotoluene	0.1 - 10	99	8.3	0.06
Dibromochloromethane	0.1 - 10	92	7.0	0.05
1,2-Dibromo-3-chloropropane	0.5 - 10	83	19.9	0.26
1,2-Dibromoethane	0.5 - 10	102	3.9	0.06
Dibromomethane	0.5 - 10	100	5.6	0.24
1,2-Dichlorobenzene	0.1 - 10	93	6.2	0.03
1,3-Dichlorobenzene	0.5 - 10	99	6.9	0.12
1,4-Dichlorobenzene	0.2 - 20	103	6.4	0.03
Dichlorodifluoromethane	0.5 - 10	90	7.7	0.10
1,1-Dichloroethane	0.5 - 10	96	5.3	0.04
1,2-Dichloroethane	0.1 - 10	95	5.4	0.06
1,1-Dichloroethene	0.1 - 10	94	6.7	0.12
cis-1,2-Dichloroethene	0.5 - 10	101	6.7	0.12
trans-1,2-Dichloroethene	0.1 - 10	93	5.6	0.06
1,2-Dichloropropane	0.1 - 10	97	6.1	0.04
1,3-Dichloropropane	0.1 - 10	96	6.0	0.04
2,2-Dichloropropane	0.5 - 10	86	16.9	0.35
1,1-Dichloropropene	0.5 - 10	98	8.9	0.10
cis-1,2-Dichloropropene				
trans-1,2-Dichloropropene				
Ethyl benzene	0.1 - 10	99	8.6	0.06
Hexachlorobutadiene	0.5 - 10	100	6.8	0.11
Isopropylbenzene	0.5 - 10	101	7.6	0.15

(a) Data obtained using column 1 (60m x 0.75mm ID with 1.5 μm film thickness VOCOL (Supelco, Inc.)), with a jet separator interface and a quadrupole mass spectrometer (sect. 11.3.1) with analytes divided among three solutions.

(b) Replicate samples at the lowest concentration listed in Column 2 of this table were analyzed. These results were used to calculate MDLs.

Method 524.2 – *Table 4*

Accuracy and Precision Data From 16-31
Determinations of the Method Analytes in Reagent Water
Using Wide Bore Capillary Column 1[a] (Continued)

Compound	True Conc. Range, (μg/L)	Mean Accuracy (% of true value)	Rel. Std. Dev. (%)	Method Det. Limit (μg/L)[b]
p-Isopropyltoluene (4-Isopropyl toluene)	0.1 - 10	99	6.7	0.12
Methylene chloride	0.1 - 10	95	5.3	0.03
Naphthalene	0.1 - 100	104	8.2	0.04
n-Propylbenzene	0.1 - 10	100	5.8	0.04
Styrene	0.1 - 100	102	7.2	0.04
1,1,1,2-Tetrachloroethane	0.5 - 10	90	6.8	0.05
1,1,2,2-Tetrachloroethane	0.1 - 10	91	6.3	0.04
Tetrachloroethene	0.5 - 10	89	6.8	0.14
Toluene	0.5 - 10	102	8.0	0.11
1,2,3-Trichlorobenzene	0.5 - 10	109	8.6	0.03
1,2,4-Trichlorobenzene	0.5 - 10	108	8.3	0.04
1,1,1-Trichloroethane	0.5 - 10	98	8.1	0.08
1,1,2-Trichloroethane	0.5 - 10	104	7.3	0.10
Trichloroethene	0.5 - 10	90	7.3	0.19
Trichlorofluoromethane	0.5 - 10	89	8.1	0.08
1,2,3-Trichloropropane	0.5 - 10	108	14.4	0.32
1,2,4-Trimethylbenzene	0.5 - 10	99	8.1	0.13
1,3,5-Trimethylbenzene	0.5 - 10	92	7.4	0.05
Vinyl chloride	0.5 - 10	98	6.7	0.17
o-Xylene	0.1 - 31	103	7.2	0.11
m-Xylene	0.1 - 10	97	6.5	0.05
p-Xylene	0.5 - 10	104	7.7	0.13

(a) Data obtained using column 1 (60m x 0.75mm ID with 1.5 μm film thickness VOCOL (Supelco, Inc.)), with a jet separator interface and a quadrupole mass spectrometer (sect. 11.3.1) with analytes divided among three solutions.

(b) Replicate samples at the lowest concentration listed in Column 2 of this table were analyzed. These results were used to calculate MDLs.

Method 524.2 – *Table 5*

Accuracy and Precision Data from Seven Determinations of the Method Analytes in Reagent Water Using The Cryogenic Trapping Option and a Narrow Bore Capillary Column 3[a]

Compound	True Conc. Range, (µg/L)	Mean Accuracy (% of true value)	Rel. Std. Dev. (%)	Method Det. Limit (µg/L)
Benzene	0.1	99	6.2	0.03
Bromobenzene	0.5	97	7.4	0.11
Bromochloromethane	0.5	97	5.8	0.07
Bromodichloromethane	0.1	100	4.6	0.03
Bromoform	0.1	99	5.4	0.20
Bromomethane	0.1	99	7.1	0.06
n-Butylbenzene	0.5	94	6.0	0.03
sec-Butylbenzene	0.5	90	7.1	0.12
tert-Butylbenzene	0.5	90	2.5	0.33
Carbon tetrachloride	0.1	92	6.8	0.08
Chlorobenzene	0.1	91	5.8	0.03
Chloroethane	0.1	100	5.8	0.02
Chloroform	0.1	95	3.2	0.02
Chloromethane	0.1	99	4.7	0.05
2-Chlorotoluene	0.1	99	4.6	0.05
4-Chlorotoluene	0.1	96	7.0	0.05
Cyanogen chloride[b]		92	10.6	0.30
Dibromochloromethane	0.1	99	5.6	0.07
1,2-Dibromo-3-chloropropane	0.1	92	10.0	0.05
1,2-Dibromoethane	0.1	97	5.6	0.02
Dibromomethane	0.1	93	6.9	0.03
1,2-Dichlorobenzene	0.1	97	3.5	0.05
1,3-Dichlorobenzene	0.1	99	6.0	0.05
1,4-Dichlorobenzene	0.1	93	5.7	0.04
Dichlorodifluoromethane	0.1	99	8.8	0.11
1,1-Dichloroethane	0.1	98	6.2	0.03
1,2-Dichloroethane	0.1	100	6.3	0.02
1,1-Dichloroethene	0.1	95	9.0	0.05
cis-1,2-Dichloroethene	0.1	100	3.7	0.06
trans-1,2-Dichloroethene	0.1	98	7.2	0.03
1,2-Dichloropropane	0.1	96	6.0	0.02
1,3-Dichloropropane	0.1	99	5.8	0.04
2,2-Dichloropropane	0.1	99	4.9	0.05
1,1-Dichloropropene	0.1	98	7.4	0.02
cis-1,2-Dichloropropene				
trans-1,2-Dichloropropene				
Ethyl benzene	0.1	99	5.2	0.03

(a) Data obtained by Caroline A. Madding using column 3 (30m x 0.32mm ID DB-5 (J&W Scientific, Inc.) with a 1µm film thickness), with a cryogenic interface and a quadrupole mass spectrometer (sect. 11.3.3).

(b) Reference 8, Flesch, J.J., P.S. Fair, "The Analysis of Cyanogen Chloride in Drinking Water," Proceeding of Water Quality Technology Conference, American Water Works Association, St. Louis, MO., November 14-16, 1988.

Appendix A

Accuracy and Precision Data from Seven Determinations of the Method Analytes in Reagent Water Using The Cryogenic Trapping Option and a Narrow Bore Capillary Column 3[a] (Continued)

Compound	True Conc. Range, (μg/L)	Mean Accuracy (% of true value)	Rel. Std. Dev. (%)	Method Det. Limit (μg/L)
Hexachlorobutadiene	0.1	100	6.7	0.04
Isopropylbenzene	0.5	98	6.4	0.10
p-Isopropyltoluene (4-Isopropyl toluene)	0.5	87	13.0	0.26
Methylene chloride	0.5	97	13.0	0.09
Naphthalene	0.1	98	7.2	0.04
n-Propylbenzene	0.1	99	6.6	0.06
Styrene	0.1	96	19.0	0.06
1,1,1,2-Tetrachloroethane	0.1	100	4.7	0.04
1,1,2,2-Tetrachloroethane	0.1	100	12.0	0.20
Tetrachloroethene	0.1	96	5.0	0.05
Toluene	0.1	100	5.9	0.08
1,2,3-Trichlorobenzene	0.1	98	8.9	0.04
1,2,4-Trichlorobenzene	0.1	91	16.0	0.20
1,1,1-Trichloroethane	0.1	100	4.0	0.04
1,1,2-Trichloroethane	0.5	98	4.9	0.03
Trichloroethene	0.1	96	2.0	0.02
Trichlorofluoromethane	0.1	97	4.6	0.07
1,2,3-Trichloropropane	0.1	96	6.5	0.03
1,2,4-Trimethylbenzene	0.1	96	6.5	0.04
1,3,5-Trimethylbenzene	0.1	99	4.2	0.02
Vinyl chloride	0.1	96	0.2	0.04
o-Xylene	0.1	94	7.5	0.06
m-Xylene	0.1	94	4.6	0.03
p-Xylene	0.1	97	6.1	0.06

(a) Data obtained by Caroline A. Madding using column 3 (30m x 0.32mm ID DB-5 (J&W Scientific, Inc.) with a 1μm film thickness), with a cryogenic interface and a quadrupole mass spectrometer (sect. 11.3.3).

(b) Reference 8, Flesch, J.J., P.S. Fair, "The Analysis of Cyanogen Chloride in Drinking Water," Proceeding of Water Quality Technology Conference, American Water Works Association, St. Louis, MO., November 14-16, 1988.

Method 524.2 – *Table 6*

Accuracy and Precision Data from Seven Determinations of the Method Analytes in Reagent Water Using Wide-Bore Capillary Column 2[a]

Compound	No.[b]	Mean Accuracy (% of true value 2 μg/L Conc.)	Rel. Std. Dev. (%)	Mean Accuracy (% of true value 0.2 μg/L Conc.)	Rel. Std. Dev. (%)
Internal Standard					
Fluorobenzene	1	-	-	-	-
Surrogates					
4-Bromofluorobenzene	2	98	1.8	96	1.3
1,2-Dichlorobenzene-d$_4$	3	97	3.2	95	1.7
Target Analytes					
Benzene	37	97	4.4	113	1.8
Bromobenzene	38	102	3.0	101	1.9
Bromochloromethane	4	99	5.2	102	2.9
Bromodichloromethane	5	96	1.8	100	1.8
Bromoform	6	89	2.4	90	2.2
Bromomethane	7	55	27.	52	6.7
n-Butylbenzene	39	89	4.8	87	2.3
sec-Butylbenzene	40	102	3.5	100	2.8
tert-Butylbenzene	41	101	4.5	100	2.9
Carbon tetrachloride	8	84	3.2	92	2.6
Chlorobenzene	42	104	3.1	103	1.6
Chloroethane [c]					
Chloroform	9	97	2.0	95	2.1
Chloromethane	10	110	5.0	(d)	
2-Chlorotoluene	43	91	2.4	108	3.1
4-Chlorotoluene	44	89	2.0	108	4.4
Dibromochloromethane	11	95	2.7	100	3.0
1,2-Dibromo-3-Chloropropane [c]					
1,2-Dibromoethane [c]					
Dibromomethane	13	99	2.1	95	2.2
1,2-Dichlorobenzene	45	93	2.7	94	5.1
1,3-Dichlorobenzene	46	100	4.0	87	2.3
1,4-Dichlorobenzene	47	98	4.1	94	2.8
Dichlorodifluoromethane	14	38	25.	(d)	
1,1-Dichloroethane	15	97	2.3	85	3.6
1,2-Dichloroethane	16	102	3.8	100	2.1
1,1-Dichloroethene	17	90	2.2	87	3.8
cis-1,2-Dichloroethene	18	100	3.4	89	2.9

(a)   Data obtained using Column 2 (30m x 0.53 mm ID DB-624 (J&W Scientific, Inc.) fused silica capillary with a 3 μm film thickness) with the open split interface and an ion trap mass spectrometer (Section 11.3.2) with all method analytes in the same reagent water solution.
(b)   Designation in Figures 1 and 2.
(c)   Not measured; authentic standards were not available.
(d)   Not found at 0.2 μg/L.

Method 524.2 – *Table 6*

Accuracy and Precision Data from Seven Determinations of the Method Analytes in Reagent Water Using Wide-Bore Capillary Column 2[a] (Continued)

Compound	No.[b]	Mean Accuracy (% of true value 2 μg/L Conc.)	Rel. Std. Dev. (%)	Mean Accuracy (% of true value 0.2 μg/L Conc.)	Rel. Std. Dev. (%)
trans-1,2-Dichloroethene	19	92	2.1	85	2.3
1,2-Dichloropropane	20	102	2.2	103	2.9
1,3-Dichloropropane	21	92	3.7	93	3.2
2,2-Dichloropropane[c]					
1,1-Dichloropropene[c]					
cis-1,2-Dichloropropene[c]					
trans-1,2-Dichloropropene	25	96	1.7	99	2.1
Ethyl benzene	48	96	9.1	100	4.0
Hexachlorobutadiene	26	91	5.3	88	2.4
Isopropylbenzene	49	103	3.2	101	2.1
p-Isopropyltoluene (4-Isopropyl toluene)	50	95	3.6	95	3.1
Methylene chloride	27	(e)		(e)	
Naphthalene	51	93	7.6	78	8.3
n-Propylbenzene	52	102	4.9	97	2.1
Styrene	53	95	4.4	104	3.1
1,1,1,2-Tetrachloroethane	28	99	2.7	95	3.8
1,1,2,2-Tetrachloroethane	29	101	4.6	84	3.6
Tetrachloroethene	30	97	4.5	92	3.3
Toluene	54	105	2.8	126	1.7
1,2,3-Trichlorobenzene	55	90	5.7	78	2.9
1,2,4-Trichlorobenzene	56	92	5.2	83	5.9
1,1,1-Trichloroethane	31	94	3.9	94	2.5
1,1,2-Trichloroethane	32	107	3.4	109	2.8
Trichloroethene	33	99	2.9	106	2.5
Trichlorofluoromethane	34	81	4.6	48	13.
1,2,3-Trichloropropane	35	97	3.9	91	2.8
1,2,4-Trimethylbenzene	57	93	3.1	106	2.2
1,3,5-Trimethylbenzene	58	88	2.4	97	3.2
Vinyl chloride	36	104	3.5	115	14.
o-Xylene	59	97	1.8	98	1.7
m-Xylene	60	(f)		(f)	
p-Xylene	61	98	2.3	103	1.4

(a) Data obtained using Column 2 (30m x 0.53 mm ID DB-624 (J&W Scientific, Inc.) fused silica capillary with a 3 μm film thickness) with the open split interface and an ion trap mass spectrometer (Section 11.3.2) with all method analytes in the same reagent water solution.
(b) Designation in Figures 1 and 2.
(c) Not measured; authentic standards were not available.
(d) Not found at 0.2 μg/L.
(e) Not measured; methylene chloride was in the laboratory reagent blank.
(f) m-xylene coelutes with and cannot be distinguished from its isomer p-xylene, No 61.

Method 525.2 – *Table 1*

Appendix A

Ion Abundance Criteria for Bis(Perfluorophenyl)Phenyl Phosphine
(Decafluorotriphenylphosphine, DFTPP)

Mass (M/z)	Relative Abundance Criteria	Purpose of Checkpoint[1]
51	10-80% of the base peak	low mass sensitivity
68	<2% of mass 69	low mass resolution
70	<2% of mass 69	low mass resolution
127	10-80% of the base peak	low-mid mass sensitivity
197	<2% of mass 198	mid-mass resolution
198	base peak or >50% of 442	mid-mass resolution and sensitivity
199	5-9% of mass 198	mid-mass resolution and isotope ratio
275	10-60% of the base peak	mid-high mass sensitivity
365	>1% of the base peak	baseline threshold
441	Present and < mass 443	high mass resolution
442	base peak or >50% of 198	high mass resolution and sensitivity
443	15-24% of mass 442	high mass resolution and isotope ratio

(1) All ions are used primarily to check the mass measuring accuracy of the mass spectrometer and data system, and this is the most important part of the performance test. The three resolution checks, which include natural abundance isotope ratios, constitute the next most important part of the performance test. The correct setting of the base line threshold, as indicated by the presence of low intensity ions, is the next most important part of the performance test. Finally, the ion abundance ranges are designed to encourage some standarization to fragmentation patterns.

Method 525.2 – *Table 2*

Retention Time Data, Quantitation Ions, and Internal Standard References for Method Analytes

Compounds	Retention Time (min:sec)		Quantitation Ion	IS Reference #
	A(a)	B(b)		
Internal Standards				
Acenaphthene-d10 (#1)	7:47	7:01	164	
Chrysene-d12 (#2)	21:33	18:09	240	
Phenanthrene-d10 (#3)	11:37	10:13	188	
Surrogates				
1,3-Dimethyl-2-Nitrobenzene	5:16	4:33	134	1
Perylene-d12	26:60	21:31	264	3
Triphenylphosphate	20:25	17:25	326/325	3
Target Analytes				
Acenaphthylene	7:30	6:46	152	1
Alachlor	12:59	11:24	160	2
Aldrin	14:24	12:31	66	2
Ametryn	13:11	11:35	227/170	2
Anthracene	10:50	10:24	178	2
Aroclor 1016		7:30-14:00	152/256/292	2
Aroclor 1221		6:38-11:25	152/222/256	2
Aroclor 1232		6:38-13:54	152/256/292	2
Aroclor 1242		6:38-15:00	152/256/292	2
Aroclor 1248		8:47-15:00	152/256/292	2
Aroclor 1254		11:00-18:00	220/326/360	2
Aroclor 1260		13:10-21:00	326/360/394	2
Atraton	10:31	9:25	196/169	1
Atrazine	10:49	9:38	200/215	1/2
Benz[a]anthracene	21:31	18:08	228	3
Benzo[b]fluoranthene	25:33	20:44	252	3
Benzo[k]fluoranthene	25:45	20:48	252	3
Benzo[g,h,i]pyrene	31:16	24:18	276	3
Benzo[a]pyrene	25:24	21:25	252	3
Bromacil	13:46	12:03	205	2
Butachlor	16:25	14:16	176/160	2
Butylate	6:60	6:23	57/146	1
Butylbenzylphthalate	19:39	16:53	149	2/3
Carboxin	17:37	15:13	143	2
Chlordane, (alpha-Chlordane	16:43	14:28	375/373	2/3
Chlordane, (gamma-Chlordane)	16:19	14:05	373	2/3
Chlordane, (*trans*-Nonachlor)	16:47	14:30	409	2/3

(a) Single-ramp linear temperature program conditions (Section 10.2.3.2).
(b) Multi-ramp linear temperature program conditions (Section 10.2.3.1).

Method 525.2 − *Table 2*

Retention Time Data, Quantitation Ions, and Internal Standard References for Method Analytes (Continued)

Compounds	Retention Time (min:sec)		Quantitation Ion	IS Reference #
	A[a]	B[b]		
Chlorneb	7:47	7:05	191	1
Chlorobenzilate	18:22	15:52	139	2
2-Chlorobiphenyl	7:53	7:08	188	1
Chlorpropham	9:33	8:36	127	1
Chlorpyrifos	14:10	12:23	197/97	2
Chlorothalonil	11:38	10:15	266	2
Chrysene	21:39	18:13	228	3
Cyanazine	14:14	12:28	225/68	2
Cycloate	9:23	8:26	83/154	1
DCPA	14:20	12:30	301	2
4,4′-DDD	18:40	16:05	235/165	2
4,4′-DDE	17:20	14:59	246	2
4,4′-DDT	19:52	17:00	235/165	2
DEF	17:24	15:05	57/169	2
Diazinon	11:19	10:05	137/179	2
Dibenz[a,h]anthracene	30:32	23:47	278	3
Di-n-Butylphthalate	13:49	12:07	149	2
2,3-Dichlorobiphenyl	10:20	9:12	222/152	1
Dichlorvos	5:31	4:52	109	1
Dieldrin	17:35	15:09	79	2
Di(2-Ethylhexyl)adipate	20:11	17:19	129	2/3
Di(2-Ethylhexyl)phthalate	22:11	18:39	149	2/3
Diethylphthalate	8:68	7:53	149	1
Dimethylphthalate	7:13	6:34	163	1
2,4-Dinitrotoluene	8:08	7:22	165	1
2,6-Dinitrotoluene	7:19	6:40	165	1
Diphenamid	14:52	12:58	72/167	2
Disulfoton	11:43	10:22	88	2
Disulfoton Sulfone	16:28	14:17	213/153	2
Disulfoton Sulfoxide	6:09	5:31	97	1
Endosulfan I	16:44	14:26	195	2
Endosulfan II	18:35	15:59	195	2
Endosulfan Sulfate	19:47	16:54	272	2
Endrin	18:15	15:42	67/81	2
Endrin Aldehyde	19:02	16:20	67	2
EPTC	6:23	5:46	128	1
Ethoprop	9:19	8:23	158	1
Etridiazole	7:14	6:37	211/183	1
Fenamiphos	16:48	14:34	303/154	2
Fenarimol	23:26	19:24	139	3
Fluorene	8:59	8:03	166	1

(a) Single-ramp linear temperature program conditions (Section 10.2.3.2).
(b) Multi-ramp linear temperature program conditions (Section 10.2.3.1).

Method 525.2 – *Table 2*

Retention Time Data, Quantitation Ions, and Internal Standard References for Method Analytes (Continued)

Compounds	Retention Time (min:sec)		Quantitation Ion	IS Reference #
	A[(a)]	B[(b)]		
Fluridone	26:51	21:26	328	3
HCH, alpha	10:19	9:10	181	1
HCH, beta	10:57	9:41	181	2
HCH, delta	11:57	10:32	181	2
HCH, gamma (Lindane)	11:13	9:54	181	2
Heptachlor	13:19	11:37	100	2
Heptachlor epoxide	15:34	13:29	81	2
2,2′,3,3′,4,4′,6-Heptachlorobiphenyl	21:23	18:04	394/396	3
Hexachlorobenzene	10:27	9:15	284	1
2,2′4,4′,5,6′-Hexachlorobiphenyl	17:32	15:09	360	2
Hexachlorocyclo-pentadiene	5:16	5:38	237	1
Hexazinone	20:00	17:06	171	2
Indeno[1,2,3-cd]pyrene	30:26	23:43	276	3
Isophorone	4:54	4:10	82	1
Merphos	15:38	13:35	209/153	2
Methoxychlor	21:36	18:14	227	3
Methyl Paraoxon	11:57	10:22	109	2
Metolachlor	14:07	12:20	162	2
Metribuzin	12:46	11:13	198	2
Mevinphos	5:54	6:19	127	1
MGK 264 - Isomer a	15:18	13:00	164/66	2
MGK 264 - Isomer b	14:55	13:19	164	2
Molinate	8:19	7:30	126	1
Napropamide	16:53	14:37	72	2
Norflurazon	19:31	16:46	145	2
2,2′,3,3′,4,5′,6,6′-Octachlorobiphenyl	21:33	18:11	430/428	3
Pebulate	7:18	6:40	128	1
2,2′,3′,4,6-Pentachlorobiphenyl	15:37	13:33	326	2
Pentachlorophenol	11:01	9:45	266	2
Permethrin, *cis*	24:25	20:01	183	3
Permethrin, *trans*	24:39	20:10	183	3
Phenanthrene	11:41	10:16	178	2
Prometon	10:39	9:32	225/168	2
Prometryn	13:15	11:39	241/184	2
Pronamide	11:19	10:02	173	2
Propachlor	9:00	8:07	120	1
Propazine	10:54	9:43	214/172	2

(a)   Single-ramp linear temperature program conditions (Section 10.2.3.2).
(b)   Multi-ramp linear temperature program conditions (Section 10.2.3.1).

Method 525.2 — *Table 2*

**Retention Time Data, Quantitation Ions, and Internal Standard
References for Method Analytes (Continued)**

Compounds	Retention Time (min:sec)		Quantitation Ion	IS Reference #
	A[(a)]	B[(b)]		
Pyrene	16:41	14:24	202	2
Simazine	10:41	9:33	201/186	2
Simetryn	13:04	11:29	213	2
Stirofos	16:20	14:11	109	2
Tebuthiuron	8:00	7:16	156	1
Terbacil	11:44	10:24	161	2
Terbufos	11:14	9:58	57	2
Terbutryn	13:39	11:58	226/185	2
2,2',4,4',-Tetrchlorobiphenyl	14:02	12:14	292	2
Toxaphene		13:00-21:00	159	2
Triademefon	14:30	12:40	57	2
2,4,5-Trichlorobiphenyl	12:44	10:53	256	2
Tricyclazole	17:15	14:51	186	2
Trifluralin	9:31	8:37	306	1
Vernolate	7:10	6:32	128	1

(a) Single-ramp linear temperature program conditions (Section 10.2.3.2).
(b) Multi-ramp linear temperature program conditions (Section 10.2.3.1).

Method 525.2 – *Table 3*

Accuracy and Precision Data from Eight Determinations of the Method Analytes in Reagent Water Using Liquid-Solid C-18 Cartridge Extraction and the Quadrupole Mass Spectrometer

Compound	True Conc. (µg/L)	Mean Observed Conc. (µg/L)	%RSD	Mean Method Accuracy (% of True Conc.)	MDL (µg/L)
Surrogates					
1,3-Dimethyl-2-Nitrobenzene	5.0	4.7	3.9	94	
Perylene-d12	5.0	4.9	4.8	98	
Triphenylphosphate	5.0	5.5	6.3	110	
Target Analytes					
Acenaphthylene	0.50	0.45	8.2	91	0.11
Alachlor	0.50	0.47	12	93	0.16
Aldrin	0.50	0.40	9.3	80	0.11
Ametryn	0.50	0.44	6.9	88	0.092
Anthracene	0.50	0.53	4.3	106	0.68
Aroclor 1016	ND	ND	ND	ND	ND
Aroclor 1221	ND	ND	ND	ND	ND
Aroclor 1232	ND	ND	ND	ND	ND
Aroclor 1242	ND	ND	ND	ND	ND
Aroclor 1248	ND	ND	ND	ND	ND
Aroclor 1254	ND	ND	ND	ND	ND
Aroclor 1260	ND	ND	ND	ND	ND
Atraton[a]	0.50	0.35	15	70	0.16
Atrazine	0.50	0.54	4.8	109	0.178
Benz[a]anthracene	0.50	0.41	16	82	0.20
Benzo[b]fluoranthene	0.50	0.49	20	98	0.30
Benzo[k]fluoranthene	0.50	0.51	35	102	0.54
Benzo[g,h,i]pyrene	0.50	0.72	2.2	144	0.047
Benzo[a]pyrene	0.50	0.58	1.9	116	0.032
Bromacil	0.50	0.54	6.4	108	0.10
Butachlor	0.50	0.62	4.1	124	0.076
Butylate	0.50	0.52	4.1	105	0.064
Butylbenzylphthalate	0.50	0.77	11	154	0.25
Carboxin	5.0	3.8	12	76	1.4
Chlordane, (alpha-Chlordane	0.50	0.36	11	72	0.12
Chlordane, (gamma-Chlordane)	0.50	0.40	8.8	80	0.11
Chlordane, (*trans*-Nonachlor)	0.50	0.43	17	87	0.22
Chlorneb	0.50	0.51	5.7	102	0.088
Chlorobenzilate	5.0	6.5	6.9	130	1.3
2-Chlorobiphenyl	0.50	0.40	7.2	80	0.086
Chlorpropham	0.50	0.61	6.2	121	0.11

Accuracy and Precision Data from Eight Determinations of the Method Analytes in Reagent Water Using Liquid-Solid C-18 Cartridge Extraction and the Quadrupole Mass Spectrometer (Continued)

Compound	True Conc. (µg/L)	Mean Observed Conc. (µg/L)	%RSD	Mean Method Accuracy (% of True Conc.)	MDL (µg/L)
Chlorpyrifos	0.50	0.55	2.7	110	0.044
Chlorothalonil	0.50	0.57	6.9	113	0.12
Chrysene	0.50	0.39	7.0	78	0.082
Cyanazine	0.50	0.71	8.0	141	0.17
Cycloate	0.50	0.52	6.1	104	0.095
DCPA	0.50	0.55	5.8	109	0.094
4,4′-DDD	0.50	0.54	4.4	107	0.071
4,4′-DDE	0.50	0.40	6.3	80	0.075
4,4′-DDT	0.50	0.79	3.5	159	0.083
Diazinon	0.50	0.41	8.8	83	0.11
Dibenz[a,h]anthracene	0.50	0.53	0.5	106	0.010
Di-n-Butylphthalate	ND	ND	ND	ND	ND
2,3-Dichlorobiphenyl	0.50	0.40	11	80	0.14
Dichlorvos	0.50	0.55	9.1	110	0.15
Dieldrin	0.50	0.48	3.7	96	0.053
Di(2-Ethylhexyl)adipate	0.50	0.42	7.1	84	0.090
Di(2-Ethylhexyl)phthalate	ND	ND	ND	ND	ND
Diethylphthalate	0.50	0.59	9.6	118	0.17
Dimethylphthalate	0.50	0.60	3.2	120	0.058
2,4-Dinitrotoluene	0.50	0.60	5.6	119	0.099
2,6-Dinitrotoluene	0.50	0.60	8.8	121	0.16
Diphenamid	0.50	0.54	2.5	107	0.041
Disulfoton	5.0	3.99	5.1	80	0.62
Disulfoton Sulfone	0.50	0.74	3.2	148	0.070
Disulfoton Sulfoxide	0.50	0.58	12	116	0.20
Endosulfan I	0.50	0.55	18	110	0.30
Endosulfan II	0.50	0.50	29	99	0.44
Endosulfan Sulfate	0.50	0.62	7.2	124	0.13
Endrin	0.50	0.54	18	108	0.29
Endrin Aldehyde	0.50	0.43	15	87	0.19
EPTC	0.50	0.50	7.2	100	0.11
Ethoprop	0.50	0.62	6.1	123	0.11
Etridiazole	0.50	0.69	7.6	139	0.16
Fenamiphos	5.0	5.2	6.1	103	0.95
Fenarimol	5.0	6.3	6.5	126	1.2
Fluorene	0.50	0.46	4.2	93	0.059
Fluridone	5.0	5.1	3.6	102	0.55
HCH, alpha	0.50	0.51	13	102	0.20
HCH, beta	0.50	0.51	20	102	0.31
HCH, delta	0.50	0.56	13	112	0.21

Accuracy and Precision Data from Eight Determinations of the Method Analytes in Reagent Water Using Liquid-Solid C-18 Cartridge Extraction and the Quadrupole Mass Spectrometer (Continued)

Compound	True Conc. (μg/L)	Mean Observed Conc. (μg/L)	%RSD	Mean Method Accuracy (% of True Conc.)	MDL (μg/L)
HCH, gamma (Lindane)	0.50	0.63	8.0	126	0.15
Heptachlor	0.50	0.41	12	83	0.15
Heptachlor epoxide	0.50	0.35	5.5	70	0.058
2,2′,3,3′,4,4′,6-Heptachlorobiphenyl	0.50	0.35	10	71	0.11
Hexachlorobenzene	0.50	0.39	11	78	0.13
2,2′4,4′,5,6′-Hexachlorobiphenyl	0.50	0.37	9.6	73	0.11
Hexachlorocyclo-pentadiene	0.50	0.43	5.6	86	0.072
Hexazinone	0.50	0.70	5.0	140	0.11
Indeno[1,2,3-cd]pyrene	0.50	0.69	2.7	139	0.057
Isophorone	0.50	0.44	3.2	88	0.042
Methoxychlor	0.50	0.62	4.2	123	0.077
Methyl Paraoxon	0.50	0.57	10	115	0.17
Metolachlor	0.50	0.37	8.0	75	0.090
Metribuzin	0.50	0.49	11	97	0.16
Mevinphos	0.50	0.57	12	114	0.20
MGK 264 - Isomer a	0.33	0.39	3.4	116	0.040
MGK 264 - Isomer b	0.17	0.16	6.4	96	0.030
Molinate	0.50	0.53	5.5	105	0.087
Napropamide	0.50	0.58	3.5	116	0.060
Norflurazon	0.50	0.63	7.1	126	0.13
2,2′,3,3′,4,5′,6,6′-Octachlorobiphenyl	0.50	0.50	8.7	101	0.13
Pebulate	0.50	0.49	5.4	98	0.080
2,2′,3′,4,6-Pentachlorobiphenyl	0.50	0.30	16	61	0.15
Pentachlorophenol	ND	ND	ND	ND	ND
Permethrin, *cis*	0.25	0.30	3.7	121	0.034
Permethrin, *trans*	0.75	0.82	2.7	109	0.067
Phenanthrene	0.50	0.46	4.3	92	0.059
Prometon[a]	0.50	0.30	42	60	0.38
Prometryn	0.50	0.46	5.6	92	0.078
Pronamide	0.50	0.54	5.9	108	0.095
Propachlor	0.50	0.49	7.5	98	0.11
Propazine	0.50	0.54	7.1	108	0.12

ND = Not determined
[a]Data from samples extracted at pH 2. For accurate determination of this analyte, a separate sample must be extracted at ambient pH.

Method 525.2 – *Table 3*

Accuracy and Precision Data from Eight Determinations of the Method Analytes in Reagent Water Using Liquid-Solid C-18 Cartridge Extraction and the Quadrupole Mass Spectrometer (Continued)

Compound	True Conc. (µg/L)	Mean Observed Conc. (µg/L)	%RSD	Mean Method Accuracy (% of True Conc.)	MDL (µg/L)
Pyrene	0.50	0.38	5.7	77	0.066
Simazine	0.50	0.55	9.1	109	0.15
Simetryn	0.50	0.52	8.2	105	0.13
Stirofos	0.50	0.75	5.8	149	0.13
Tebuthiuron	5.0	6.8	14	136	2.8
Terbacil	5.0	4.9	14	97	2.1
Terbufos	0.50	0.53	6.1	106	0.096
Terbutryn	0.50	0.47	7.6	95	0.11
2,2′,4,4′,-Tetrchloro-biphenyl	0.50	0.36	4.1	71	0.044
Toxaphene	ND	ND	ND	ND	ND
Triademefon	0.50	0.57	20	113	0.33
2,4,5-Trichlorobiphenyl	0.50	0.38	6.7	75	0.075
Tricyclazole	5.0	4.6	19	92	2.6
Trifluralin	0.50	0.63	5.1	127	0.096
Vernolate	0.50	0.51	5.5	102	0.084

ND = Not determined

[a]Data from samples extracted at pH 2. For accurate determination of this analyte, a separate sample must be extracted at ambient pH.

Method 525.2 – *Table 4*

Accuracy and Precision Data from Eight Determinations of the Method Analytes in Reagent Water Using Liquid-Solid C-18 Disk Extraction and the Quadrupole Mass Spectrometer

Compound	True Conc. (μg/L)	Mean Observed Conc. (μg/L)	%RSD	Mean Method Accuracy (% of True Conc.)	MDL (μg/L)
Surrogates					
1,3-Dimethyl-2-Nitrobenzene	5.0	4.6	2.6	93	
Perylene-d12	5.0	4.8	1.6	95	
Triphenylphosphate	5.0	5.0	2.5	101	
Target Analytes					
Acenaphthylene	0.50	0.47	8.4	94	0.12
Alachlor	0.50	0.50	5.8	100	0.087
Aldrin	0.50	0.39	13	78	0.16
Ametryn	0.50	0.38	28	76	0.32
Anthracene	0.50	0.49	13	98	0.18
Aroclor 1016	ND	ND	ND	ND	ND
Aroclor 1221	ND	ND	ND	ND	ND
Aroclor 1232	ND	ND	ND	ND	ND
Aroclor 1242	ND	ND	ND	ND	ND
Aroclor 1248	ND	ND	ND	ND	ND
Aroclor 1254	ND	ND	ND	ND	ND
Aroclor 1260	ND	ND	ND	ND	ND
Atraton[a]	0.50	0.07	139	19	0.29
Atrazine	0.50	0.60	3.7	119	0.065
Benz[a]anthracene	0.50	0.38	6.1	76	0.070
Benzo[b]fluoranthene	0.50	0.61	2.5	121	0.046
Benzo[k]fluoranthene	0.50	0.61	27	122	0.50
Benzo[g,h,i]pyrene	0.50	0.69	1.4	138	0.029
Benzo[a]pyrene	0.50	0.58	6.1	116	0.11
Bromacil	0.50	0.49	23	99	0.34
Butachlor	0.50	0.63	2.1	127	0.039
Butylate	0.50	0.50	4.9	99	0.073
Butylbenzylphthalate	0.50	0.78	5.5	156	0.13
Carboxin	5.0	2.7	12	54	0.98
Chlordane, (alpha-Chlordane	0.50	0.37	5.5	74	0.061
Chlordane, (gamma-Chlordane)	0.50	0.40	4.2	80	0.050
Chlordane, (*trans*-Nonachlor)	0.50	0.45	7.8	90	0.11
Chlorneb	0.50	0.51	7.3	100	0.11

ND   Not determined

(a)   Data from samples extracted at pH 2. For accurate determination of this analyte, a separate sample must be extracted at ambient pH.

Method 525.2 – *Table 4*

**Accuracy and Precision Data from Eight Determinations of the Method Analytes
in Reagent Water Using Liquid-Solid C-18 Disk Extraction
and the Quadrupole Mass Spectrometer (Continued)**

Compound	True Conc. (µg/L)	Mean Observed Conc. (µg/L)	%RSD	Mean Method Accuracy (% of True Conc.)	MDL (µg/L)
Chlorobenzilate	5.0	7.9	8.4	156	2.0
2-Chlorobiphenyl	0.50	0.42	1.9	84	0.023
Chlorpropham	0.50	0.68	5.4	134	0.11
Chlorpyrifos	0.50	0.61	6.5	119	0.12
Chlorothalonil	0.50	0.59	6.5	116	0.11
Chrysene	0.50	0.35	3.6	71	0.038
Cyanazine	0.50	0.68	15	136	0.31
Cycloate	0.50	0.53	4.9	106	0.077
DCPA	0.50	0.55	4.5	110	0.073
4,4'-DDD	0.50	0.67	14	137	0.28
4,4'-DDE	0.50	0.48	4.9	96	0.070
4,4'-DDT	0.50	0.93	3.2	187	0.090
Diazinon	0.50	0.56	6.8	109	0.11
Dibenz[a,h]anthracene	0.50	0.61	15	122	0.28
Di-n-Butylphthalate	ND	ND	ND	ND	ND
2,3-Dichlorobiphenyl	0.50	0.46	8.1	93	0.11
Dichlorvos	0.50	0.54	5.6	108	0.092
Dieldrin	0.50	0.52	7.8	104	0.12
Di(2-Ethylhexyl)adipate	ND	ND	ND	ND	ND
Di(2-Ethylhexyl)phthalate	ND	ND	ND	ND	ND
Diethylphthalate	0.50	0.66	10	132	0.20
Dimethylphthalate	0.50	0.57	8.3	114	0.14
2,4-Dinitrotoluene	0.50	0.54	5.7	109	0.093
2,6-Dinitrotoluene	0.50	0.48	4.9	96	0.071
Diphenamid	0.50	0.60	3.8	118	0.067
Disulfoton	5.0	4.8	9.4	96	1.3
Disulfoton Sulfone	0.50	0.82	2.8	164	0.070
Disulfoton Sulfoxide	0.50	0.68	8.9	136	0.18
Endosulfan I	0.50	0.65	10	132	0.20
Endosulfan II	0.50	0.60	21	122	0.38
Endosulfan Sulfate	0.50	0.67	6.1	133	0.12
Endrin	0.50	0.58	18	116	0.31
Endrin Aldehyde	0.50	0.51	16	101	0.24
EPTC	0.50	0.50	3.8	100	0.056
Ethoprop	0.50	0.69	2.3	138	0.048
Etridiazole	0.50	0.74	4.0	149	0.090
Fenamiphos	5.0	6.3	8.8	124	1.6
Fenarimol	5.0	7.5	5.5	150	1.2
Fluorene	0.50	0.47	8.1	94	0.11
Fluridone	5.0	5.7	4.5	114	0.77

ND Not determined

Method 525.2 – *Table 4*

Accuracy and Precision Data from Eight Determinations of the Method Analytes in Reagent Water Using Liquid-Solid C-18 Disk Extraction and the Quadrupole Mass Spectrometer (Continued)

Compound	True Conc. (μg/L)	Mean Observed Conc. (μg/L)	%RSD	Mean Method Accuracy (% of True Conc.)	MDL (μg/L)
HCH, alpha	0.50	0.54	12	107	0.20
HCH, beta	0.50	0.57	17	112	0.28
HCH, delta	0.50	0.61	8.2	120	0.15
HCH, gamma (Lindane)	0.50	0.62	6.6	124	0.12
Heptachlor	0.50	0.40	12	80	0.14
Heptachlor epoxide	0.50	0.36	8.7	71	0.093
2,2',3,3',4,4',6-Heptachlorobiphenyl	0.50	0.36	13	71	0.14
Hexachlorobenzene	0.50	0.47	8.3	95	0.12
2,2'4,4',5,6'-Hexachlorobiphenyl	0.50	0.41	11	83	0.13
Hexachlorocyclo-pentadiene	0.50	0.42	12	84	0.16
Hexazinone	0.50	0.85	5.6	169	0.14
Indeno[1,2,3-cd]pyrene	0.50	0.69	2.4	138	0.050
Isophorone	0.50	0.41	4.2	83	0.052
Methoxychlor	0.50	0.58	1.9	117	0.033
Methyl Paraoxon	0.50	0.62	14	122	0.25
Metolachlor	0.50	0.38	7.5	75	0.084
Metribuzin	0.50	0.54	3.9	107	0.062
Mevinphos	0.50	0.72	3.7	143	0.079
MGK 264 - Isomer a	0.33	0.40	8.8	119	0.10
MGK 264 - Isomer b	0.17	0.17	5.9	103	0.030
Molinate	0.50	0.53	3.2	105	0.050
Napropamide	0.50	0.64	5.9	126	0.11
Norflurazon	0.50	0.70	4.2	141	0.089
2,2',3,3',4,5',6,6'-Octachlorobiphenyl	0.50	0.51	4.2	102	0.064
Pebulate	0.50	0.48	5.8	96	0.084
2,2',3',4,6-Pentachlorobiphenyl	0.50	0.35	4.2	70	0.044
Pentachlorophenol	2.0	1.9	16	95	.89
Permethrin, *cis*	0.25	0.32	3.3	126	0.031
Permethrin, *trans*	0.75	0.89	1.9	118	0.051
Phenanthrene	0.50	0.48	5.0	95	0.071
Prometon[a]	0.50	0.21	66	45	0.44
Prometryn	0.50	0.46	24	93	0.33
Pronamide	0.50	0.58	7.1	113	0.12
Propachlor	0.50	0.49	5.4	98	0.079

ND Not determined
(a) Data from samples extracted at pH 2 - for accurate determination of this analyte, a separate sample must be extracted at ambient pH.

Method 525.2 – *Table 4*

Accuracy and Precision Data from Eight Determinations of the Method Analytes in Reagent Water Using Liquid-Solid C-18 Disk Extraction and the Quadrupole Mass Spectrometer (Continued)

Compound	True Conc. (μg/L)	Mean Observed Conc. (μg/L)	%RSD	Mean Method Accuracy (% of True Conc.)	MDL (μg/L)
Pyrene	0.50	0.40	3.2	79	0.038
Simazine	0.50	0.60	10	120	0.18
Simetryn	0.50	0.41	15	83	0.19
Stirofos	0.50	0.84	3.2	168	0.081
Tebuthiuron	5.0	9.3	8.6	187	2.4
Terbacil	5.0	5.0	11	100	1.7
Terbufos	0.50	0.62	4.2	123	0.077
Terbutryn	0.50	0.46	23	94	0.32
2,2′,4,4′,-Tetrchlorobiphenyl	0.50	0.40	7.4	79	0.088
Toxaphene	ND	ND	ND	ND	ND
Triademefon	0.50	0.73	7.2	145	0.16
2,4,5-Trichlorobiphenyl	0.50	0.44	5.3	89	0.071
Tricyclazole	5.0	6.8	12	137	2.4
Trifluralin	0.50	0.62	2.6	124	0.048
Vernolate	0.50	0.51	3.4	100	0.051

ND   Not determined
(a)   Data from samples extracted at pH 2. For accurate determination of this analyte, a separate sample must be extracted at ambient pH.

Method 525.2 – *Table 5*

Accuracy and Precision Data from Eight Determinations of the Method Analytes in Reagent Water Using Liquid-Solid C-18 Cartridge Extraction and the Ion Trap Mass Spectrometer

Compound	True Conc. (µg/L)	Mean Observed Conc. (µg/L)	%RSD	Mean Method Accuracy (% of True Conc.)	MDL (µg/L)
Surrogates					
1,3-Dimethyl-2-Nitrobenzene	5.0	4.9	8.4	98	
Perylene-d12	5.0	4.3	18	86	
Triphenylphosphate	5.0	4.8	13	96	
Target Analytes					
Acenaphthylene	0.50	0.50	8.8	100	0.13
Alachlor	0.50	0.58	4.0	115	0.069
Aldrin	0.50	0.42	3.5	85	0.045
Ametryn	0.50	0.46	3.3	91	0.045
Anthracene	0.50	0.42	3.8	84	0.048
Aroclor 1016	1.0	1.1	4.4	113	0.15
Aroclor 1221	ND	ND	ND	ND	ND
Aroclor 1232	ND	ND	ND	ND	ND
Aroclor 1242	ND	ND	ND	ND	ND
Aroclor 1248	ND	ND	ND	ND	ND
Aroclor 1254 [a]	1.0	1.1	17	110	0.56
Aroclor 1260	1.0	0.96	9.3	96	0.27
Atraton [c]	0.50	0.35	11	70	0.12
Atrazine	0.50	0.55	5.0	109	0.081
Benz[a]anthracene	0.50	0.43	7.3	85	0.093
Benzo[b]fluoranthene	0.50	0.44	16	88	0.21
Benzo[k]fluoranthene	0.50	0.34	22	68	0.23
Benzo[g,h,i]pyrene	0.50	0.38	31	76	0.35
Benzo[a]pyrene	0.50	0.36	21	73	0.23
Bromacil	0.50	0.45	9.1	90	0.12
Butachlor	0.50	0.67	12	133	0.24
Butylate	0.50	0.52	5.2	104	0.082
Butylbenzylphthalate [b]	5.0	5.7	7.7	114	1.4
Carboxin	0.50	0.58	22	117	0.38
Chlordane, (alpha-Chlordane	0.50	0.47	12	95	0.17
Chlordane, (gamma-Chlordane)	0.50	0.50	10	99	0.16

(a) Seven replicates.
(b) Seven replicates in fortified tap water.
(c) Data from samples extracted at pH 2 - for accurate determination of this analyte, a separate sample must be extracted at ambient pH.

ND Not determined

Method 525.2 – *Table 5*

Accuracy and Precision Data from Eight Determinations of the Method Analytes in Reagent Water Using Liquid-Solid C-18 Cartridge Extraction and the Ion Trap Mass Spectrometer (Continued)

Compound	True Conc. (μg/L)	Mean Observed Conc. (μg/L)	%RSD	Mean Method Accuracy (% of True Conc.)	MDL (μg/L)
Chlordane, (*trans*-Nonachlor)	0.50	0.48	11	96	0.16
Chlorne [b]	0.50	0.51	8.1	103	0.13
Chlorobenzilate	0.50	0.61	9.7	123	0.17
2-Chlorobiphenyl	0.50	0.47	4.8	94	0.068
Chlorpropham	0.50	0.55	8.1	109	0.13
Chlorpyrifos	0.50	0.50	2.4	99	0.35
Chlorothalonil	0.50	0.62	5.3	123	0.098
Chrysene	0.50	0.50	9.2	99	0.14
Cyanazine	0.50	0.49	13	97	0.19
Cycloate	0.50	0.52	7.6	103	0.12
DCPA	0.50	0.55	7.2	109	0.12
4,4'-DDD	0.50	0.52	3.6	103	0.055
4,4'-DDE	0.50	0.41	5.8	81	0.070
4,4'-DDT	0.50	0.54	2.4	108	0.039
Diazinon	0.50	0.37	2.7	75	0.030
Dibenz[a,h]anthracene	0.50	0.37	29	74	0.32
Di-n-Butylphthalate [b]	5.0	6.2	4.6	124	0.89
2,3-Dichlorobiphenyl	0.50	0.45	5.8	90	0.079
Dichlorvos	0.50	0.53	8.0	106	0.13
Dieldrin	0.50	0.50	10	100	0.15
Di(2-Ethylhexyl)adipate	0.50	0.59	18	117	0.31
Di(2-Ethylhexyl) phthalate [b]	5.0	6.5	6.6	130	1.3
Diethylphthalate	0.50	0.63	15	126	0.28
Dimethylphthalate	0.50	0.51	9.5	102	0.14
2,4-Dinitrotoluene	0.50	0.45	18	91	0.24
2,6-Dinitrotoluene	0.50	0.40	17	80	0.20
Diphenamid	0.50	0.55	6.5	111	0.11
Disulfoton	0.50	0.62	9.8	124	0.18
Disulfoton Sulfone	0.50	0.64	3.5	128	0.068
Disulfoton Sulfoxide	0.50	0.57	8.6	114	0.15
Endosulfan I	0.50	0.60	6.1	121	0.11
Endosulfan II	0.50	0.64	3.9	128	0.074
Endosulfan Sulfate	0.50	0.58	5.4	116	0.093
Endrin	0.50	0.62	18	124	0.34
Endrin Aldehyde	0.50	0.58	8.7	116	0.15
EPTC	0.50	0.53	7.7	105	0.12
Ethoprop	0.50	0.62	10	124	0.19

[b] Seven replicates in fortified tap water.

Method 525.2 – *Table 5*

Accuracy and Precision Data from Eight Determinations of the Method Analytes in Reagent Water Using Liquid-Solid C-18 Cartridge Extraction and the Ion Trap Mass Spectrometer (Continued)

Compound	True Conc. (μg/L)	Mean Observed Conc. (μg/L)	%RSD	Mean Method Accuracy (% of True Conc.)	MDL (μg/L)
Etridiazole	0.50	0.61	6.5	122	0.12
Fenamiphos	0.50	0.67	12	133	0.24
Fenarimol	0.50	0.74	11	148	0.25
Fluorene	0.50	0.49	9.0	98	0.13
Fluridone	5.0	5.2	2.5	105	0.39
HCH, alpha	0.50	0.55	6.8	109	0.11
HCH, beta	0.50	0.54	5.3	107	0.085
HCH, delta	0.50	0.52	3.1	105	0.049
HCH, gamma (Lindane)	0.50	0.53	5.3	105	0.084
Heptachlor	0.50	0.50	4.1	100	0.061
Heptachlor epoxide	0.50	0.54	8.2	108	0.13
2,2',3,3',4,4',6-Heptachlorobiphenyl	0.50	0.45	11	90	0.15
Hexachlorobenzene	0.50	0.41	6.0	82	0.074
2,2'4,4',5,6'-Hexachlorobiphenyl	0.50	0.40	15	80	0.18
Hexachlorocyclopenta-diene	0.50	0.34	13	68	0.13
Hexazinone	0.50	0.80	5.6	159	0.14
Indeno[1,2,3-cd]pyrene	0.50	0.36	28	71	0.30
Isophorone	0.50	0.54	7.9	107	0.13
Methoxychlor	0.50	0.58	7.7	115	0.13
Methyl Paraoxon	0.50	0.85	3.7	170	0.094
Metolachlor	0.50	0.58	4.8	117	0.085
Metribuzin	0.50	0.54	14	108	0.22
Mevinphos	0.50	0.47	12	95	0.17
MGK 264 - Isomer a	0.33	0.38	9.5	113	0.11
MGK 264 - Isomer b	0.16	0.18	5.4	105	0.029
Molinate	0.50	0.55	5.2	111	0.086
Napropamide	0.50	0.63	10	127	0.20
Norflurazon	0.50	0.82	3.8	165	0.093
2,2',3,3',4,5',6,6'-Octachlorobiphenyl	0.50	0.49	19	99	0.28
Pebulate	0.50	0.56	6.1	112	0.10
2,2',3',4,6-Pentachlorobiphenyl	0.50	0.43	8.7	86	0.11
Pentachlorophenol	2.0	2.4	10	119	0.72
Permethrin, *cis*	0.25	0.45	3.2	179	0.043
Permethrin, *trans*	0.75	1.1	2.2	153	0.074
Phenanthrene	0.50	0.48	4.8	96	0.069

Method 525.2 – *Table 5*

Accuracy and Precision Data from Eight Determinations of the Method Analytes in Reagent Water Using Liquid-Solid C-18 Cartridge Extraction and the Ion Trap Mass Spectrometer (Continued)

Compound	True Conc. (μg/L)	Mean Observed Conc. (μg/L)	%RSD	Mean Method Accuracy (% of True Conc.)	MDL (μg/L)
Prometon[c]	0.50	0.24	27	48	0.20
Prometryn	0.50	0.46	3.0	92	0.041
Pronamide	0.50	0.56	5.3	113	0.089
Propachlor	0.50	0.56	8.6	112	0.14
Propazine	0.50	0.52	4.3	103	0.066
Pyrene	0.50	0.47	11	95	0.16
Simazine	0.50	0.48	8.8	96	0.13
Simetryn	0.50	0.48	2.9	96	0.042
Stirofos	0.50	0.80	3.9	160	0.093
Tebuthiuron	0.50	0.67	7.4	134	0.15
Terbacil	0.50	0.59	12	119	0.22
Terbufos	0.50	0.46	11	92	0.15
Terbutryn	0.50	0.48	2.6	97	0.038
2,2′,4,4′,-Tetrchloro-biphenyl	0.50	0.40	6.4	81	0.077
Toxaphene	10	11	4.9	118	1.7
Triademefon	0.50	0.73	6.4	146	0.14
2,4,5-Trichlorobiphenyl	0.50	0.44	3.3	88	0.043
Tricyclazole	0.50	0.63	16	127	0.31
Trifluralin	0.50	0.62	13	124	0.24
Vernolate	0.50	0.50	9.3	101	0.14

(c)   Data from samples extracted at pH 2 - for accurate determination of this analyte, a separate sample must be extracted at ambient pH.

Method 525.2 – *Table 6*

Accuracy and Precision Data from Eight Determinations of the Method Analytes in Reagent Water Using Liquid-Solid C-18 Disk Extraction and the Ion Trap Mass Spectrometer

Compound	True Conc. (μg/L)	Mean Observed Conc. (μg/L)	%RSD	Mean Method Accuracy (% of True Conc.)	MDL (μg/L)
Surrogates					
1,3-Dimethyl-2-Nitrobenzene	5.0	4.9	10	98	
Perylene-d12	5.0	4.9	4.5	98	
Triphenylphosphate	5.0	5.9	8.1	117	
Target Analytes					
Acenaphthylene	0.50	0.51	4.5	102	0.068
Alachlor	0.50	0.54	6.6	108	0.11
Aldrin	0.50	0.45	6.3	90	0.085
Ametryn	0.50	0.41	23	82	0.29
Anthracene	0.50	0.39	15	79	0.18
Aroclor 1016	0.20	0.25	4.7	123	0.040
Aroclor 1221	0.20	0.26	6.1	130	0.054
Aroclor 1232	0.20	0.24	4.7	121	0.042
Aroclor 1242	0.20	0.26	4.9	129	0.043
Aroclor 1248	0.20	0.24	4.1	118	0.038
Aroclor 1254	0.20	0.22	3.7	110	0.028
Aroclor 1260 [a]	0.20	0.21	2.2	108	0.018
Atraton [d]	0.50	0.10	46	21	0.14
Atrazine	0.50	0.56	4.6	111	0.076
Benz[a]anthracene	0.50	0.44	7.4	88	0.098
Benzo[b]fluoranthene	0.50	0.50	9.1	100	0.14
Benzo[k]fluoranthene	0.50	0.46	2.2	91	0.031
Benzo[g,h,i]pyrene	0.50	0.47	7.9	95	0.11
Benzo[a]pyrene	0.50	0.44	12	89	0.16
Bromacil	0.50	0.49	4.4	99	0.066
Butachlor	0.50	0.66	5.1	132	0.10
Butylate	0.50	0.50	5.4	100	0.082
Butylbenzylphthalate [b]	5.0	5.7	7.7	114	1.4
Carboxin	0.50	0.40	38.1	79	0.45
Chlordane, (alpha-Chlordane	0.50	0.50	4.3	101	0.065
Chlordane, (gamma-Chlordane)	0.50	0.51	7.2	102	0.11
Chlordane, (trans-Nonachlor)	0.50	0.52	6.2	104	0.097

(a)   Six replicates.
(b)   Seven replicates in fortified tap water.

Method 525.2 – *Table 6*

Accuracy and Precision Data from Eight Determinations of the Method Analytes in Reagent Water Using Liquid-Solid C-18 Disk Extraction and the Ion Trap Mass Spectrometer (Continued)

Compound	True Conc. (μg/L)	Mean Observed Conc. (μg/L)	%RSD	Mean Method Accuracy (% of True Conc.)	MDL (μg/L)
Chlorneb	0.50	0.54	6.3	108	0.10
Chlorobenzilate	0.50	0.59	9.7	117	0.17
2-Chlorobiphenyl	0.50	0.50	4.7	100	0.070
Chlorpropham	0.50	0.55	4.7	111	0.079
Chlorpyrifos	0.50	0.54	11	109	0.18
Chlorothalonil	0.50	0.59	4.4	119	0.079
Chrysene	0.50	0.48	6.1	96	0.088
Cyanazine	0.50	0.52	8.3	105	0.13
Cycloate	0.50	0.51	4.1	102	0.063
DCPA	0.50	0.53	3.2	105	0.051
4,4'-DDD	0.50	0.63	16	127	0.31
4,4'-DDE	0.50	0.48	3.7	96	0.054
4,4'-DDT	0.50	0.58	7.2	117	0.13
Diazinon	0.50	0.50	4.5	101	0.068
Dibenz[a,h]anthracene	0.50	0.47	9.9	94	0.14
Di-n-Butylphthalate [b]	5.0	5.7	3.3	115	0.59
2,3-Dichlorobiphenyl	0.50	0.50	2.6	100	0.039
Dichlorvos	0.50	0.50	8.7	99	0.13
Dieldrin	0.50	0.53	7.0	106	0.11
Di(2-Ethylhexyl)adipate [b]	5.0	5.4	7.5	107	1.3
Di(2-Ethylhexyl) phthalate [b]	5.0	5.7	2.6	114	0.46
Diethylphthalate	0.50	0.68	5.0	137	0.10
Dimethylphthalate	0.50	0.51	5.0	102	0.077
2,4-Dinitrotoluene	0.50	0.30	8.1	59	0.072
2,6-Dinitrotoluene	0.50	0.28	6.4	56	0.054
Diphenamid	0.50	0.56	6.4	112	0.11
Disulfoton	0.50	0.70	5.3	139	0.11
Disulfoton Sulfone	0.50	0.64	5.9	128	0.11
Disulfoton Sulfoxide	0.50	0.60	3.8	119	0.068
Endosulfan I	0.50	0.61	4.9	122	0.089
Endosulfan II	0.50	0.66	6.1	131	0.12
Endosulfan Sulfate	0.50	0.57	9.0	115	0.16
Endrin	0.50	0.68	7.9	137	0.16
Endrin Aldehyde	0.50	0.57	2.8	114	0.048
EPTC	0.50	0.48	5.2	97	0.076

(b) Seven replicates in fortified tap water.

Method 525.2 – *Table 6*

Accuracy and Precision Data from Eight Determinations of the Method Analytes in Reagent Water Using Liquid-Solid C-18 Disk Extraction and the Ion Trap Mass Spectrometer (Continued)

Compound	True Conc. (μg/L)	Mean Observed Conc. (μg/L)	%RSD	Mean Method Accuracy (% of True Conc.)	MDL (μg/L)
Ethoprop	0.50	0.61	7.5	122	0.14
Etridiazole	0.50	0.54	4.2	108	0.067
Fenamiphos	0.50	0.67	10	133	0.20
Fenarimol	0.50	0.59	5.8	118	0.10
Fluorene	0.50	0.53	3.4	106	0.054
Fluridone	5.0	5.2	2.3	104	0.16
HCH, alpha	0.50	0.55	5.0	110	0.083
HCH, beta	0.50	0.54	4.1	109	0.068
HCH, delta	0.50	0.53	3.6	106	0.058
HCH, gamma (Lindane)	0.50	0.50	3.2	100	0.047
Heptachlor	0.50	0.49	4.0	98	0.059
Heptachlor epoxide	0.50	0.50	3.2	100	0.048
2,2',3,3',4,4',6-Heptachlorobiphenyl	0.50	0.46	7.3	92	0.10
Hexachlorobenzene	0.50	0.49	3.4	97	0.049
2,2'4,4',5,6'-Hexachlorobiphenyl	0.50	0.50	5.3	99	0.079
Hexachlorocyclopenta-diene	0.50	0.37	9.3	73	0.10
Hexazinone	0.50	0.75	4.2	150	0.094
Indeno[1,2,3-cd]pyrene	0.50	0.48	7.3	96	0.10
Isophorone	0.50	0.51	4.3	102	0.066
Methoxychlor	0.50	0.52	6.7	104	0.10
Methyl Paraoxon	0.50	0.75	4.5	151	0.10
Metolachlor	0.50	0.57	3.2	114	0.054
Metribuzin	0.50	0.53	5.7	107	0.090
Mevinphos	0.50	0.56	6.2	112	0.10
MGK 264 - Isomer a	0.33	0.38	6.7	113	0.076
MGK 264 - Isomer b	0.16	0.18	5.3	110	0.029
Molinate	0.50	0.53	3.8	105	0.060
Napropamide	0.50	0.58	7.9	116	0.14
Norflurazon	0.50	0.71	4.3	142	0.091
2,2',3,3',4,5',6,6'-Octachlorobiphenyl	0.50	0.47	5.3	94	0.076
Pebulate	0.50	0.56	7.1	112	0.11
2,2',3',4,6-Pentachlorobiphenyl	0.50	0.49	4.0	97	0.059
Pentachlorophenol	2.0	2.2	15	111	1.0

Method 525.2 − *Table 6*

Accuracy and Precision Data from Eight Determinations of the Method Analytes in Reagent Water Using Liquid-Solid C-18 Disk Extraction and the Ion Trap Mass Spectrometer (Continued)

Compound	True Conc. (μg/L)	Mean Observed Conc. (μg/L)	%RSD	Mean Method Accuracy (% of True Conc.)	MDL (μg/L)
Permethrin, *cis*	0.25	0.37	3.1	149	0.035
Permethrin, *trans*	0.75	0.84	1.6	112	0.039
Phenanthrene	0.50	0.49	6.3	97	0.092
Prometon[(d)]	0.50	0.16	63	32	0.30
Prometryn	0.50	0.46	23	91	0.32
Pronamide	0.50	0.56	3.9	111	0.064
Propachlor	0.50	0.58	5.7	115	0.098
Propazine	0.50	0.53	4.7	106	0.074
Pyrene	0.50	0.52	5.2	104	0.080
Simazine	0.50	0.54	2.8	107	0.045
Simetryn	0.50	0.36	20	71	0.22
Stirofos	0.50	0.72	3.7	144	0.080
Tebuthiuron	0.50	0.67	7.9	133	0.16
Terbacil	0.50	0.64	12	129	0.23
Terbufos	0.50	0.57	6.8	113	0.11
Terbutryn	0.50	0.46	24	93	0.34
2,2′,4,4′,- Tetrchlorobiphenyl	0.50	0.46	7.4	91	0.10
Toxaphene [(c)]	10	12	2.7	122	1.0
Triademefon	0.50	0.71	7.3	142	0.16
2,4,5-Trichlorobiphenyl	0.50	0.48	4.5	97	0.066
Tricyclazole	0.50	0.65	14	130	0.27
Trifluralin	0.50	0.59	7.8	117	0.14
Vernolate	0.50	0.50	3.2	99	0.047

(c) Seven replicates.
(d) Data from samples extracted at pH 2. For accurate determination of this analyte, a separate sample must be extracted at ambient pH.

Accuracy and Precision Data from Eight Determinations at 5 µg/L in Reagent Water of Poorly Chromatographed Nitrogen and Phosphorous Containing Pesticides

Compound	Ion Trap Mass Spectrometer				Quadrupole Mass Spectrometer			
	Cartridge		Disk		Cartridge		Disk	
	% RSD	Mean Method Accuracy (% of True Conc.)	%RSD	Mean Method Accuracy (% of True Conc.)	% RSD	Mean Method Accuracy (% of True Conc.)	% RSD	Mean Method Accuracy (% of True Conc.)
Fenamiphos	7.7	99	4.5	108	6.1	103	8.8	124
Fenarimol	2.0	104	10	110	6.5	126	5.5	150
Fluridone	2.5	105	2.3	104	3.6	102	4.5	114
Hexazinone	4.2	106	9.7	116	5.3	104	8.3	127
Norflurazon	4.1	111	9.6	119	3.2	98	11.1	113
Stirofos	8.2	114	12	124	4.1	110	11.1	125
Tebuthiuron	9.5	119	5.3	145	13	136	8.6	182
Triademeton	7.8	113	10	128	3.7	100	9.8	118
Tricyclazole	16	81	9.5	99	19	92	12	137

Method 525.2 – *Table 8*

Accuracy and Precision Data from Seven Determinations
of the Method Analytes in Tap Water using Liquid-Solid
C-18 Cartridge Extraction and the Ion Trap Mass Spectrometer

Compound	True Conc.	Mean	% RSD	% REC
Acenaphthylene	5.0	5.2	5.3	104
Alachlor	5.0	5.5	6.9	110
Aldrin	5.0	4.4	14	88
Ametryn	5.0	4.2	3.4	83
Anthracene	5.0	4.3	5.2	87
Aroclor 1016	ND	ND	ND	ND
Aroclor 1221	ND	ND	ND	ND
Aroclor 1232	ND	ND	ND	ND
Aroclor 1242	ND	ND	ND	ND
Aroclor 1248	ND	ND	ND	ND
Aroclor 1254	ND	ND	ND	ND
Aroclor 1260	ND	ND	ND	ND
Atraton[a]	5.0	2.2	28	43
Atrazine	5.0	5.6	6.2	111
Benz[a]anthracene	5.0	4.9	8.8	97
Benzo[b]fluoranthene	5.0	5.7	7.5	114
Benzo[k]fluoranthene	5.0	5.7	2.9	113
Benzo[g,h,i]pyrene	5.0	5.6	7.1	113
Benzo[a]pyrene	5.0	6.1	4.6	121
Bromacil	5.0	3.5	5.1	69
Butachlor	5.0	5.4	7.5	109
Butylate	5.0	5.1	4.5	102
Butylbenzylphthalate	5.0	7.2	8.3	144
Carboxin	5.0	1.0	23	20
Chlordane, (alpha-Chlordane	5.0	5.2	8.9	104
Chlordane, (gamma-Chlordane)	5.0	5.1	8.0	102
Chlordane, (trans-Nonachlor)	5.0	5.6	7.4	111
Chlorneb	5.0	5.2	3.0	105
Chlorobenzilate	5.0	5.7	4.4	114
2-Chlorobiphenyl	5.0	5.8	5.4	115
Chlorpropham	5.0	6.3	4.9	127
Chlorpyrifos	5.0	5.3	7.2	107
Chlorothalonil	5.0	5.4	9.9	108
Chrysene	5.0	5.5	3.9	110
Cyanazine	5.0	6.1	13	122
Cycloate	5.0	5.6	1.5	112
DCPA	5.0	5.4	5.0	107
4,4'-DDD	5.0	5.3	6.5	105
4,4'-DDE	5.0	5.2	6.6	104
4,4'-DDT	5.0	5.6	9.6	111

(a) Data from samples extracted at pH 2. For accurate determination of this analyte, a separate sample must be extracted at ambient pH.

Method 525.2 – *Table 8*

Accuracy and Precision Data from Seven Determinations of the Method Analytes in Tap Water using Liquid-Solid C-18 Cartridge Extraction and the Ion Trap Mass Spectrometer (Continued)

Compound	True Conc.	Mean	% RSD	% REC
Diazinon	5.0	4.9	8.7	98
Dibenz[a,h]anthracene	5.0	5.9	7.5	118
Di-n-Butylphthalate	5.0	6.2	4.6	124
2,3-Dichlorobiphenyl	5.0	5.3	7.4	106
Dichlorvos	5.0	2.8	7.3	56
Dieldrin	5.0	5.3	7.2	105
Di(2-Ethylhexyl)adipate	5.0	6.7	10	134
Di(2-Ethylhexyl)phthalate	5.0	6.5	6.6	130
Diethylphthalate	5.0	6.4	7.4	127
Dimethylphthalate	5.0	5.8	7.1	116
2,4-Dinitrotoluene	5.0	4.2	8.7	84
2,6-Dinitrotoluene	5.0	4.1	8.5	82
Diphenamid	5.0	5.2	7.7	104
Disulfoton	5.0	2.5	33	50
Disulfoton Sulfone	5.0	5.5	7.4	110
Disulfoton Sulfoxide	5.0	9.4	11	188
Endosulfan I	5.0	5.5	11	109
Endosulfan II	5.0	5.3	9.6	106
Endosulfan Sulfate	5.0	5.3	7.8	106
Endrin	5.0	6.1	3.9	121
Endrin Aldehyde	5.0	5.1	9.1	102
EPTC	5.0	5.1	2.1	102
Ethoprop	5.0	6.3	4.2	125
Etridiazole	5.0	5.8	7.5	117
Fenamiphos	5.0	5.9	22	119
Fenarimol	5.0	7.1	3.3	141
Fluorene	5.0	5.7	5.2	114
Fluridone	5.0	6.2	9.0	125
HCH, alpha	5.0	5.9	2.6	118
HCH, beta	5.0	5.3	8.4	106
HCH, delta	5.0	5.3	5.2	106
HCH, gamma (Lindane)	5.0	5.3	6.9	107
Heptachlor	5.0	4.7	8.7	93
Heptachlor epoxide	5.0	5.2	7.7	105
2,2',3,3',4,4',6-Heptachlorobiphenyl	5.0	5.1	6.9	103
Hexachlorobenzene	5.0	4.6	7.4	93
2,2'4,4',5,6'-Hexachlorobiphenyl	5.0	5.6	8.1	112
Hexachlorocyclopentadiene	5.0	6.0	4.8	120
Hexazinone	5.0	6.9	6.3	138
Indeno[1,2,3-cd]pyrene	5.0	6.8	7.7	135

Method 525.2 – *Table 8*

Accuracy and Precision Data from Seven Determinations of the Method Analytes in Tap Water using Liquid-Solid C-18 Cartridge Extraction and the Ion Trap Mass Spectrometer (Continued)

Compound	True Conc.	Mean	% RSD	% REC
Isophorone	5.0	4.9	12	99
Methoxychlor	5.0	5.6	4.9	112
Methyl Paraoxon	5.0	5.6	11	111
Metolachlor	5.0	5.6	7.7	111
Metribuzin	5.0	2.1	5.8	42
Mevinphos	5.0	3.3	1.6	67
MGK 264 - Isomer a	3.3	3.6	6.2	107
MGK 264 - Isomer b	1.7	1.8	7.6	110
Molinate	5.0	5.5	1.5	110
Napropamide	5.0	5.3	8.9	106
Norflurazon	5.0	6.7	7.2	135
2,2′,3,3′,4,5′,6,6′-Octachlorobiphenyl	5.0	4.9	6.9	97
Pebulate	5.0	5.3	3.1	106
2,2′,3′,4,6-Pentachlorobiphenyl	5.0	5.3	8.1	107
Pentachlorophenol	20.	33	4.9	162
Permethrin, cis	5.0	3.3	3.5	130
Permethrin, trans	5.0	8.5	2.2	113
Phenanthrene	5.0	5.5	4.0	109
Prometona [a]	5.0	2.0	25	40
Prometryn	5.0	4.5	4.3	89
Pronamide	5.0	5.7	5.3	115
Propachlor	5.0	6.2	4.0	124
Propazine	5.0	5.6	4.9	113
Pyrene	5.0	5.2	6.7	104
Simazine	5.0	6.0	9.0	120
Simetryn	5.0	3.9	7.0	78
Stirofos	5.0	6.1	12	121
Tebuthiuron	5.0	6.5	9.7	130
Terbacil	5.0	4.0	5.5	79
Terbufos	5.0	4.5	8.4	90
Terbutryn	5.0	4.3	6.5	86
2,2′,4,4′,-Tetrchlorobiphenyl	5.0	5.3	4.3	106
Toxaphene	ND	ND	ND	ND
Triademefon	5.0	6.0	12	121
2,4,5-Trichlorobiphenyl	5.0	5.2	5.1	103
Tricyclazole	5.0	4.8	5.2	96
Trifluralin	5.0	5.9	7.8	119
Vernolate	5.0	5.4	3.3	108

(a) Data from samples extracted at pH 2 - for accurate determination of this analyte, a separate sample must be extracted at ambient pH.

Method 601 – *Table 1*

Appendix A

Chromatographic Conditions and Method Detection Limits

Compound	Retention Time (min)		Method Detection Limit (µg/L)
	Col.1	Col.2	
Chloromethane	1.50	5.28	0.08
Bromomethane	2.17	7.05	1.18
Dichlorodifluoromethane	2.62	ND	1.81
Vinyl chloride	2.67	5.28	0.18
Chloroethane	3.33	8.68	0.52
Methylene chloride	5.25	10.1	0.25
Trichlorofluoromethane	7.18	ND	ND
1,1-Dichloroethene	7.93	7.72	0.13
1,1-Dichloroethane	9.30	12.6	0.07
trans-1,2-Dichloroethene	10.1	9.38	0.10
Chloroform	10.7	12.1	0.05
1,2-Dichloroethane	11.4	15.4	0.03
1,1,1-Trichloroethane	12.6	13.1	0.03
Carbon tetrachloride	13.0	14.4	0.12
Bromodichloromethane	13.7	14.6	0.10
1,2-Dichloropropane	14.9	16.6	0.04
cis-1,3-Dichloropropene	15.2	16.6	0.34
Trichloroethene	15.8	13.1	0.12
Dibromochloromethane	16.5	16.6	0.09
1,1,2-Trichloroethane	16.5	18.1	0.02
trans-1,3-Dichloropropene	16.5	18.0	0.20
2-Chloroethylvinyl ether	18.0	ND	0.13
Bromoform	19.2	19.2	0.20
1,1,2,2-Tetrachloroethane	21.6	ND	0.03
Tetrachloroethene	21.7	15.0	0.03
Chlorobenzene	24.2	18.8	0.25
1,3-Dichlorobenzene	34.0	22.4	0.32
1,2-Dichlorobenzene	34.9	23.5	0.15
1,4-Dichlorobenzene	35.4	22.3	0.24

Column 1 conditions: Carbopack B (60/80 mesh) coated with 1% SP-1000 packed in an 8 ft x 0.1 in. ID stainless steel or glass column with helium carrier gas at 40 mL/min flow rate. Column temperature held at 45 °C for 3 min then programmed at 8 °C/min to 220 °C and held for 15 min.

Column 2 conditions: Porisil-C (100/120 mesh) coated with n-octane packed in a 6 ft x 0.1 in. ID stainless steel or glass column with helium carrier gas at 40 mL/min flow rate. Column temperature held at 50 °C for 3 min then programmed a 6 °C/min to 170 °C and held for 4 min.

ND = not determined

Method 601 - *Table 2*

Calibration and QC Acceptance Criteria - Method 601[a]

Parameter	Range for Q (μg/L)	Limit for s (μg/L)	Range for $\overline{x}$ (μg/L)	Range for P, P_S (%)
Bromodichloromethane	15.2-24.8	4.3	10.7-32.0	42-172
Bromoform	14.7-25.3	4.7	5.0-29.3	13-159
Bromomethane	11.7-28.3	7.6	3.4-24.5	D-144
Carbon tetrachloride	13.7-26.3	5.6	11.8-25.3	43-143
Chlorobenzene	14.4-25.6	5.0	10.2-27.4	38-150
Chloroethane	15.4-24.6	4.4	11.3-25.2	46-137
2-Chloroethylvinyl ether	12.0-28.0	8.3	4.5-35.5	14-186
Chloroform	15.0-25.0	4.5	12.4-24.0	49-133
Chloromethane	11.9-28.1	7.4	D-34.9	D-193
Dibromochloromethane	13.1-26.9	6.3	7.9-35.1	24-191
1,2-Dichlorobenzene	14.0-26.0	5.5	1.7-38.9	D-208
1,3-Dichlorobenzene	9.9-30.1	9.1	6.2-32.6	7-187
1,4-Dichlorobenzene	13.9-26.1	5.5	11.5-25.5	42-143
1,1-Dichloroethane	16.8-23.2	3.2	11.2-24.6	47-132
1,2-Dichloroethane	14.3-25.7	5.2	13.0-26.5	51-147
1,1-Dichloroethene	12.6-27.4	6.6	10.2-27.3	28-167
trans-1,2-Dichloroethene	12.8-27.2	6.4	11.4-27.1	38-155
1,2-Dichloropropane	14.8-25.2	5.2	10.1-29.9	44-156
cis-1,3-Dichloropropene	12.8-27.2	7.3	6.2-33.8	22-178
trans-1,3-Dichloropropene	12.8-27.2	7.3	6.2-33.8	22-178
Methylene chloride	15.5-24.5	4.0	7.0-27.6	25-162
1,1,2,2-Tetrachloroethane	9.8-30.2	9.2	6.6-31.8	8-184
Tetrachloroethene	14.0-26.0	5.4	8.1-29.6	26-162
1,1,1-Trichloroethane	14.2-25.8	4.9	10.8-24.8	41-138
1,1,2-Trichloroethane	15.7-24.3	3.9	9.6-25.4	39-136
Trichloroethene	15.4-24.6	4.2	9.2-26.6	35-146
Trichlorofluoromethane	13.3-26.7	6.0	7.4-28.1	21-156
Vinyl chloride	13.7-26.3	5.7	8.2-29.9	28-163

Q   = Concentration measured in QC check sample, in μg/L (Section 7.5.3)
s   = Standard deviation of four recovery measurements, in μg/L (Section 8.2.4)
$\overline{x}$ = Average recovery for four recovery measurements, in μg/L (Section 8.2.4)
P, P_S = Percent recovery measured (Section 8.3.2, Section 8.4.2)
D   = Detected; result must be greater than zero.
(a) Criteria were calculated assuming a QC check sample concentration of 20 μg/L.

Note: These criteria are based directly upon the method performance data in Table 3. Where necessary, the limits for recovery have been broadened to assure applicability of the limits to concentrations below those used to develop Table 3.

Method 602 – *Table 1*

Chromatographic Conditions and Method Detection Limits

Compound	Retention time (min)		Method Detection Limit (μg/L)
	Col.1	Col.2	
Benzene	3.33	2.75	0.2
Toluene	5.75	4.25	0.2
Ethylbenzene	8.25	6.25	0.2
Chlorobenzene	9.17	8.02	0.2
1,4-Dichlorobenzene	16.8	16.2	0.3
1,3-Dichlorobenzene	18.2	15	0.4
1,2-Dichlorobenzene	25.9	19.4	0.4

Column 1 conditions: Supelcoport (100/120 mesh) coated with 5% SP-1200/1.75% Bentone-34 packed in a 6 ft by 0.085 in. ID stainless steel column with helium carrier gas at 36 mL/min flow rate. Column temperature held at 50 °C for 2 min then programmed at 6 °C/min to 90 °C for a final hold.

Column 2 conditions: Chromosorb W-AW (60/80 mesh) coated with 5% 1,2,3-Tris(2-cyanoethyoxy)-propane packed in a 6 ft by 0.085 in. ID stainless steel column with helium carrier gas at 30 mL/min flow rate. Column temperature held at 40 °C for 2 min then programmed at 2 °C/min to 100 °C for a final hold.

Method 602 – *Table 2*

Calibration and QC Acceptance Criteria - Method 602[a]

Parameter	Range for Q (μg/L)	Limit for s (μg/L)	Range for $\bar{x}$ (μg/L)	Range for P, P_s (%)
Benzene	15.4-24.6	4.1	10.0-27.9	39-150
Chlorobenzene	16.1-23.9	3.5	12.7-25.4	55-135
1,2-Dichlorobenzene	13.6-26.4	5.8	10.6-27.6	37-154
1,3-Dichlorobenzene	14.5-25.5	5.0	12.8-25.5	50-141
1,4-Dichlorobenzene	13.9-26.1	5.5	11.6-25.5	42-143
Ethylbenzene	12.6-27.4	6.7	10.0-28.2	32-160
Toluene	15.5-24.5	4.0	11.2-27.7	46-148

Q = Concentration measured in QC check sample, in μg/L (Section 7.5.3)
s = Standard deviation of four recovery measurements, in μg/L (Section 8.2.4)
$\bar{x}$ = Average recovery for four recovery measurements, in μg/L (Section 8.2.4)
P, P_s = Percent recovery measured (Section 8.3.2, Section 8.4.2)

(a) Criteria were calculated assuming a QC check sample concentration of 20 μg/L.

Note: These criteria are based directly upon the method performance data in Table 3. Where necessary, the limits for recovery have been broadened to assure applicability of the limits to concentrations below those used to develop Table 3.

Method 608 – *Table 1*

Chromatographic Conditions and Method Detection Limits

Parameter	Retention Time (min)		Method Detection Limit (μg/L)
	Col.1	**Col.2**	
α-BHC	1.35	1.82	0.003
γ-BHC (Lindane)	1.70	2.13	0.004
β-BHC	1.90	1.97	0.006
Heptachlor	2.00	3.35	0.003
δ-BHC	2.15	2.20	0.009
Aldrin	2.40	4.10	0.004
Heptachlor epoxide	3.50	5.00	0.083
Endosulfan I	4.50	6.20	0.014
4,4'-DDE	5.13	7.15	0.004
Dieldrin	5.45	7.23	0.002
Endrin	6.55	8.10	0.006
4,4'-DDD	7.83	9.08	0.011
Endosulfan II	8.00	8.28	0.004
4,4'-DDT	9.40	11.75	0.012
Endrin aldehyde	11.82	9.30	0.023
Endosulfan sulfate	14.22	10.70	0.066
Chlordane	MR	MR	0.014
Toxaphene	MR	MR	0.24
PCB-1016	MR	MR	ND
PCB-1221	MR	MR	ND
PCB-1232	MR	MR	ND
PCB-1242	MR	MR	0.065
PCB-1248	MR	MR	ND
PCB-1254	MR	MR	ND
PCB-1260	MR	MR	ND

Column 1 conditions: Supelcoport (100/120 mesh) coated with 1.5% SP-2250/1.95% SP-2401 packed in a 1.8 m long by 4 mm ID glass column with 5% methane/95% argon carrier gas at 60 mL/min flow rate. Column temperature held isothermal at 200 °C, except for PCB-1016 through PCB-1248, should be measured at 160 °C.

Column 2 conditions: Supelcoport (100/120 mesh) coated with 3% OV-1 packed in a 1.8 m long by 4 mm ID glass column with 5% methane/95% argon carrier gas at 60 mL/min flow rate. Column temperature held isothermal at 200 °C for the pesticides; at 140 °C for PCB-1221 and 1232; and at 170 °C for PCB-1016 and 1242 to 1268.

MR = Multiple peak response
ND = not determined

Method 608 —*Table 3*

QC Acceptance Criteria

Parameter	Test conc. (μg/L)	Limit for s (μg/L)	Range for $\overline{X}$ (μg/L)	Range for P, P$_S$ (%)
Aldrin	2.0	0.042	1.08-2.24	42-122
α-BHC	2.0	0.48	0.98-2.44	37-134
β-BHC	2.0	0.64	0.78-2.60	17-147
δ-BHC	2.0	0.72	1.01-2.37	19-140
γ-BHC	2.0	0.46	0.86-2.32	32-127
Chlordane	50	10.0	27.6-54.3	45-119
4,4'-DDD	10	2.8	4.8-12.6	31-141
4,4'-DDE	2.0	0.55	1.08-2.60	30-145
4,4'-DDT	10	3.6	4.6-13.7	25-160
Dieldrin	2.0	0.76	1.15-2.49	36-146
Endosulfan I	2.0	0.49	1.14-2.82	45-153
Endosulfan II	10	6.1	2.2-17.1	D-202
Endosulfan sulfate	10	2.7	3.8-13.2	26-144
Endrin	10	3.7	5.1-12.6	30-147
Heptachlor	2.0	0.4	0.86-2.00	34-111
Heptachlor epoxide	2.0	0.41	1.13-2.63	37-142
Toxaphene	50	12.7	27.8-55.6	41-126
PCB-1016	50	10.0	30.5-51.5	50-114
PCB-1221	50	24.4	22.1-75.2	15-178
PCB-1232	50	17.9	14.0-98.5	10-215
PCB-1242	50	12.2	24.8-69.6	39-150
PCB-1248	50	15.9	29.0-70.2	38-158
PCB-1254	50	13.8	22.2-57.9	29-131
PCB-1260	50	10.4	18.7-54.9	8-127

s = Standard deviation of four recovery measurements, in μg/L (Section 8.2.4)
$\overline{X}$ = Average recovery for four recovery measurements, in μg/L (Section 8.2.4)
P, P$_S$ = Percent recovery measured (Section 8.3.2, Section 8.4.2)
D = Detected; result must be greater than zero.

Note: These criteria are based directly upon the method performance data in Table 4. Where necessary, the limits for recovery have been broadened to assure applicability of the limits to concentrations below those used to develop Table 4.

Method 624 – *Table 1*

Chromatographic Conditions and Method Detection Limits

Parameter	Retention Time (min)	Method Detection Limit (µg/L)
Chloromethane	2.3	ND
Bromomethane	3.1	ND
Vinyl chloride	3.8	ND
Chloroethane	4.6	ND
Methylene chloride	6.4	2.8
Trichlorofluoromethane	8.3	ND
1,1-Dichloroethene	9.0	2.8
1,1-Dichloroethane	10.1	4.7
trans-1,2-Dichloroethene	10.8	1.6
Chloroform	11.4	1.6
1,2-Dichloroethane	12.1	2.8
1,1,1-Trichloroethane	13.4	3.8
Carbon tetrachloride	13.7	2.8
Bromodichloromethane	14.3	2.2
1,2-Dichloropropane	15.7	6.0
cis-1,3-Dichloropropene	15.9	5.0
Trichloroethene	16.5	1.9
Benzene	17.0	4.4
Dibromochloromethane	17.1	3.1
1,1,2-Trichloroethane	17.2	5.0
trans-1,3-Dichloropropene	17.2	ND
2-Chloroethylvinyl ether	18.6	ND
Bromoform	19.8	4.7
1,1,2,2-Tetrachloroethane	22.1	6.9
Tetrachloroethene	22.2	4.1
Toluene	23.5	6.0
Chlorobenzene	24.6	6.0
Ethyl benzene	26.4	7.2
1,3-Dichlorobenzene	33.9	ND
1,2-Dichlorobenzene	35.0	ND
1,4-Dichlorobenzene	35.4	ND

Column conditions: Carbopak B (60/80 mesh) coated with 1% SP-1000 packed in a 6 ft by .1 in. ID glass column with helium carrier gas at 30 mL/min. flow rate. Column temperature held at 45 °C for 3 min., then programmed at 8 °C/ min. to 220 °C and held for 15 min.

ND = not determined

Method 624 – *Table 2*

BFB Key M/Z Abundance Criteria

Mass	m/z Abundance Criteria
50	15 to 40% of mass 95
75	30 to 60% of mass 95
95	Base Peak, 100% Relative Abundance
96	5 to 9% of mass 95
173	less than 2% of mass 174
174	greater than 50% of mass 95
175	5 to 9% of mass 174
176	greater than 95% but less than 101% of mass 174
177	5 to 9% of mass 176

Method 624 – *Table 3*

Suggested Surrogate and Internal Standards

Compound	Retention Time (min)[a]	Primary m/z	Secondary masses
Benzene d_6	17.0	84	
4-Bromofluorobenzene	28.3	95	174, 176
1,2-Dichloroethane d_4	12.1	102	
1,4-Difluorobenzene	19.6	114	63, 88
Ethylbenzene d_5	26.4	111	
Ethylbenzene d_{10}	26.4	98	
Fluorobenzene	18.4	96	70
Pentafluorobenzene	23.5	168	
Bromochloromethane	9.3	128	49, 130, 51
2-Bromo-1-chloropropane	19.2	77	79, 156
1,4-Dichlorobutane	25.8	55	90, 92

(a)   For chromatographic conditions, see Table 1.

Method 624 – *Table 5*

Calibration and QC Acceptance Criteria[a]

Parameter	Range for Q (µg/L)	Limit for s (µg/L)	Range for $\bar{x}$ (µg/L)	Range for P, P_S (%)
Benzene	12.8-27.2	6.9	15.2-26.0	37-151
Bromodichloromethane	13.1-26.9	6.4	10.1-28.0	35-155
Bromoform	14.2-25.8	5.4	11.4-31.1	45-169
Bromomethane	2.8-37.2	17.9	D-41.2	D-242
Carbon tetrachloride	14.6-25.4	5.2	17.2-23.5	70-140
Chlorobenzene	13.2-26.8	6.3	16.4-27.4	37-160
Chloroethane	7.6-32.4	11.4	8.4-40.4	14-230
2-Chloroethylvinyl ether	D-44.8	25.9	D-50.4	D-305
Chloroform	13.5-26.5	6.1	13.7-24.2	51-138
Chloromethane	D-40.8	19.8	D-45.9	D-273
Dibromochloromethane	13.5-26.5	6.1	13.8-26.6	53-149
1,2-Dichlorobenzene	12.6-27.4	7.1	11.8-34.7	18-190
1,3-Dichlorobenzene	14.6-25.4	5.5	17.0-28.8	59-156
1,4-Dichlorobenzene	12.6-27.4	7.1	11.8-34.7	18-190
1,1-Dichloroethane	14.5-25.5	5.1	14.2-28.5	59-155
1,2-Dichloroethane	13.6-26.4	6.0	14.3-27.4	49-155
1,1-Dichloroethene	10.1-29.9	9.1	3.7-42.3	D-234
trans-1,2-Dichloroethene	13.9-26.1	5.7	13.6-28.4	54-156
1,2-Dichloropropane	6.8-33.2	13.8	3.8-36.2	D-210
cis-1,3-Dichloropropene	4.8-35.2	15.8	1.0-39.0	D-227
trans-1,3-Dichloropropene	10.0-30.0	10.4	7.6-32.4	17-183
Ethyl benzene	11.8-28.2	7.5	17.4-26.7	37-162
Methylene chloride	12.1-27.9	7.4	D-41.0	D-221
1,1,2,2-Tetrachloroethane	12.1-27.9	7.4	13.5-27.2	46-157
Tetrachloroethene	14.7-25.3	5.0	17.0-26.6	64-148
Toluene	14.9-25.1	4.8	16.6-26.7	47-150
1,1,1-Trichloroethane	15.0-25.0	4.6	13.7-30.1	52-162
1,1,2-Trichloroethane	14.2-25.8	5.5	14.3-27.1	52-150
Trichloroethene	13.3-26.7	6.6	18.6-27.6	71-157
Trichlorofluoromethane	9.6-30.4	10.0	8.9-31.5	17-181
Vinyl chloride	0.8-39.2	20.0	D-43.5	D-251

Q = Concentration measured in QC check sample, in µg/L (Section 7.5.3)
s = Standard deviation of four recovery measurements, in µg/L (Section 8.2.4)
$\bar{x}$ = Average recovery for four recovery measurements, in µg/L (Section 8.2.4)
P, P_S = Percent recovery measured, (Section 8.3.2, Section 8.4.2)
D = Detected; result must be greater than zero.
(a) Criteria were calculated assuming a QC check sample concentration of 20 µg/L.

Note: These criteria are based directly upon the method performance data in Table 6. Where necessary, the limits for recovery have been broadened to assure applicability of the limits to concentrations below those used to develop Table 6.

Method 625 – *Table 4*

Chromatographic Conditions, Method Detection Limits, and Characteristic Masses for Base/Neutral Extractables

Parameter	Retention Time (min)	Method Detection Limit (mg/L)	Characteristic Masses					
			Electron Impact			Chemical Ionization		
			Primary	Secondary	Secondary	Methane	Methane	Methane
1,3-Dichlorobenzene	7.4	1.9	146	148	113	146	148	150
1,4-Dichlorobenzene	7.8	4.4	146	148	113	146	148	150
Hexachloroethane	8.4	1.6	117	201	199	199	201	203
Bis(2-chloroethyl)ether	8.4	5.7	93	63	95	63	107	109
1,2-Dichlorobenzene	8.4	1.9	146	148	113	146	148	150
Bis(2-chloroisopropyl)ether[a]	9.3	5.7	45	77	79	77	135	137
N-Nitrosodi-n-propylamine			130	42	101			
Nitrobenzene	11.1	1.9	77	123	65	124	152	164
Hexachlorobutadiene	11.4	0.9	225	223	227	223	225	227
1,2,4-Trichlorobenzene	11.6	1.9	180	182	145	181	183	209
Isophorone	11.9	2.2	82	95	138	139	167	178
Naphthalene	12.1	1.6	128	129	127	129	157	169
Bis(2-chloroethoxy)methane	12.2	5.3	93	95	123	65	107	137
Hexachlorocyclopentadiene[a]	13.9		237	235	272	235	237	239
2-Chloronaphthalene	15.9	1.9	162	164	127	163	191	203
Acenaphthylene	17.4	3.5	152	151	153	152	153	181
Acenaphthene	17.8	1.9	154	153	152	154	155	183
Dimethyl phthalate	18.3	1.6	163	194	164	151	163	164
2,6-Dinitrotoluene	18.7	1.9	165	89	121	183	211	223
Fluorene	19.5	1.9	166	165	167	166	167	195
4-Chlorophenyl phenyl ether	19.5	4.2	204	206	141			
2,4-Dinitrotoluene	19.8	5.7	165	63	182	183	211	223
Diethyl phthalate	20.1	1.9	149	177	150	177	223	251
N-Nitrosodiphenylamine[b]	20.5	1.9	169	168	167	169	170	198
Hexachlorobenzene	21.0	1.9	284	142	249	284	286	288
β-BHC[b]	21.1		183	181	109			
4-Bromophenyl phenyl ether	21.2	1.9	248	250	141	249	251	277
δ-BHC[b]	22.4		183	181	109			
Phenanthrene	22.8	5.4	178	179	176	178	179	207
Anthracene	22.8	1.9	178	179	176	178	179	207
β-BHC	23.4	4.2	181	183	109			
Heptachlor	23.4	1.9	100	272	274			
δ-BHC	23.7	3.1	183	109	181			
Aldrin	24.0	1.9	66	263	220			
Dibutyl phthalate	24.7	2.5	149	150	104	149	205	279
Heptachlor epoxide	25.6	2.2	353	355	351			
Endosulfan I[b]	26.4		237	339	341			
Fluoranthene	26.5	2.2	202	101	100	203	231	243
Dieldrin	27.2	2.5	79	263	279			
4,4'-DDE	27.2	5.6	246	248	176			

Method 625 – *Table 4*

Chromatographic Conditions, Method Detection Limits, and Characteristic Masses for Base/Neutral Extractables (Continued)

Parameter	Reten-tion Time (min)	Method Detection Limit (mg/L)	Characteristic Masses					
			Electron Impact			Chemical Ionization		
			Pri-mary	Second-ary	Second-ary	Meth-ane	Meth-ane	Meth-ane
Pyrene	27.3	1.9	202	101	100	203	231	243
Endrin[b]	27.9		81	263	82			
Endosulfan II[b]	28.6		237	339	341			
4,4'-DDD	28.6	2.8	235	237	165			
Benzidine[b]	28.8	44	184	92	185	185	213	225
4,4'-DDT	29.3	4.7	235	237	165			
Endosulfan sulfate	29.8	5.6	272	387	422			
Endrin aldehyde			67	345	250			
Butyl benzyl phthalate	29.9	2.5	149	91	206	149	299	327
Bis(2-ethylhexyl)phthalate	30.6	2.5	149	167	279	149		
Chrysene	31.5	2.5	228	226	229	228	229	257
Benzo(a)anthracene	31.5	7.8	228	229	226	228	229	257
3,3'-Dichlorobenzidine	32.2	16.5	252	254	126			
Di-n-octyl phthalate	32.5	2.5	149					
Benzo(b)fluoranthene	34.9	4.8	252	253	125	252	253	281
Benzo(k)fluoranthene	34.9	2.5	252	253	125	252	253	281
Benzo(a)pyrene	36.4	2.5	252	253	125	252	253	281
Indeno(1,2,3-cd)pyrene	42.7	3.7	276	138	277	276	277	305
Dibenzo(a,h)anthracene	43.2	2.5	278	139	279	278	279	307
Benzo(ghi)perylene	45.1	4.1	276	138	277	276	277	305
N-Nitrosodimethylamine[b]			42	74	44			
Chlordane[c]	19-30		373	375	377			
Toxaphene[c]	25-34		159	231	233			
PCB-1016[c]	18-30		224	260	294			
PCB-1221[c]	15-30	30	190	224	260			
PCB-1232[c]	15-32		190	224	260			
PCB-1242[c]	15-32		224	260	294			
PCB-1248[c]	12-34		294	330	262			
PCB-1254[c]	22-34	36	294	330	362			
PCB-1260[c]	23-32		330	362	394			

(a) The proper chemical name is 2,2'-oxybis(1-chloropropane).
(b) See Section 1.2.
(c) These compounds are mixtures of various isomers. Column conditions: Supelcoport (100/120 mesh) coated with 3% SP-2250 packed in a 1.8 m long by 2 mm ID glass column with helium carrier gas at 30 mL/min. flow rate. Column temperature held isothermal at 50 °C for 4 min., then programmed at 8 °C/min. to 270 °C and held for 30 min.

Method 625 – *Table 5*

Chromatographic Conditions, Method Detection Limits, and Characteristic Masses for Acid Extractables

Parameter	Retention Time (min)	Method Detection Limit (mg/L)	Characteristic Masses					
			Electron Impact			Chemical Ionization		
			Primary	Secondary	Secondary	Methane	Methane	Methane
2-Chlorophenol	5.9	3.3	128	64	130	129	131	157
2-Nitrophenol	6.5	3.6	139	65	109	140	168	122
Phenol	8.0	1.5	94	65	66	95	123	135
2,4-Dimethylphenol	9.4	2.7	122	107	121	123	151	163
2,4-Dichlorophenol	9.8	2.7	162	164	98	163	165	167
2,4,6-Trichlorophenol	11.8	2.7	196	198	200	197	199	201
4-Chloro-3-methylphenol	13.2	3.0	142	107	144	143	171	183
2,4-Dinitrophenol	15.9	42	184	63	154	185	213	225
2-Methyl-4,6-dinitrophenol	16.2	24	198	182	77	199	227	239
Pentachlorophenol	17.5	3.6	266	264	268	267	265	269
4-Nitrophenol	20.3	2.4	65	139	109	140	168	122

Column conditions: Supelcoport (100/120 mesh) coated with 1% SP-1240DA packed in a 1.8 m long by 2 mm ID glass column with helium carrier gas at 30 mL/min. flow rate. Column temperature held isothermal at 70 °C for 2 min. then programmed at 8 °C/min. to 200 °C.

Method 625 – *Table 6*

QC Acceptance Criteria

Parameter	Test conc. (μg/L)	Limit for s (μg/L)	Range for $\overline{x}$ (μg/L)	Range for P, P$_S$ (%)
Acenaphthene	100	27.6	60.1-132.3	47-145
Acenaphthylene	100	40.2	53.5-126.0	33-145
Aldrin	100	39.0	7.2-152.2	D-166
Anthracene	100	32.0	43.4-118.0	27-133
Benzo(a)anthracene	100	27.6	41.8-133.0	33-143
Benzo(b)fluoranthene	100	38.8	42.0-140.4	24-159
Benzo(k)fluoranthene	100	32.3	25.2-145.7	11-162
Benzo(a)pyrene	100	39.0	31.7-148.0	17-163
Benzo(ghi)perylene	100	58.9	D-195.0	D-219
Benzyl butyl phthalate	100	23.4	D-139.9	D-152
β-BHC	100	31.5	41.5-130.6	24-149
δ-BHC	100	21.6	D-100.0	D-110
Bis(2-chloroethyl)ether	100	55.0	42.9-126.0	12-158
Bis(2-chloroethoxy)methane	100	34.5	49.2-164.7	33-184
Bis(2-chloroisopropyl)ether[a]	100	46.3	62.8-138.6	36-166
Bis(2-ethylhexyl)phthalate	100	41.1	28.9-136.8	8-158
4-Bromophenyl phenyl ether	100	23.0	64.9-114.4	53-127
2-Chloronaphthalene	100	13.0	64.5-113.5	60-118
4-Chlorophenyl phenyl ether	100	33.4	38.4-144.7	25-158
Chrysene	100	48.3	44.1-139.9	17-168
4,4'-DDD	100	31.0	D-134.5	D-145
4,4'-DDE	100	32.0	19.2-119.7	4-136
4,4'-DDT	100	61.6	D-170.6	D-203
Dibenzo(a,h)anthracene	100	70.0	D-199.7	D-227
Di-n-butyl phthalate	100	16.7	8.4-111.0	1-118
1,2-Dichlorobenzene	100	30.9	48.6-112.0	32-129
1,3-Dichlorobenzene	100	41.7	16.7-153.9	D-172
1,4-Dichlorobenzene	100	32.1	37.3-105.7	20-124
3,3'-Dichlorobenzidine	100	71.4	8.2-212.5	D-262
Dieldrin	100	30.7	44.3-119.3	29-136
Diethyl phthalate	100	26.5	D-100.0	D-114
Dimethyl phthalate	100	23.2	D-100.0	D-112
2,4-Dinitrotoluene	100	21.8	47.5-126.9	39-139
2,6-Dinitrotoluene	100	29.6	68.1-136.7	50-158
Di-n-octyl phthalate	100	31.4	18.6-131.8	4-146
Endosulfan sulfate	100	16.7	D-103.5	D-107
Endrin aldehyde	100	32.5	D-188.8	D-209

Method 625 – *Table 6*

QC Acceptance Criteria (Continued)

Parameter	Test conc. (μg/L)	Limit for s (μg/L)	Range for $\overline{x}$ (μg/L)	Range for P, P_s (%)
Fluoranthene	100	32.8	42.9-121.3	26-137
Fluorene	100	20.7	71.6-108.4	59-121
Heptachlor	100	37.2	D-172.2	D-192
Heptachlor epoxide	100	54.7	70.9-109.4	26-155
Hexachlorobenzene	100	24.9	7.8-141.5	D-152
Hexachlorobutadiene	100	26.3	37.8-102.2	24-116
Hexachloroethane	100	24.5	55.2-100.0	40-113
Indeno(1,2,3-cd)pyrene	100	44.6	D-150.9	D-171
Isophorone	100	63.3	46.6-180.2	21-196
Naphthalene	100	30.1	35.6-119.6	21-133
Nitrobenzene	100	39.3	54.3-157.6	35-180
N-Nitrosodi-n-propylamine	100	55.4	13.6-197.9	D-230
PCB-1260	100	54.2	19.3-121.0	D-164
Phenanthrene	100	20.6	65.2-108.7	54-120
Pyrene	100	25.2	69.6-100.0	52-115
1,2,4-Trichlorobenzene	100	28.1	57.3-129.2	44-142
4-Chloro-3-methylphenol	100	37.2	40.8-127.9	22-147
2-Chlorophenol	100	28.7	36.2-120.4	23-134
2,4-Dichlorophenol	100	26.4	52.5-121.7	39-135
2,4-Dimethylphenol	100	26.1	41.8-109.0	32-119
2,4-Dinitrophenol	100	49.8	D-172.9	D-191
2-Methyl-4,6-dinitrophenol	100	93.2	53.0-100.0	D-181
2-Nitrophenol	100	35.2	45.0-166.7	29-182
4-Nitrophenol	100	47.2	13.0-106.5	D-132
Pentachlorophenol	100	48.9	38.1-151.8	14-176
Phenol	100	22.6	16.6-100.0	5-112
2,4,6-Trichlorophenol	100	31.7	52.4-129.2	37-144

s = Standard deviation of four recovery measurements, in μg/L. (Section 8.2.4)
$\overline{x}$ = Average recovery for four recovery measurements, in μg/L. (Section 8.2.4)
P, P_s = Percent recovery measured. (Section 8.3.2, Section 8.4.2)
D = Detected; result must be greater than zero.

(a) The proper chemical name is 2,2'-oxybis(1-chloropropane).

Note: These criteria are based directly upon the method performance data in Table 7. Where necessary, the limits for recovery have been broadened to assure applicability of the limits to concentrations below those used to develop Table 7.

Method 625 – *Table 8*

Suggested Internal and Surrogate Standards

Base/neutral Fraction	Acid Fraction
Aniline-d_5	2-Fluorophenol
Anthracene-d_{10}	Pentafluorophenol
Benzo(a)anthracene-d_{12}	Phenol-d_5
4,4'-Dibromobiphenyl	2-Perfluoromethyl phenol
4,4'-Dibromoctafluorobiphenyl	
Decafluorobiphenyl	
2,2'-Difluorobiphenyl	
4-Fluoroaniline	
1-Fluoronaphthalene	
2-Fluoronaphthalene	
Naphthalene-d_8	
Nitrobenzene-d_5	
2,3,4,5,6-Pentafluorobiphenyl	
Phenanthrene-d_{10}	
Pyridine-d_5	

Method 625 – *Table 9*

DFTPP Key Masses and Ion Abundance Criteria

Mass	m/z Abundance Criteria
51	30-60 percent of mass 198
68	Less than 2 percent of mass 69
70	Less than 2 percent of mass 69
127	40-60 percent of mass 198
197	Less than 1 percent of mass 198
198	Base peak, 100 percent relative abundance
199	5-9 percent of mass 198
275	10-30 percent of mass 198
365	Greater than 1 percent of mass 198
441	Present but less than mass 443
442	Greater than 40 percent of mass 198
443	17-23 percent of mass 442

Method 3111 – *Table 3111:I*

Atomic Absorption Concentration Ranges with Direct Aspiration Atomic Absorption

Element	Wavelength nm	Flame Gases*	Instrument Detection Limit mg/L	Sensitivity mg/L	Optimum Concentration Range mg/L
Ag	328.1	A-Ac	0.01	0.06	0.1-4
Al	309.3	N-Ac	0.1	1	5-100
Au	242.8	A-Ac	0.01	0.25	0.5-20
Ba	553.6	N-Ac	0.03	0.4	1-20
Be	234.9	N-Ac	0.005	0.03	0.05-2
Bi	223.1	A-Ac	0.06	0.4	1-50
Ca	422.7	A-Ac	0.003	0.08	0.2-20
Cd	228.8	A-Ac	0.002	0.025	0.05-2
Co	240.7	A-Ac	0.03	0.2	0.5-10
Cr	357.9	A-Ac	0.02	0.1	0.2-10
Cs	852.1	A-Ac	0.02	0.3	0.5-15
Cu	324.7	A-Ac	0.01	0.1	0.2-10
Fe	248.3	A-Ac	0.02	0.12	0.3-10
Ir	264.0	A-Ac	0.6	8	-
K	766.5	A-Ac	0.005	0.04	0.1-2
Li	670.8	A-Ac	0.002	0.04	0.1-2
Mg	285.2	A-Ac	0.0005	0.007	0.02-2
Mn	279.5	A-Ac	0.01	0.05	0.1-10
Mo	313.3	N-Ac	0.1	0.5	1-20
Na	589.0	A-Ac	0.002	0.015	0.03-1
Ni	232.0	A-Ac	0.02	0.15	0.3-10
Os	290.9	N-Ac	0.08	1	-
Pb**	283.3	A-Ac	0.05	0.5	1-20
Pt	265.9	A-Ac	0.1	2.0	5-75
Rh	343.5	A-Ac	0.5	0.3	-
Ru	349.9	A-Ac	0.07	0.5	-
Sb	217.6	A-Ac	0.07	0.5	1-40
Si	251.6	N-Ac	0.3	2	5-150
Sn	224.6	A-Ac	0.8	4	10-200
Sr	460.7	A-Ac	0.03	0.15	0.3-5
Ti	365.3	N-Ac	0.3	2	5-100
V	318.4	N-Ac	0.2	1.5	2-100
Zn	213.9	A-Ac	0.005	0.02	0.05-2

* A-Ac = air-acetylene; N-Ac = nitrous oxide-acetylene

** The more sensitive 217.0 nm wavelength is recommended for instruments with background correction capabilities.

Copyright © ASTM. Reprinted with permission.

Method 3113 – *Table 3113:II*

Detection Levels and Concentration Ranges for Electrothermal Atomization Atomic Absorption Spectrometry

Element	Wavelength nm	Estimated Detection Limit μg/L	Optimum Concentration Range μg/L
Al	309.3	3	20-200
Sb	217.6	3	20-300
As*	193.7	1	5-100
Ba**	553.6	2	10-200
Be	234.9	0.2	1-30
Cd	228.8	0.1	0.5-10
Cr	357.9	2	5-100
Co	240.7	1	5-100
Cu	324.7	1	5-100
Fe	248.3	1	5-100
Pb***	283.3	1	5-100
Mn	279.5	0.2	1-30
Mo**	313.3	1	3-60
Ni**	232.0	1	5-100
Se*	196.0	2	5-100
Ag	328.1	0.2	1-25
Sn	224.6	5	20-300

* Gas interrupt utilized

** Pyrolytic graphite tubes utilized

*** The more sensitive 217.0-nm wavelength is recommended for instruments with background correction capabilities.

Method 3120 − *Table 3120:I*

Suggested Wavelengths, Estimated Detection Limits, Alternate Wavelengths, Calibration Concentrations, and Upper Limits

Element	Suggested Wavelength nm	Estimated Detection Limit μg/L	Alternate Wavelength* nm	Calibration Concentration mg/L	Upper Limit Concentration mg/L
Aluminum	308.22	40	237.32	10.0	100
Antimony	206.83	30	217.58	10.0	100
Arsenic	193.70	50	189.04**	10.0	100
Barium	455.40	2	493.41	1.0	50
Beryllium	313.04	0.3	234.86	1.0	10
Boron	249.77	5	249.68	1.0	50
Cadmium	226.50	4	214.44	2.0	50
Calcium	317.93	10	315.89	10.0	100
Chromium	267.72	7	206.15	5.0	50
Cobalt	228.62	7	230.79	2.0	50
Copper	324.75	6	219.96	1.0	50
Iron	259.94	7	238.20	10.0	100
Lead	220.35	40	217.00	10.0	100
Lithium	670.78	4***	-	5.0	100
Magnesium	279.08	30	279.55	10.0	100
Manganese	257.61	2	294.92	2.0	50
Molybdenum	202.03	8	203.84	10.0	100
Nickel	231.60	15	221.65	2.0	50
Potassium	766.49	100***	769.90	10.0	100
Selenium	196.03	75	203.99	5.0	100
Silica (SiO_2)	212.41	20	251.61	21.4	100
Silver	328.07	7	338.29	2.0	50
Sodium	589.00	30***	589.59	10.0	100
Strontium	407.77	0.5	421.55	1.0	50
Thallium	190.86**	40	377.57	10.0	100
Vanadium	292.40	8	-	1.0	50
Zinc	213.86	2	206.20	5.0	100

* Other wavelengths may be substituted if they provide the needed sensitivity and are corrected for spectral interference.

** Available with vacuum or inert gas purged optical path

*** Sensitive to operating conditions

Method 6210A – *Table 6210:I*

Chromatographic Conditions and Method Detection Limits (MDL)

Compound	Retention Time min	Method Detection Limit µg/L
Chloromethane	2.3	ND
Bromomethane	3.1	ND
Vinyl chloride	3.8	ND
Chloroethane	4.6	ND
Methylene chloride	6.4	2.8
Trichlorofluoromethane	8.3	ND
1,1-Dichloroethene	9.0	2.8
1,1-Dichloroethane	10.1	4.7
trans-1,2-Dichloroethene	10.8	1.6
Chloroform	11.4	1.6
1,2-Dichloroethane	12.1	2.8
1,1,1-Trichloroethane	13.4	3.8
Carbon tetrachloride	13.7	2.8
Bromodichloromethane	14.3	2.2
1,2-Dichloropropane	15.7	6.0
cis-1,3-Dichloropropene	15.9	5.0
Trichloroethene	16.5	1.9
Benzene	17.0	4.4
Dibromochloromethane	17.1	3.1
1,1,2-Trichloroethane	17.2	5.0
trans-1,3-Dichloropropene	17.2	ND
2-Chloroethylvinyl ether	18.6	ND
Bromoform	19.8	4.7
1,1,2,2-Tetrachloroethane	22.1	6.9
Tetrachloroethene	22.2	4.1
Toluene	23.5	6.0
Chlorobenzene	24.6	6.0
Ethyl benzene	26.4	7.2
1,3-Dichlorobenzene	33.9	ND
1,2-Dichlorobenzene	35.0	ND
1,4-Dichlorobenzene	35.4	ND

Column conditions: Carbopack B (60/80 mesh) coated with 1% SP-1000 packed in a 1.8 m by 3 mm ID glass column with helium carrier gas at 30 mL/min flow rate. Column temperature held at 45 °C for 3 min, then programmed at 8 °C/min to 220 °C and held for 15 min.

ND = not determined

Method 6210A – *Table 6210:II*

BFB KEY m/z Abundance Criteria

Mass	m/z Abundance Criteria
50	15 to 40% of mass 95
75	30 to 60% of mass 95
95	Base peak, 100% relative abundance
96	5 to 9% of mass 95
173	<2% of mass 174
174	>50% of mass 95
175	5 to 9% of mass 174
176	>95% but <101% of mass 174
177	5 to 9% of mass 176

Method 6210A – *Table 6210:III*

Suggested Surrogate and Internal Standards

Compound	Retention Time* min	Primary m/z	Secondary Masses
Benzene-d$_6$	17.0	84	
4-Bromofluorobenzene	28.3	95	174, 176
1,2-Dichloroethane-d$_4$	12.1	102	
1,4-Difluorobenzene	19.6	114	63, 88
Ethylbenzene-d$_5$	26.4	111	
Ethylbenzene-d$_{10}$	26.4	98	
Fluorobenzene	18.4	96	70
Pentafluorobenzene	23.5	168	
Bromochloromethane	9.3	128	49, 130, 51
2-Bromo-1-chloropropane	19.2	77	79, 156
1,4-Dichlorobutane	25.8	55	90, 92

*For chromatographic conditions, see Table 6210:I.

Method 6210A – Table 6210:IV

Calibration and QC Acceptance Criteria

Compound	Range for Q µg/L	Limit for s µg/L	Range for $\bar{x}$ g/L	Range for P, P_S %
Benzene	12.8-27.2	6.9	15.2-26.0	37-151
Bromodichloromethane	13.1-26.9	6.4	10.1-28.0	35-155
Bromoform	14.2-25.8	5.4	11.4-31.1	45-169
Bromomethane	2.8-37.2	17.9	D-41.2	D-242
Carbon tetrachloride	14.6-25.4	5.2	17.2-23.5	70-140
Chlorobenzene	13.2-26.8	6.3	16.4-27.4	37-160
Chloroethane	7.6-32.4	11.4	8.4-40.4	14-230
2-Chloroethylvinyl ether	D-44.8	25.9	D-50.4	D-305
Chloroform	13.5-26.5	6.1	13.7-24.2	51-138
Chloromethane	D-40.8	19.8	D-45.9	D-273
Dibromochloromethane	13.5-26.5	6.1	13.8-26.6	53-149
1,2-Dichlorobenzene	12.6-27.4	7.1	11.8-34.7	18-190
1,3-Dichlorobenzene	14.6-25.4	5.5	17.0-28.8	59-156
1,4-Dichlorobenzene	12.6-27.4	7.1	11.8-34.7	18-190
1,1-Dichloroethane	14.5-25.5	5.1	14.2-28.5	59-155
1,2-Dichloroethane	13.6-26.4	6.0	14.3-27.4	49-155
1,1-Dichloroethene	10.1-29.9	9.1	3.7-42.3	D-234
trans-1,2-Dichloroethene	13.9-26.1	5.7	13.6-28.5	54-156
1,2-Dichloropropane	6.8-33.2	13.8	3.8-36.2	D-210
cis-1,3-Dichloropropene	4.8-35.2	15.8	1.0-39.0	D-227
trans-1,3-Dichloropropene	10.0-30.0	10.4	7.6-32.4	17-183
Ethyl benzene	11.8-28.2	7.5	17.4-26.7	37-162
Methylene chloride	12.1-27.9	7.4	D-41.0	D-221
1,1,2,2-Tetrachloroethane	12.1-27.9	7.4	13.5-27.2	46-157
Tetrachloroethene	14.7-25.3	5.5	17.0-26.6	64-148
Toluene	14.9-25.1	4.8	16.6-26.7	47-150
1,1,1-Trichloroethane	15.0-25.0	4.6	13.7-30.1	52-162
1,1,2-Trichloroethane	14.2-25.8	5.5	14.3-27.1	52-150
Trichloroethene	13.3-26.7	6.6	18.6-27.6	71-157
Trichlorofluoromethane	9.6-30.4	10.0	8.9-31.5	17-181
Vinyl chloride	0.8-39.2	20.0	D-43.5	D-251

Q = concentration measured in QC check sample

s = standard deviation of four recovery measurements

$\bar{x}$ = average recovery of four recovery measurements

P, P_S = percent recovery measured

D = Detected; result must be greater than zero.

Criteria calculated assuming a QC check sample concentration of 20 µg/L

Note: These criteria are based directly on the method performance data in Table 6210:VI. Where necessary, the limits for recovery were broadened to assure applicability of the limits to concentrations below those used to develop Table 6210:VI.

Method 6210A – *Table 6210:V*

Characteristic Masses for Purgeable Organic Compounds

Compound	Primary Ion	Secondary Ions
Chloromethane	50	52
Bromomethane	94	96
Vinyl chloride	62	64
Chloroethane	64	66
Methylene chloride	84	49, 51, 86
Trichlorofluoromethane	101	103
1,1-Dichloroethene	96	61, 98
1,1-Dichloroethane	63	65, 83, 85, 98, 100
trans-1,2-Dichloroethene	96	61, 98
Chloroform	83	85
1,2-Dichloroethane	98	62, 64, 100
1,1,1-Trichloroethane	97	99, 117, 119
Carbon tetrachloride	117	119, 121
Bromodichloromethane	127	83, 85, 129
1,2-Dichloropropane	112	63, 65, 114
trans-1,3-Dichloropropene	75	77
Trichloroethene	130	95, 97, 132
Benzene	78	
Dibromochloromethane	127	129, 208, 206
1,1,2-Trichloroethane	97	83, 85, 99, 132, 134
cis-1,3-Dichloropropene	75	77
2-Chloroethylvinyl ether	106	63, 65
Bromoform	173	171, 175, 250, 252, 254, 256
1,1,2,2-Tetrachloroethane	168	83, 85, 131, 133, 166
Tetrachloroethene	164	129, 131, 166
Toluene	92	91
Chlorobenzene	112	114
Ethyl benzene	106	91
1,3-Dichlorobenzene	146	148, 113
1,2-Dichlorobenzene	146	148, 113
1,4-Dichlorobenzene	146	148, 113

Method 6210A – *Table 6210:VIII*

Characteristic Masses (m/z) for Purgeable Organic Compounds

Compound	Primary Ion	Secondary Ions
Benzene	78	-
Bromobenzene	156	77, 158
Bromochloromethane	128	49, 130
Bromodichloromethane	83	85, 127
Bromoform	173	175, 254
Bromomethane	94	96
n-Butylbenzene	91	92, 134
sec-Butylbenzene	105	134
tert-Butylbenzene	119	91, 134
Carbon tetrachloride	117	119
Chlorobenzene	112	77, 114
Chloroethane	64	66
Chloroform	83	85
Chloromethane	50	52
2-Chlorotoluene	91	126
4-Chlorotoluene	91	126
Dibromochloromethane	129	127
1,2-Dibromo-3-Chloropropane	75	155, 157
1,2-Dibromoethane	107	109, 188
Dibromomethane	93	95, 174
1,2-Dichlorobenzene	146	111, 148
1,3-Dichlorobenzene	146	111, 148
1,4-Dichlorobenzene	146	111, 148
Dichlorodifluoromethane	85	87
1,1-Dichloroethane	63	65, 83
1,2-Dichloroethane	62	98
1,1-Dichloroethene	96	61, 63
cis-1,2-Dichloroethene	96	61, 98
trans-1,2-Dichloroethene	96	61, 98
1,2-Dichloropropane	63	112
1,3-Dichloropropane	76	78
2,2-Dichloropropane	77	97
1,1-Dichloropropene	75	110, 77
Ethylbenzene	91	106
Hexachlorobutadiene	225	223, 227

Appendix A

Characteristic Masses (m/z) for Purgeable Organic Compounds (Continued)

Compound	Primary Ion	Secondary Ions
Isopropylbenzene	105	120
p-Isopropyltoluene	119	134, 91
Methylene chloride	84	86, 49
Naphthalene	128	—
n-Propylbenzene	91	120
Styrene	104	78
1,1,1,2-Tetrachloroethane	131	133, 119
1,1,2,2-Tetrachloroethane	83	131, 85
Tetrachloroethene	166	168, 129
Toluene	92	91
1,1,1-Trichloroethane	97	99, 61
1,1,2-Trichloroethane	83	97, 85
Trichlorethene	95	130, 132
Trichlorofluoromethane	101	103
1,2,3-Trichloropropane	75	77
Vinyl chloride	62	64
m-Xylene	106	91
o-Xylene	106	91
p-Xylene	106	91
Internal Standards/Surrogates:		
Fluorobenzene	96	70
1,2-Dichlorobenzene-d$_4$	150	115, 152
p-Bromofluorobenzene	95	174, 176

Method 6220A – *Table 6220:I*

Chromatographic Conditions and Method Detection Limits

| Compound | Retention Time (min) | | Method Detection |
	Column 1	Column 2	Limit μg/L
Benzene	3.33	2.75	0.2
Toluene	5.75	4.25	0.2
Ethylbenzene	8.25	6.25	0.2
Chlorobenzene	9.17	8.02	0.2
1,4-Dichlorobenzene	16.8	16.2	0.3
1,3-Dichlorobenzene	18.2	15.0	0.4
1,2-Dichlorobenzene	25.9	19.4	0.4

Column 1 conditions: Supelcoport (100/120 mesh) coated with 5% SP-1200/1, 75% Bentone-34 packed in a 1.8 m by 2.2 mm ID stainless steel column with helium carrier gas at 36 mL/min flow rate. Column temperature held at 50 °C for 2 min, then programmed at 6 °C/min to 90°C for a final hold.

Column 2 conditions: Chromosorb W-AW (60/80 mesh) coated with 5% 1,2,3-tris(2-cyanoethoxy)propane packed in a 1.8 m by 2.2 mm ID stainless steel column with helium carrier gas at 30 mL/min flow rate. Column temperature held at 40 °C for 2 min, then programmed at 2 °C/min to 100 °C for a final hold.

Method 6220A – *Table 6220:II*

Calibration and QC Acceptance Criteria

Compound	Range for Q μg/L	Limit for s μg/L	Range for $\bar{X}$ μg/L	Range for P, P_s %
Benzene	15.4-24.6	4.1	10.0-27.9	39-150
Chlorobenzene	16.1-23.9	3.5	12.7-25.4	55-135
1,2-Dichlorobenzene	13.6-26.4	5.8	10.6-27.6	37-154
1,3-Dichlorobenzene	14.5-25.5	5	12.8-25.5	50-141
1,4-Dichlorobenzene	13.9-26.1	5.5	11.6-25.5	42-143
Ethylbenzene	12.6-27.4	6.7	10.0-28.2	32-160
Toluene	15.5-24.5	4	11.2-27.7	46-148

Q   =   concentration measured in QC check sample
s   =   standard deviation of four recovery measurements
$\bar{X}$ =   average recovery for four recovery measurements
P, P_s =   recovery measured

Criteria calculated assuming a QC check sample concentration of 20 μg/L

NOTE: These criteria are based directly on the method performance data in Table 6220:III. Where necessary, the limits for recovery were broadened to assure applicability of the limits to concentrations below those used to develop Table 6220:III.

Method 6230A – *Table 6230:I*

Chromatographic Conditions and Method Detection Limits (MDL)

Compound	Retention Time (min)		Method Detection Limit µg/L
	Column 1	Column 2	
Chloromethane	1.50	5.28	0.08
Bromomethane	2.17	7.05	1.18
Dichlorodifluoromethane	2.62	ND	1.81
Vinyl chloride	2.67	5.28	0.18
Chloroethane	3.33	8.68	0.52
Methylene chloride	5.25	10.1	0.25
Trichlorofluoromethane	7.18	ND	ND
1,1-Dichloroethene	7.93	7.72	0.13
1,1-Dichloroethane	9.30	12.6	0.07
trans-1,2-Dichloroethene	10.1	9.38	0.10
Chloroform	10.7	12.1	0.05
1,2-Dichloroethane	11.4	15.4	0.03
1,1,1-Trichloroethane	12.6	13.1	0.03
Carbon tetrachloride	13.0	14.4	0.12
Bromodichloromethane	13.7	14.6	0.10
1,2-Dichloropropane	14.9	16.6	0.04
cis-1,3-Dichloropropene	15.2	16.6	0.34
Trichloroethene	15.8	13.1	0.12
Dibromochloromethane	16.5	16.6	0.09
1,1,2-Trichloroethane	16.5	18.1	0.02
trans-1,3-Dichloropropene	16.5	18	0.20
2-Chloroethylvinyl ether	18.0	ND	0.13
Bromoform	19.2	19.2	0.20
1,1,2,2-Tetrachloroethane	21.6	ND	0.03
Tetrachloroethene	21.7	15	0.03
Chlorobenzene	24.2	18.8	0.25
1,3-Dichlorobenzene	34.0	22.4	0.32
1,2-Dichlorobenzene	34.9	23.5	0.15
1,4-Dichlorobenzene	35.4	22.3	0.24

Column 1 conditions: Carbopack B (60/80 mesh) coated with 1% SP-1000 packed in a 2.4 m by 3 mm ID stainless steel or glass column with helium carrier gas at 40 mL/min flow rate. Column temperature held at 45 °C for 3 min, then programmed at 8 °C/min to 220 °C and held for 15 min.

Column 2 conditions: Porisil-C (100/120 mesh) coated with n-octane packed in a 1.8 m by 3 mm ID stainless steel or glass column with helium carrier gas at 40 mL/min flow rate. Column temperature held at 50 °C for 3 min, then programmed at 6 °C/min to 170 °C and held for 4 min.

ND = Not determined

Method 6230A – *Table 6230:II*

Calibration and QC Acceptance Criteria

Compounds	Range for Q µg/L	Limit for s µg/L	Range for $\bar{x}$ µg/L	Range for P,P_s %
Bromodichloromethane	15.2-24.8	4.3	10.7-32.0	42-172
Bromoform	14.7-25.3	4.7	5.0-29.3	13-159
Bromomethane	11.7-28.3	7.6	3.4-24.5	D-144
Carbon tetrachloride	13.7-26.3	5.6	11.8-25.3	43-143
Chlorobenzene	14.4-25.6	5.0	10.2-27.4	38-150
Chloroethane	15.4-24.6	4.4	11.3-25.2	46-137
2-Chloroethylvinyl ether	12.0-28.0	8.3	4.5-35.5	14-186
Chloroform	15.0-25.0	4.5	12.4-24.0	49-133
Chloromethane	11.9-28.1	7.4	D-34.9	D-193
Dibromochloromethane	13.1-26.9	6.3	7.9-35.1	24-191
1,2-Dichlorobenzene	14.0-26.0	5.5	1.7-38.9	D-208
1,3-Dichlorobenzene	9.9-30.1	9.1	6.2-32.6	7-187
1,4-Dichlorobenzene	13.9-26.1	5.5	11.5-25.5	42-143
1,1-Dichloroethane	16.8-23.2	3.2	11.2-24.6	47-132
1,2-Dichloroethane	14.3-25.7	5.2	13.0-26.5	51-147
1,1-Dichloroethene	12.6-27.4	6.6	10.2-27.3	28-167
trans-1,2-Dichloroethene	12.8-27.2	6.4	11.4-27.1	38-155
1,2-Dichloropropane	14.8-25.2	5.2	10.1-29.9	44-156
cis-1,3-Dichloropropene	12.8-27.2	7.3	6.2-33.8	22-178
trans-1,3-Dichloropropene	12.8-27.2	7.3	6.2-33.8	22-178
Methylene chloride	15.5-24.5	4.0	7.0-27.6	25-162
1,1,2,2-Tetrachloroethane	9.8-30.2	9.2	6.6-31.8	8-184
Tetrachloroethene	14.0-26.0	5.4	8.1-29.6	26-162
1,1,1-Trichloroethane	14.2-25.8	4.9	10.8-24.8	41-138
1,1,2-Trichloroethane	15.7-24.3	3.9	9.6-25.4	39-136
Trichloroethene	15.4-24.6	4.2	9.2-26.6	35-146
Trichlorofluoromethane	13.3-26.7	6.0	7.4-28.1	21-156
Vinyl chloride	13.7-26.3	5.7	8.2-29.9	28-163

Q = concentration measured in QC check sample

s = standard deviation of four recovery measurements

$\bar{x}$ = average recovery for four recovery measurements

P, P_s = recovery measured

D = detected; result must be greater than zero.

Criteria calculated assuming a QC check sample concentration of 20 µg/L

NOTE: These criteria are based directly on the method performance data in Table 6230:III. Where necessary, the limits for recovery were broadened to assure applicability of the limits to concentrations below those used to develop Table 6230:III.

Method 6230A – *Table 6230:III*

Method Bias and Precision as Functions of Concentration

Compound	Bias, as Recovery, X′ µg/L	Single-Analyst Precision, s′ µg/L	Overall Precision, S′ µg/L
Bromodichloromethane	1.12C-1.02	$0.11\,\overline{x} + 0.04$	$0.20\,\overline{x} + 1.00$
Bromoform	0.96C-2.05	$0.12\,\overline{x} + 0.58$	$0.21\,\overline{x} + 2.41$
Bromomethane	0.76C-1.27	$0.28\,\overline{x} + 0.27$	$0.36\,\overline{x} + 0.94$
Carbon tetrachloride	0.98C-1.04	$0.15\,\overline{x} + 0.38$	$0.20\,\overline{x} + 0.39$
Chlorobenzene	1.00C-1.23	$0.15\,\overline{x} - 0.02$	$0.18\,\overline{x} + 1.21$
Chloroethane	0.99C-1.53	$0.14\,\overline{x} - 0.13$	$0.17\,\overline{x} + 0.63$
2-Chloroethylvinyl ether**	1.00C	$0.20\,\overline{x}$	$0.35\,\overline{x}$
Chloroform	0.93C-0.39	$0.13\,\overline{x} + 0.15$	$0.19\,\overline{x} - 0.02$
Chloromethane	0.77C+0.18	$0.28\,\overline{x} - 0.31$	$0.52\,\overline{x} + 1.31$
Dibromochloromethane	0.94C+2.72	$0.11\,\overline{x} + 1.10$	$0.24\,\overline{x} + 1.68$
1,2-Dichlorobenzene	0.93C+1.70	$0.20\,\overline{x} + 0.97$	$0.13\,\overline{x} + 6.13$
1,3-Dichlorobenzene	0.95C+0.43	$0.14\,\overline{x} + 2.33$	$0.26\,\overline{x} + 2.34$
1,4-Dichlorobenzene	0.93C-0.09	$0.15\,\overline{x} + 0.29$	$0.20\,\overline{x} + 0.41$
1,1-Dichloroethane	0.95C-1.08	$0.08\,\overline{x} + 0.17$	$0.14\,\overline{x} + 0.94$
1,2-Dichloroethane	1.04C-1.06	$0.11\,\overline{x} + 0.70$	$0.15\,\overline{x} + 0.94$
1,1-Dichloroethene	0.98C-0.87	$0.21\,\overline{x} - 0.23$	$0.29\,\overline{x} - 0.40$
trans-1,2-Dichloroethene	0.97C-0.16	$0.11\,\overline{x} + 1.46$	$0.17\,\overline{x} + 1.46$
1,2-Dichloropropane**	1.00C	$0.13\,\overline{x}$	$0.23\,\overline{x}$
cis-1,3-Dichloropropene**	1.00C	$0.18\,\overline{x}$	$0.32\,\overline{x}$
trans-1,3-Dichloropropene**	1.00C	$0.18\,\overline{x}$	$0.32\,\overline{x}$
Methylene chloride	0.91C-0.93	$0.11\,\overline{x} + 0.33$	$0.21\,\overline{x} + 1.43$
1,1,2,2-Tetrachloroethene	0.95C+0.19	$0.14\,\overline{x} + 2.41$	$0.23\,\overline{x} + 2.79$
Tetrachloroethene	0.94C+0.06	$0.14\,\overline{x} + 0.38$	$0.18\,\overline{x} + 2.21$
1,1,1-Trichloroethane	0.90C-0.16	$0.15\,\overline{x} + 0.04$	$0.20\,\overline{x} + 0.37$
1,1,2-Trichloroethane	0.86C+0.30	$0.13\,\overline{x} - 0.14$	$0.19\,\overline{x} + 0.67$
Trichloroethene	0.87C+0.48	$0.13\,\overline{x} - 0.03$	$0.23\,\overline{x} + 0.30$
Trichlorofluoromethane	0.89C-0.07	$0.15\,\overline{x} + 0.67$	$0.26\,\overline{x} + 0.91$
Vinyl chloride	0.97C-0.36	$0.13\,\overline{x} + 0.65$	$0.27\,\overline{x} + 0.40$

X′ = expected recovery for one or more measurements of a sample containing a concentration of C

s′ = expected single-analyst standard deviation of measurements at an average concentration found of $\overline{x}$

S′ = expected interlaboratory standard deviation of measurements at an average concentration found of $\overline{x}$

C = true value for the concentration, µg/L

$\overline{x}$ = average recovery found for measurements of samples containing a concentration of C, µg/L

**Estimates based on performance in a single laboratory

Method 6410 – *Table 6410:I*

Chromatographic Conditions, Method Detection Limits, and Characteristic Masses for Base/Neutral Extractables

Compound	Retention Time min	Method Detection Limit mg/L	Characteristic Masses					
			Electron Impact			Chemical Ionization		
			Primary	Secondary	Secondary	Methane	Methane	Methane
1,3-Dichlorobenzene	7.4	1.9	146	148	113	146	148	150
1,4-Dichlorobenzene	7.8	4.4	146	148	113	146	148	150
Hexachloroethane	8.4	1.6	117	201	199	199	201	203
bis(2-Chloroethyl) ether	8.4	5.7	93	63	95	63	107	109
1,2-Dichlorobenzene	8.4	1.9	146	148	113	146	148	150
bis(2-Chloroisopropyl) ether[a]	9.3	5.7	45	77	79	77	135	137
N-Nitrosodi-n-propylamine			130	42	101			
Nitrobenzene	11.1	1.9	77	123	65	124	152	164
Hexachlorobutadiene	11.4	0.9	225	223	227	223	225	227
1,2,4-Trichlorobenzene	11.6	1.9	180	182	145	181	183	209
Isophorone	11.9	2.2	82	95	138	139	167	178
Naphthalene	12.1	1.6	128	129	127	129	157	169
bis(2-Chloroethoxy) methane	12.2	5.3	93	95	123	65	107	137
Hexachlorocyclopentadiene [b]	13.9		237	235	272	235	237	239
2-Chloronaphthalene	15.9	1.9	162	164	127	163	191	203
Acenaphthylene	17.4	3.5	152	151	153	152	153	181
Acenaphthene	17.8	1.9	154	153	152	154	155	183
Dimethyl phthalate	18.3	1.6	163	194	164	151	163	164
2,6-Dinitrotoluene	18.7	1.9	165	89	121	183	211	223
Fluorene	19.5	1.9	166	165	167	166	167	195
4-Chlorophenyl phenyl ether	19.5	4.2	204	206	141			
2,4-Dinitrotoluene	19.8	5.7	165	63	182	183	211	223
Diethyl phthalate	20.1	1.9	149	177	150	177	223	251
N-Nitrosodiphenylamine[b]	20.5	1.9	169	168	167	169	170	198
Hexachlorobenzene	21.0	1.9	284	142	249	284	286	288
α-BHC[b]	21.1		183	181	109			
4-Bromophenyl phenyl ether	21.2	1.9	248	250	141	249	251	277
γ-BHC[b]	22.4		183	181	109			
Phenanthrene	22.8	5.4	178	179	176	178	179	207
Anthracene	22.8	1.9	178	179	176	178	179	207
β-BHC	23.4	4.2	181	183	109			
Heptachlor	23.4	1.9	100	272	274			
δ-BHC	23.7	3.1	183	109	181			
Aldrin	24.0	1.9	66	263	220			
Dibutyl phthalate	24.7	2.5	149	150	104	149	205	279
Heptachlor epoxide	25.6	2.2	353	355	351			
Endosulfan I[b]	26.4		237	338	341			
Fluoranthene	26.5	2.2	202	101	100	203	231	243

(a) The proper chemical name is 2,2'-oxybis(1-chloropropane).

(b) See introductory section of text.

Column conditions: Supelcoport (100/120 mesh) coated with 3% SP-2250 packed in a 1.8 m long x 2 mm ID glass column with helium carrier gas at 30 mL/min flow rate. Column temperature held isothermal at 50 °C for 4 min, then programmed at 8 °C/min to 270 °C and held for 30 min.

Chromatographic Conditions, Method Detection Limits, and Characteristic Masses for Base/Neutral Extractables (Continued)

Compound	Retention Time min	Method Detection Limit mg/L	Characteristic Masses					
			Electron Impact			Chemical Ionization		
			Primary	Secondary	Secondary	Methane	Methane	Methane
Dieldrin	27.2	2.5	79	263	279			
4,4'-DDE	27.2	5.6	246	248	176			
Pyrene	27.3	1.9	202	101	100	203	231	243
Endrin[b]	27.9		81	263	82			
Endosulfan II [b]	28.6		237	339	341			
4,4'-DDD	28.6	2.8	235	237	165			
Benzidine [b]	28.8	44	184	92	185	185	213	225
4,4'-DDT	29.3	4.7	235	237	165			
Endosulfan sulfate	29.8	5.6	272	387	422			
Endrin aldehyde			67	345	250			
Butyl benzyl phthalate	29.9	2.5	149	91	206	149	299	327
bis(2-Ethylhexyl) phthalate	30.6	2.5	149	167	279	149		
Chrysene	31.5	2.5	228	226	229	228	229	257
Benzo(a)anthracene	31.5	7.8	228	229	226	228	229	257
3,3'–Dichlorobenzidine	32.2	16.5	252	254	126			
Di-n-octyl phthalate	32.5	2.5	149					
Benzo(b)fluoranthene	34.9	4.8	252	253	125	252	253	281
Benzo(k)fluoranthene	34.9	2.5	252	253	125	252	253	281
Benzo(a)pyrene	36.4	2.5	252	253	125	252	253	281
Indeno(1,2,3-cd)pyrene	42.7	3.7	276	138	277	276	277	305
Dibenzo(a,h)anthracene	43.2	2.5	278	139	279	278	279	307
Benzo(ghi)perylene	45.1	4.1	276	138	277	276	277	305
N-Nitrosodimethylamine[b]			42	74	44			
Chlordane[c]	19-30		373	375	377			
Toxaphene[c]	25-34		159	231	233			
PCB 1016[c]	18-30		224	260	294			
PCB 1221[c]	15-30	30	190	224	260			
PCB 1232[c]	15-32		190	224	260			
PCB 1242[c]	15-32		224	260	294			
PCB 1248[c]	12-34		294	330	262			
PCB 1254[c]	22-34	36	294	330	362			
PCB 1260[c]	23-32		330	362	394			

(a)   The proper chemical name is 2,2'-oxybis(1-chloropropane).
(b)   See introductory section of text.
(c)   These compounds are mixtures of various isomers.

Column conditions: Supelcoport (100/120 mesh) coated with 3% SP-2250 packed in a 1.8 m long by 2 mm ID glass column with helium carrier gas at 30 mL/min flow rate. Column temperature held isothermal at 50 °C for 4 min, then programmed at 8 °C/min to 270 °C and held for 30 min.

Method 6410 – *Table 6410:II*

Chromatographic Conditions, Method Detection Limits, and Characteristic Masses for Acid Extractables

Compound	Retention Time min	Method Detection Limit mg/L	Characteristic Masses					
			Electron Impact			Chemical Ionization		
			Primary	Secondary	Secondary	Methane	Methane	Methane
2-Chlorophenol	5.9	3.3	128	64	130	129	131	157
2-Nitrophenol	6.5	3.6	139	65	109	140	168	122
Phenol	8.0	1.5	94	65	66	95	123	135
2,4-Dimethylphenol	9.4	2.7	122	107	121	123	151	163
2,4-Dichlorophenol	9.8	2.7	162	164	98	163	165	167
2,4,6-Trichlorophenol	11.8	2.7	196	198	200	197	199	201
4-Chloro-3-methylphenol	13.2	3.0	142	107	144	143	171	183
2,4-Dinitrophenol	15.9	42	184	63	154	185	213	225
2-Methyl-4,6-dinitrophenol	16.2	24	198	182	77	199	227	239
Pentachlorophenol	17.5	3.6	266	264	268	267	265	269
4-Nitrophenol	20.3	2.4	65	139	109	140	168	122

Column conditions: Supelcoport (100/120 mesh) coated with 1% SP-1240DA packed in a 1.8 m long by 2 mm ID glass column with helium carrier gas at 30 mL/min flow rate. Column temperature held isothermal at 70 °C for 2 min then programmed at 8 °C/min to 200 °C.

Method 6410 – *Table 6410:III*

DFTPP Key Masses and Abundance Criteria

Mass	m/z Abundance Criteria
51	30-60% of mass 198
68	Less than 2% of mass 69
70	Less than 2% of mass 69
127	40-60% of mass 198
197	Less than 1% of mass 198
198	Base peak, 100% relative abundance
199	5-9% of mass 198
275	10-30% of mass 198
365	Greater than 1% of mass 198
441	Present but less than mass 443
442	Greater than 40% of mass 198
443	17-23% of mass 442

Method 6410 – *Table 6410:IV*

Suggested Internal and Surrogate Standards

Base/Neutral Fraction	Acid Fraction
Aniline-d_5	2-Fluorophenol
Anthracene-d_{10}	Pentafluorophenol
Benzo(a)anthracene-d_{12}	Phenol-d_5
4,4′-Dibromobiphenyl	2-Perfluoromethyl phenol
4,4′-Dibromooctafluorobiphenyl	
Decafluorobiphenyl	
2,2′-Difluorobiphenyl	
4-Fluoroaniline	
1-Fluoronaphthylene	
2-Fluoronaphthylene	
Naphthalene-d_8	
Nitrobenzene-d_5	
2,3,4,5,6-Pentafluorobiphenyl	
Phenanthrene-d_{10}	
Pyridine-d_5	

Method 6410 – *Table 6410:V*

QC Acceptance Criteria

Compound	Test Concentration μg/L	Limits for s μg/L	Range for $\bar{x}$ μg/L	Range for P, P_S %
Acenaphthene	100	27.6	60.1-132.3	47-145
Acenaphthylene	100	40.2	53.5-126.0	33-145
Aldrin	100	39.0	7.2-152.2	D-166
Anthracene	100	32.0	43.4-118.0	27-133
Benzo(a)anthracene	100	27.6	41.8-133.0	33-143
Benzo(b)fluoranthene	100	38.8	42.0-140.4	24-159
Benzo(k)fluoranthene	100	32.3	25.2-145.7	11-162
Benzo(a)pyrene	100	39.0	31.7-148.0	17-163
Benzo(g,h,i)perylene	100	58.9	D-195.0	D-219
Benzyl butyl phthalate	100	23.4	D-139.9	D-152
δ-BHC	100	31.5	41.5-130.6	24-149
β-BHC	100	21.6	D-100.0	D-110
bis(2-Chloroethyl) ether	100	55.0	42.9-126.0	12-158
bis(2-Chloroethoxy) methane	100	34.5	49.2-164.7	33-184
bis(2-Chloroisopropyl) ether**	100	46.3	62.8-138.6	36-166
bis(2-Ethylhexyl) phthalate	100	41.1	28.9-136.8	8-158
4-Bromophenyl phenyl ether	100	23.0	64.9-114.4	53-127
2-Chloronapthalene	100	13.0	64.5-113.5	60-118
4-Chlorophenyl phenyl ether	100	33.4	38.4-144.7	25-158
Chrysene	100	48.3	44.1-139.9	17-168
4,4'-DDD	100	31.0	D-134.5	D-145
4,4'-DDE	100	32.0	19.2-119.7	4-136
4,4'-DDT	100	61.6	D-170.6	D-203
Dibenzo(a,h)anthracene	100	70.0	D-199.7	D-227
Di-n-butyl phthalate	100	16.7	8.4-111.0	1-118
1,2-Dichlorobenzene	100	30.9	48.6-112.0	32-129
1,3-Dichlorobenzene	100	41.7	16.7-153.9	D-172
1,4-Dichlorobenzene	100	32.1	37.3-105.7	20-124
3,3'-Dichlorobenzidine	100	71.4	8.2-212.5	D-262
Dieldrin	100	30.7	44.3-119.3	29-136
Diethyl phthalate	100	26.5	D-100.0	D-114
Dimethyl phthalate	100	23.2	D-100.0	D-112
2,4-Dinitrotoluene	100	21.8	47.5-126.9	39-139

s = standard deviation for four recovery measurements

$\bar{x}$ = average recovery for four recovery measurements

P, P_S = percent recovery measured

D = detected; results must be greater than zero.

*The proper chemical name is 2,2'-oxybis(1-chloropropane).

NOTE: These criteria are based directly upon the method performance data in Table 6410:VI. Where necessary, the limits for recovery were broadened to assure applicability of the limits to concentrations below those used to develop Table 6410:VI.

Method 6410 – *Table 6410:V*

QC Acceptance Criteria (Continued)

Compound	Test Concentration µg/L	Limits for s µg/L	Range for $\overline{x}$ µg/L	Range for P,P$_S$ %
2,6-Dinitrotoluene	100	29.6	68.1-136.7	50-158
Di-n-octylphthalate	100	31.4	18.6-131.8	4-146
Endosulfan sulfate	100	16.7	D-103.5	D-107
Endrin aldehyde	100	32.5	D-188.8	D-209
Fluoranthene	100	32.8	42.9-121.3	26-137
Fluorene	100	20.7	71.6-108.4	59-121
Heptachlor	100	37.2	D-172.2	D-192
Heptachlor epoxide	100	54.7	70.9-109.4	26-155
Hexachlorobenzene	100	24.9	7.8-141.5	D-152
Hexachlorobutadiene	100	26.3	37.8-102.2	24-116
Hexachloroethane	100	24.5	55.2-100.0	40-113
Indeno(1,2,3-cd)pyrene	100	44.6	D-150.9	D-171
Isophorone	100	63.3	46.6-180.2	21-196
Naphthalene	100	30.1	35.6-119.6	21-133
Nitrobenzene	100	39.3	54.3-157.6	35-180
N-Nitrosodi-n-propylamine	100	55.4	13.6-197.9	D-230
PCB-1260	100	54.2	19.3-121.0	D-164
Phenanthrene	100	20.6	65.2-108.7	54-120
Pyrene	100	25.2	69.6-100.0	52-115
1,2,4-Trichlorobenzene	100	28.1	57.3-129.2	44-142
4-Chloro-3-methylphenol	100	37.2	40.8-127.9	22-147
2-Chlorophenol	100	28.7	36.2-120.4	23-134
2,4-Dichlorophenol	100	26.4	52.5-121.7	39-135
2,4-Dimethylphenol	100	26.1	41.8-109.0	32-119
2,4-Dinitrophenol	100	49.8	D-172.9	D-191
2-Methyl-4,6-dinitrophenol	100	93.2	53.0-100.0	D-181
2-Nitrophenol	100	35.2	45.0-166.7	29-182
4-Nitrophenol	100	47.2	13.0-106.5	D-132
Pentachlorophenol	100	48.9	38.1-151.8	14-176
Phenol	100	22.6	16.6-100.0	5-112
2,4,6-Trichlorophenol	100	31.7	52.4-129.2	37-144

s   =   standard deviation for four recovery measurements
$\overline{x}$ =   average recovery for four recovery measurements
P, P$_S$ =   percent recovery measured
D   =   detected; results must be greater than zero.
NOTE: These criteria are based directly upon the method performance data in Table 6410:VI. Where necessary, the limits for recovery were broadened to assure applicability of the limits to concentrations below those used to develop Table 6410:VI.

Method 6630 – *Table 6630:III*

Chromatographic Conditions and Method Detection Limits

Parameter	Retention time (min) Col. 1	Col. 2	Method Detection Limit (μg/L)
α-BHC	1.35	1.82	0.003
γ-BHC (Lindane)	1.70	2.13	ND
β-BHC	1.90	1.97	ND
Heptachlor	2.00	3.35	0.003
δ-BHC	2.15	2.20	0.009
Aldrin	2.40	4.10	0.004
Heptachlor epoxide	3.50	5.00	0.083
Endosulfan I	4.50	6.20	0.014
4,4′-DDE	5.13	7.15	0.004
Dieldrin	5.45	7.23	0.002
Endrin	6.55	8.10	0.006
4,4′-DDD	7.83	9.08	0.011
Endosulfan II	8.00	8.28	0.004
4,4′-DDT	9.40	11.75	0.012
Endrin aldehyde	11.82	9.30	0.023
Endosulfan sulfate	14.22	10.70	0.066
Chlordane	MR	MR	0.014
Toxaphene	MR	MR	0.24
PCB-1016	MR	MR	ND
PCB-1221	MR	MR	ND
PCB-1232	MR	MR	ND
PCB-1242	MR	MR	0.065
PCB-1248	MR	MR	ND
PCB-1254	MR	MR	ND
PCB-1260	MR	MR	ND

Column 1 conditions: Supelcoport (100/120 mesh) coated with 1.5% SP-2250/1.95% SP-2401 packed in a 1.8 m long by 4 mm ID glass column with 5% methane/95% argon carrier gas at 60 mL/min flow rate. Column temperature held isothermal at 200 °C, except for PCB-1016 through PCB-1248, should be measured at 160 °C.

Column 2 conditions: Supelcoport (100/120 mesh) coated with 3% OV-1 packed in a 1.8 m long by 4 mm ID glass column with 5% methane/95% argon carrier gas at 60 mL/min flow rate. Column temperature held isothermal at 200 °C for the pesticides; at 140 °C for PCB-1221 and 1232; and at 170 °C for PCB-1016 and 1242 to 1268.

MR = Multiple peak response

ND = Not determined

Method 6630 – *Table 6630:V*

QC Acceptance Criteria

Parameter	Test conc. (µg/L)	Limit for s (µg/L)	Range for $\bar{x}$ (µg/L)	Range for P, P_s (%)
Aldrin	2.0	0.042	1.08-2.24	42-122
α-BHC	2.0	0.48	0.98-2.44	37-134
β-BHC	2.0	0.64	0.78-2.60	17-147
δ-BHC	2.0	0.72	1.01-2.37	19-140
γ-BHC	2.0	0.46	0.86-2.32	32-127
Chlordane	50	10.0	27.6-54.3	45-119
4,4′-DDD	10	2.8	4.8-12.6	31-141
4,4′-DDE	2.0	0.55	1.08-2.60	30-145
4,4′-DDT	10	3.6	4.6-13.7	25-160
Dieldrin	2.0	0.76	1.15-2.49	36-146
Endosulfan I	2.0	0.49	1.14-2.82	45-153
Endosulfan II	10	6.1	2.2-17.1	D-202
Endosulfan sulfate	10	2.7	3.8-13.2	26-144
Endrin	10	3.7	5.1-12.6	30-147
Heptachlor	2.0	0.4	0.86-2.00	34-111
Heptachlor epoxide	2.0	0.41	1.13-2.63	37-142
Toxaphene	50	12.7	27.8-55.6	41-126
PCB-1016	50	10.0	30.5-51.5	50-114
PCB-1221	50	24.4	22.1-75.2	15-178
PCB-1232	50	17.9	14.0-98.5	10-215
PCB-1242	50	12.2	24.8-69.6	39-150
PCB-1248	50	15.9	29.0-70.2	38-158
PCB-1254	50	13.8	22.2-57.9	29-131
PCB-1260	50	10.4	18.7-54.9	8-127

s = Standard deviation of four recovery measurements, in µg/L (Section 8.2.4)
$\bar{x}$ = Average recovery for four recovery measurements, in µg/L (Section 8.2.4)
P, P_s = Percent recovery measured (Section 8.3.2, Section 8.4.2)
D = Detected; result must be greater than zero.

Note: These criteria are based directly upon the method performance data in 6630:IV. Where necessary, the limits for recovery have been broadened to assure applicability of the limits to concentrations below those used to develop Table 6630:IV.

Method 6640 – *Table 6640: I*

Method Detection Limits*

Compound	Amount Added µg/L	Average Recovery %	Mean Amount Found µg/L	Standard Deviation µg/L	Relative Standard Deviation %	Method Detection Limit µg/L
Dalapon	2.0	100	2.00	0.031	1.56	0.1
Dicamba	0.4	122	0.48	0.007	1.55	0.02
2,4-D	2.0	98.1	1.9	0.036	1.85	0.1
Pentachlorophenol	0.2	101	0.20	0.004	2.40	0.02
2,4,5-TP	0.2	119	0.24	0.003	1.58	0.01
2,4,5-T	0.2	97.8	0.20	0.005	2.73	0.02
Dinoseb	0.4	111	0.44	0.012	2.86	0.04
Bentazon	0.4	94.6	0.38	0.032	8.68	0.01
Picloram	0.4	123	0.49	0.011	2.29	0.04
Surrogate				11.3	5.92	
Internal Standard				3.74	1.18	

*Based on the analysis of seven portions with known additions.

Method 6640 – *Table 6640: IV*

Short-Range Calibration Standards*

Compound	Level 1 µg/L	Level 2 µg/L	Level 3 µg/L
Dalapon	1.0	2.0	4.0
Dicamba	0.2	0.4	0.8
2,4-D	1.0	2.0	4.0
Pentachlorophenol	0.1	0.2	0.4
2,4,5-TP	0.1	0.2	0.4
2,4,5-T	0.1	0.2	0.4
Dinoseb	0.2	0.4	0.8
Bentazon	0.2	0.4	0.8
Picloram	0.2	0.4	0.8

*Levels 1, 2, and 3 are prepared by adding 2 µL, 4 µL, and 8 µL of herbicide addition mix, respectively, into 2 mL MtBE.

Method 6640 – *Table 6640: VI*

Single-Laboratory Bias and Precision Data,* Level 1 Addition Reagent Water

Compound	Added Amount µg/L	Mean Recovery %	Mean Amount Found µg/L	Standard Deviation µg/L	Relative Standard Deviation %
Dalapon	10	114	11	0.72	6.4
Dicamba	1.0	84	0.83	0.02	2.6
2,4-D	5.0	75	4.0	0.22	5.5
Pentachlorophenol	2.0	107	2.1	0.04	1.7
2,4,5-TP	0.5	107	0.5	0.02	3.0
2,4,5-T	0.5	104	0.50	0.02	3.7
Dinoseb	2.0	90	1.8	0.1	5.4
Bentazon	1.0	84	0.85	0.020	2.3
Picloram	2.0	80	1.60	0.11	7.0

Method 6640 – *Table 6640: VII*

Single-Laboratory Bias and Precision Data,* Level 2 Addition Reagent Water

Compound	Added Amount µg/L	Mean Recovery %	Mean Amount Found µg/L	Standard Deviation µg/L	Relative Standard Deviation %
Dalapon	30	99.3	29.8	0.21	0.71
Dicamba	3.0	92.2	2.77	0.06	2.2
2,4-D	15	90.5	13.6	0.15	1.10
Pentachlorophenol	6.0	95.8	5.75	0.09	1.5
2,4,5-TP	1.5	98.5	1.48	0.02	1.3
2,4,5-T	1.5	86.4	1.30	0.02	1.3
Dinoseb	6.0	90.0	5.40	0.03	5.1
Bentazon	3.0	103	3.09	0.05	1.6
Picloram	6.0	86.0	5.16	0.18	3.5

* Based on analysis of four portions of reagent water with known additions.

Method 6640 – *Table 6640: VIII*

Laboratory Performance Check Solution

Test	Test Compound	Concentration μg/mL	Requirements
Sensitivity	Dinoseb	0.004	Detection of compound S/N > 3
Chromatographic performance	4-Nitrophenol	1.6	0.70 <PGF<1.05[a]
Column performance	3,5-Dichlorobenzoic acid	0.6	Resolution >0.04 [b]
	4-Nitrophenol	1.6	Resolution >0.04 [b]

(a) PGF = peak Gaussian factor. Calculated using the equation: $PGF = \dfrac{1.83 \times W(1/2)}{W(1/10)}$, where W(1/2) is the peak width at half height, and W(1/10) is the peak width at tenth height.

(b) Resolution between the two peaks as defined by the equation: R = t/w, where t is the difference in elution times between the two peaks, and W is the average peak width, at the baseline, of the two peaks. Source: U.S. ENVIRONMENTAL PROTECTION AGENCY. 1989. Determination of Chlorinated Acids by Gas Chromatography with an Electron Capture Detector. Method 515.1, rev. 4.0. Environmental Monitoring Systems Lab., Off. Research & Development, Cincinnati, Ohio.

Method 6010B – *Table 1*

Appendix A

Recommended Wavelengths and Estimated Instrumental Detection Limits[a]

Element	Wavelength (nm)[a]	Estimated Detection Limit (μg/L)[b]
Aluminum	308.215	45
Antimony	206.833	32
Arsenic	193.696	53
Barium	455.403	2
Beryllium	313.042	0.3
Cadmium	226.502	4
Calcium	317.933	10
Chromium	267.716	7
Cobalt	228.616	7
Copper	324.754	6
Iron	259.940	7
Lead	220.353	42
Lithium	670.784	5
Magnesium	279.079	30
Manganese	257.610	2
Molybdenum	202.030	8
Nickel	231.604	15
Phosphorus	213.618	51
Potassium	766.491	See note [c]
Selenium	196.026	75
Silver	328.068	7
Sodium	588.995	29
Strontium	407.771	0.3
Thallium	190.864	40
Vanadium	292.402	8
Zinc	213.856	2

[a] The wavelengths listed are recommended because of their sensitivity and overall acceptance. Other wavelengths may be substituted if they can provide the needed sensitivity and are treated with the same corrective techniques for spectral interference (see Paragraph 3.1). In time, other elements may be added as more information becomes available and as required.

[b] The estimated instrumental detection limits shown are taken from Reference 1 in Section 10.0. They are given as a guide for an instrumental limit. The actual method detection limits are sample dependent and may vary as the sample matrix varies.

[c] Highly dependent on operating conditions and plasma position.

Method 6020 – *Table 1*

Elements Approved for ICP-MS Determination

Element	CAS Number
Aluminum	7429-90-5
Antimony	7440-36-0
Arsenic	7440-38-2
Barium	7440-39-3
Beryllium	7440-41-7
Cadmium	7440-43-9
Chromium	7440-47-3
Cobalt	7440-48-4
Copper	7440-50-8
Lead	7439-92-1
Manganese	7439-96-5
Nickel	7440-02-0
Silver	7440-22-4
Thallium	7440-28-0
Zinc	7440-66-6

Method 6020 – *Table 2*

Recommended Interference Check Sample Components and Concentrations

Solution Component	Solution A Concentration (mg/L)	Solution AB Concentration (mg/L)
Al	100.0	100.0
Ca	100.0	100.0
Fe	100.0	100.0
Mg	100.0	100.0
Na	100.0	100.0
P	100.0	100.0
K	100.0	100.0
S	100.0	100.0
C	200.0	200.0
Cl	1000.0	1000.0
Mo	2.0	2.0
Ti	2.0	2.0
As	0.0	0.0200
Cd	0.0	0.0200
Cr	0.0	0.0200
Co	0.0	0.0200
Cu	0.0	0.0200
Mn	0.0	0.0200
Ni	0.0	0.0200
Ag	0.0	0.0200
Zn	0.0	0.0200

Method 6020 – *Table 3*

Recommended Isotopes for Selected Elements

Mass	Element of Interest
27	Aluminum
121, **123**	Antimony
75	Arsenic
138, 137, 136, **135**, 134	Barium
9	Beryllium
209	Bismuth (IS)
114, 112, **111**, 110, 113, 116, 106	Cadmium
42, 43, **44**, 46, 48	Calcium (I)
35, 37, (77, 82)(a)	Chlorine (I)
52, **53**, **50**, 54	Chromium
59	Cobalt
63, **65**	Copper
165	Holmium (IS)
115, 113	Indium (IS)
56, **54**, **57**, 58	Iron (I)
139	Lanthanum (I)
208, **207**, **206**, 204	Lead
6(b), 7	Lithium (IS)
24, **25**, **26**	Magnesium (I)
55	Manganese
98, 96, 92, **97**, 94, (108)(a)	Molybdenum (I)
58, **60**, 62, **61**, 64	Nickel
39	Potassium
103	Rhodium (IS)
45	Scandium (IS)
107, **109**	Silver
23	Sodium (I)
159	Terbium (IS)
205, 203	Thallium
120, **118**	Tin (I)
89	Yttrium (IS)
64, **66**, **68**, **67**, 70	Zinc

NOTE: Method 6020 is recommended for only those analytes listed in Table 1. Other elements are included in this table because they are potential interferents (labeled I) in the determination of recommended analytes, or because they are commonly used internal standards (labeled IS). Isotopes are listed in descending order of natural abundance. The most generally useful isotopes are underlined and in Boldface, although certain matrices may require the use of alternative isotopes.

(a) These masses are useful for interference correction (section 3.2).

(b) Internal standard must be enriched in the ^{6}Li isotope. This minimizes interference from indigenous lithium.

Method 7000 Series – *Table 1*

Atomic Absorption Concentration Ranges

Metal	Direct Aspiration Detection Limit (mg/L)	Sensitivity (mg/L)	Furnace Procedure[a],[c] Detection Limit (μg/L)
Aluminum	0.1	1	
Antimony	0.2	0.5	3
Arsenic[b]	0.002		1
Barium	0.1	0.4	2
Beryllium	0.005	0.025	0.2
Cadmium	0.005	0.025	0.1
Calcium	0.01	0.08	
Chromium	0.05	0.25	1
Cobalt	0.05	0.2	1
Copper	0.02	0.1	1
Iron	0.03	0.12	1
Lead	0.1	0.5	1
Lithium	0.002	0.04	
Magnesium	0.001	0.007	
Manganese	0.01	0.05	0.2
Mercury[d]	0.0002		
Molybdenum(p)	0.1	0.4	1
Nickel	0.04	0.15	
Osmium	0.03	1	
Potassium	0.01	0.04	
Selenium[b]	0.002		2
Silver	0.01	0.06	0.2
Sodium	0.002	0.015	
Strontium	0.03	0.15	
Thallium	0.1	0.5	1
Tin	0.8	4	
Vanadium(p)	0.2	0.8	4
Zinc	0.005	0.02	0.05

NOTE: The symbol (p) indicates the use of pyrolytic graphite with the furnace procedure.

(a) For furnace sensitivity values, consult instrument operating manual.

(b) Gaseous hydride method

(c) The listed furnace values are those expected when using a 20-μL injection and normal gas flow, except in the cases of arsenic and selenium, where gas interrupt is used.

(d) Cold vapor technique

Chromatographic Retention Times and Method Detection Limits (MDL) for Volatile Organic Compounds on Photoionization Detection (PID) and Hall Electrolytic Conductivity Detector (HECD) Detectors

Analyte	PID Ret. Time[a] minute	HECD Ret. Time minute	PID MDL µg/L	HECD MDL µg/L
Dichlorodifluoromethane	(b)	8.47		0.05
Chloromethane	(b)	9.47		0.03
Vinyl chloride	9.88	9.93	0.02	0.04
Bromomethane	(b)	11.95		1.1
Chloroethane	(b)	12.37		0.1
Trichlorofluoromethane	(b)	13.49		0.03
1,1-Dichloroethene	16.14	16.18	ND	0.07
Methylene chloride	(b)	18.39		0.02
trans-1,2-Dichloroethene	19.30	19.33	0.05	0.06
1,1-Dichloroethane	(b)	20.99		0.07
2,2-Dichloropropane	(b)	22.88		0.05
cis-1,2-Dichloroethene	23.11	23.14	0.02	0.01
Chloroform	(b)	23.64		0.02
Bromochloromethane	(b)	24.16		0.01
1,1,1-Trichloroethane	(b)	24.77		0.03
1,1-Dichloropropene	25.21	25.24	0.02	0.02
Carbon tetrachloride	(b)	25.47		0.01
Benzene	26.10	(b)	0.009	
1,2-Dichloroethane	(b)	26.27		0.03
Trichloroethene	27.99	28.02	0.02	0.01
1,2-Dichloropropane	(b)	28.66		0.006
Bromodichloromethane	(b)	29.43		0.02
Dibromomethane	(b)	29.59		2.2
Toluene	31.95	(b)	0.01	
1,1,2-Trichloroethane	(b)	33.21		ND
Tetrachloroethene	33.88	33.90	0.05	0.04
1,3-Dichloropropane	(b)	34.00		0.03
Dibromochloromethane	(b)	34.73		0.03
1,2-Dibromoethane	(b)	35.34		0.8
Chlorobenzene	36.56	36.59	0.003	0.01
Ethyl benzene	36.72	(b)	0.005	
1,1,1,2-Tetrachloroethane	(b)	36.80		0.005
m-Xylene	36.98	(b)	0.01	

(a) Retention times determined on 60 m x 0.75 mm ID VOCOL capillary column. Program: Hold 10 °C for 8 minutes, then program at 4 °C/min to 180 °C, and hold until all expected compounds eluted.

(b) Detector does not respond.

ND = Not determined.

Method 8021B – *Table 1*

Chromatographic Retention Times and Method Detection Limits (MDL) for Volatile Organic Compounds on Photoionization Detection (PID) and Hall Electrolytic Conductivity Detector (HECD) Detectors (Continued)

Analyte	PID Ret. Time[a] minute	HECD Ret. Time minute	PID MDL µg/L	HECD MDL µg/L
p-Xylene	36.98	(b)	0.01	
o-Xylene	38.39	(b)	0.02	
Styrene	38.57	(b)	0.01	
Isopropylbenzene	39.58	(b)	0.05	
Bromoform	(b)	39.75		1.6
1,1,2,2-Tetrachloroethane	(b)	40.35		0.01
1,2,3-Trichloropropane	(b)	40.81		0.4
n-Propylbenzene	40.87	(b)	0.004	
Bromobenzene	40.99	41.03	0.006	0.03
1,3,5-Trimethylbenzene	41.41	(b)	0.004	
2-Chlorotoluene	41.41	41.45	ND	0.01
4-Chlorotoluene	41.60	41.63	0.02	0.01
1,2,4-Trimethylbenzene	42.71	(b)	0.05	
tert-Butylbenzene	42.92	(b)	0.06	
sec-Butylbenzene	43.31	(b)	0.02	
4-Isopropyltoluene	43.81	(b)	0.01	
1,3-Dichlorobenzene	44.08	44.11	0.02	0.02
1,4-Dichlorobenzene	44.43	44.47	0.007	0.01
n-Butylbenzene	45.20	(b)	0.02	
1,2-Dichlorobenzene	45.71	45.74	0.05	0.02
1,2-Dibromo-3-chloropropane	(b)	48.57		3.0
1,2,4-Trichlorobenzene	51.43	51.46	0.02	0.03
Hexachlorobutadiene	51.92	51.96	0.06	0.02
Naphthalene	52.38	(b)	0.06	
1,2,3-Trichlorobenzene	53.34	53.37	ND	0.03
Internal Standards				
Fluorobenzene	26.84			
2-Bromo-1-Chloropropane		33.08		

(a) Retention times determined on 60 m x 0.75 mm ID VOCOL capillary column. Program: Hold 10 °C for 8 minutes, then program at 4 °C/min to 180 °C, and hold until all expected compounds eluted.

(b) Detector does not respond.

ND = Not determined.

Method 8021B – *Table 2*

Single Laboratory Accuracy and Precision Data for Volatile Organic Compounds in Water[c]

Analyte	Photoionizaton Detector		Hall Electrolytic Conductivity Detector	
	Recovery,[a] %	Standard Deviation of Recovery	Recovery,[a] %	Standard Deviation of Recovery
Benzene	99	1.2	(b)	(b)
Bromobenzene	99	1.7	97	2.7
Bromochloromethane	(b)	(b)	96	3.0
Bromodichloromethane	(b)	(b)	97	2.9
Bromoform	(b)	(b)	106	5.5
Bromomethane	(b)	(b)	97	3.7
n-Butylbenzene	100	4.4	(b)	(b)
sec-Butylbenzene	97	2.6	(b)	(b)
tert-Butylbenzene	98	2.3	(b)	(b)
Carbon tetrachloride	(b)	(b)	92	3.3
Chlorobenzene	100	1.0	103	3.7
Chloroethane	(b)	(b)	96	3.8
Chloroform	(b)	(b)	98	2.5
Chloromethane	(b)	(b)	96	8.9
2-Chlorotoluene	ND	ND	97	2.6
4-Chlorotoluene	101	1.0	97	3.1
1,2-Dibromo-3-chloropropane	(b)	(b)	86	9.9
Dibromochloromethane	(b)	(b)	102	3.3
1,2-Dibromoethane	(b)	(b)	97	2.7
Dibromomethane	(b)	(b)	109	7.4
1,2-Dichlorobenzene	102	2.1	100	1.5
1,3-Dichlorobenzene	104	1.7	106	4.3
1,4-Dichlorobenzene	103	2.2	98	2.3
Dichlorodifluoromethane	(b)	(b)	89	5.9
1,1-Dichloroethane	(b)	(b)	100	5.7
1,2-Dichloroethane	(b)	(b)	100	3.8
1,1-Dichloroethene	100	2.4	103	2.9
cis-1,2-Dichloroethene	ND	ND	105	3.5
trans-1,2-Dichloroethene	93	3.7	99	3.7

(a) Recoveries and standard deviations were determined from seven samples and spiked at 10 µg/L of each analyte. Recoveries were determined by internal standard method. Internal standards were: Fluorobenzene for PID;
2-Bromo-1-chloropropane for HECD.

(b) Detector does not respond.

(c) This method was tested in a single laboratory using water spiked at 10 µg/L (see reference 8).

ND = Not determined.

Method 8021B – *Table 2*

Single Laboratory Accuracy and Precision Data for Volatile Organic Compounds in Water[c] (Continued)

Analyte	Photoionizaton Detector		Hall Electrolytic Conductivity Detector	
	Recovery,[a] %	Standard Deviation of Recovery	Recovery,[a] %	Standard Deviation of Recovery
1,2-Dichloropropane	(b)	(b)	103	3.8
1,3-Dichloropropane	(b)	(b)	100	3.4
2,2-Dichloropropane	(b)	(b)	105	3.6
1,1-Dichloropropene	103	3.6	103	3.4
Ethyl benzene	101	1.4	(b)	(b)
Hexachlorobutadiene	99	9.5	98	8.3
Isopropylbenzene	98	0.9	(b)	(b)
4-Isopropyltoluene	98	2.4	(b)	(b)
Methylene chloride	(b)	(b)	97	2.8
Naphthalene	102	6.3	(b)	(b)
n-Propylbenzene	103	2.0	(b)	(b)
Styrene	104	1.4	(b)	(b)
1,1,1,2-Tetrachloroethane	(b)	(b)	99	2.3
1,1,2,2-Tetrachloroethane	(b)	(b)	99	6.8
Tetrachloroethene	101	1.8	97	2.4
Toluene	99	0.8	(b)	(b)
1,2,3-Trichlorobenzene	106	1.9	98	3.1
1,2,4-Trichlorobenzene	104	2.2	102	2.1
1,1,1-Trichloroethane	(b)	(b)	104	3.4
1,1,2-Trichloroethane	(b)	(b)	109	6.2
Trichloroethene	100	0.78	96	3.5
Trichlorofluoromethane	(b)	(b)	96	3.4
1,2,3-Trichloropropane	(b)	(b)	99	2.3
1,2,4-Trimethylbenzene	99	1.2	(b)	(b)
1,3,5-Trimethylbenzene	101	1.4	(b)	(b)
Vinyl chloride	109	5.4	95	5.6
o-Xylene	99	0.8	(b)	(b)
m-Xylene	100	1.4	(b)	(b)
p-Xylene	99	0.9	(b)	(b)

(a) Recoveries and standard deviations were determined from seven samples and spiked at 10 µg/L of each analyte. Recoveries were determined by internal standard method. Internal standards were: Fluorobenzene for PID; 2-Bromo-1-chloropropane for HECD.

(b) Detector does not respond.

(c) This method was tested in a single laboratory using water spiked at 10 µg/L (see reference 8).

Determination of Estimated Quantitation Limits (EQL) for Various Matrices[a]

Matrix	Factor[b]
Ground water	10
Low-concentration soil	10
Water miscible liquid waste	500
High-concentration soil and sludge	1250
Non-water miscible waste	1250

(a)   Sample EQLs are highly matrix-dependent. The EQLs listed herein are provided for guidance and may not always be achievable.

(b)   EQL = [Method detection limit (Table 1)] X [Factor (Table 3)]. For non-aqueous samples, the factor is on a wet-weight basis.

Method 8081A– *Table 3*

Factors for Determination of Estimated Quantitation Limits (EQLs) [a] for Various Matrices

Matrix	Factor
Ground water	10
Low-concentration soil by sonication with GPC cleanup	670
High-concentration soil and sludges by sonication	10,000
Non-water miscible waste	100,000

(a)   EQL = [Method detection limit in water (see section 1.8)] times [factor found in this table].

For nonaqueous samples, the factor is on a wet-weight basis. Sample EQLs are highly matrix-dependent. The EQLs to be determined therein are provided for guidance and may not always be achievable.

Method 8082– *Table 1*

Factors for Determination of Estimated Quantitation Limits (EQLs) [a] for Various Matrices

Matrix	Factor
Ground water	10
Low-concentration soil by sonication with GPC cleanup	670
High-concentration soil and sludges by sonication	10,000
Non-water miscible waste	100,000

[a] Laboratories may estimate the quantitation limits of the target analytes in environmental and waste media by generating MDLs in organic-free reagent water and using the following equation (see Sec. 5.0 of Chapter One for information on generating MDL data):

EQL = [Method detection limit in water] times [factor found in this table].

For nonaqueous samples, the factor is on a wet-weight basis. Sample EQLs are highly matrix-dependent. The EQLs to be determined therein are provided for guidance and may not always be achievable.

Method 8151A – *Table 1*

Estimated Method Detection Limits for Method 8151A, Diazomethane Derivatization

Analyte	Aqueous Samples GC/ECD Estimated Detection Limit[a] (μg/L)	Soil Samples	
		GC/ECD Estimated Detection Limit[b] (μg/kg)	GC/MS Estimated Identification Limit[c] (ng)
Acifluorfen	0.096		
Bentazon	0.2		
Chloramben	0.093	4.0	1.7
2,4-D	0.2	0.11	1.25
Dalapon	1.3	0.12	0.5
2,4-DB	0.8		
DCPA diacid[e]	0.02		
Dicamba	0.081		
3,5-Dichlorobenzoic acid	0.061	0.38	0.65
Dichloroprop	0.26		
Dinoseb	0.19		
5-Hydroxydicamba	0.4		
MCPP	0.09[d]	66	0.43
MCPA	0.056[d]	43	0.3
4-Nitrophenol	0.13	0.34	0.44
Pentachlorophenol	0.076	0.16	1.3
Picloram	0.14		
2,4,5-T	0.08		
2,4,5-TP	0.075	0.28	4.5

(a) EDL = estimated detection limit; defined as either the MDL (40 CFR Part 136, Appendix B, Revision 1.11), or a concentration of analyte in a sample yielding a peak in the final extract with signal-to-noise ratio of approximately 5, whichever value is higher.

(b) Detection limits determined from standard solutions corrected back to 50 g samples, extracted and concentrated to 10 mL, with 5 μL injected. Chromatography using narrow bore capillary column, 0.25 μm film, 5% phenyl/95% methyl silicone.

(c) The minimum amount of analyte to give a Finnigan INCOS FIT value of 800 as the methyl derivative vs. the spectrum obtained from 50 ng of the respective free acid herbicide.

(d) 40 CFR Part 136, Appendix B (49 FR 43234). Chromatography using wide-bore capillary column.

(e) DCPA monoacid and diacid metabolites included in method scope; DCPA diacid metabolite used for validation studies. DCPA is a dimethyl ester.

Method 8151A – *Table 4*

Accuracy and Precision for Method 8151A Diazomethane Derivatization, Organic-free Reagent Water Matrix

Analyte	Spike Concentration (µg/L)	Mean[a] Percent Recovery	Standard Deviation of Percent Recovery
Acifluorfen	0.2	121	15.7
Bentazon	1	120	16.8
Chloramben	0.4	111	14.4
2,4-D	1	131	27.5
Dalapon	10	100	20.0
2,4-DB	4	87	13.1
DCPA diacid [b]	0.2	74	9.7
Dicamba	0.4	135	32.4
3,5-Dichlorobenzoic acid	0.6	102	16.3
Dichloroprop	2	107	20.3
Dinoseb	0.4	42	14.3
5-Hydroxydicamba	0.2	103	16.5
4-Nitrophenol	1	131	23.6
Pentachlorophenol	0.04	130	31.2
Picloram	0.6	91	15.5
2,4,5-TP	0.4	117	16.4
2,4,5-T	0.2	134	30.8

(a) Mean percent recovery calculated from 7-8 determinations of spiked organic-free reagent water.

(b) DCPA monoacid and diacid metabolites included in method scope; DCPA diacid metabolite used for validation studies. DCPA is a dimethyl ester.

Method 8151A – *Table 6*

Relative Recoveries of PFB Derivatives of Herbicides[a]

Analyte	Standard Concentration mg/L	Relative Recoveries, %								
		1	2	3	4	5	6	7	8	Mean
MCPP	5.1	95.6	88.8	97.1	100	95.5	97.2	98.1	98.2	96.3
Dicamba	3.9	91.4	99.2	100	92.7	84.0	93.0	91.1	90.1	92.7
MCPA	10.1	89.6	79.7	87.0	100	89.5	84.9	92.3	98.6	90.2
Dichloroprop	6.0	88.4	80.3	89.5	100	85.2	87.9	84.5	90.5	88.3
2,4-D	9.8	55.6	90.3	100	65.9	58.3	61.6	60.8	67.6	70.0
Silvex	10.4	95.3	85.8	91.5	100	91.3	95.0	91.1	96.0	93.3
2,4,5-T	12.8	78.6	65.6	69.2	100	81.6	90.1	84.3	98.5	83.5
2,4-DB	20.1	99.8	96.3	100	88.4	97.1	92.4	91.6	91.6	95.0
Mean		86.8	85.7	91.8	93.4	85.3	89.0	87.1	91.4	

(a) Percent recovery determinations made using eight spiked water samples.

Method 8260B– *Table 1*

Chromatographic Retention Times and Method Detection Limits (MDL) for Volatile Organic Compounds on Wide Bore Capillary Columns.

Analyte	Retention Time (minutes)			MDL[d] (μg/L)
	Column 1[a]	Column 2[b]	Column 2′[c]	
Dichlorodifluoromethane	1.55	0.70	3.13	0.10
Chloromethane	1.63	0.73	3.13	0.10
Vinyl chloride	1.71	0.79	3.93	0.17
Bromomethane	2.01	0.96	4.80	0.11
Chloroethane	2.09	1.02	—	0.10
Trichlorofluoromethane	2.27	1.19	6.20	0.08
1,1-Dichloroethene	2.89	1.57	7.83	0.12
Methylene chloride	3.60	2.06	9.27	0.03
trans-1,2-Dichloroethene	3.98	2.36	9.90	0.06
1,1-Dichloroethane	4.85	2.93	10.80	0.04
2,2-Dichloropropane	6.01	3.80	11.87	0.35
cis-1,2-Dichloroethene	6.19	3.90	11.93	0.12
Chloroform	6.40	4.80	12.60	0.03
Bromochloromethane	6.74	4.38	12.37	0.04
1,1,1-Trichloroethane	7.27	4.84	12.83	0.08
Carbon tetrachloride	7.61	5.26	13.17	0.21
1,1-Dichloropropene	7.68	5.29	13.10	0.10
Benzene	8.23	5.67	13.50	0.04
1,2-Dichloroethane	8.40	5.83	13.63	0.06
Trichloroethene	9.59	7.27	14.80	0.19
1,2-Dichloropropane	10.09	7.66	15.20	0.04
Bromodichloromethane	10.59	8.49	15.80	0.08
Dibromomethane	10.65	7.93	15.43	0.24
cis-1,3-Dichloropropene	—	—	17.90	—
Toluene	12.43	10.00	17.40	0.11
trans-1,3-Dichloropropene	—	—	16.70	—
1,1,2-Trichloroethane	13.41	11.05	18.30	0.10
Tetrachloroethene	13.74	11.15	18.60	0.14
1,3-Dichloropropane	14.04	11.31	18.70	0.04
Dibromochloromethane	14.39	11.85	19.20	0.05
1,2-Dibromoethane	14.73	11.83	19.40	0.06
1-Chlorohexane	15.46	13.29	—	0.05
Chlorobenzene	15.76	13.01	20.67	0.04
1,1,1,2-Tetrachloroethane	15.94	13.33	20.87	0.05
Ethyl benzene	15.99	13.39	21.00	0.06
p-Xylene	16.12	13.69	21.30	0.13
m-Xylene	16.17	13.68	21.37	0.05
o-Xylene	17.11	14.52	22.27	0.11

(a) Column 1 - 60 meter by 0.75 mm ID VOCOL capillary. Hold at 10 °C for 8 minutes, then program to 180 °C at 4 °C/min.

(b) Column 2 - 30 meter by 0.53 mm ID DB-624 wide-bore capillary using cryogenic oven. Hold at 10 °C for 5 minutes, then program to 160 °C at 6 °C/min.

(c) Column 2′ - 30 meter by 0.53 ID DB-624 wide-bore capillary, cooling GC oven to ambient temperatures. Hold at 10 °C for 6 minutes, program to 70 °C at 10 °C/min, program to 120 °C at 5 °C/min, then program to 180 °C at 8 °C/min.

(d) MDL based on a 25 mL sample volume.

Method 8260B – *Table 1*

Chromatographic Retention Times and Method Detection Limits (MDL) for Volatile Organic Compounds on Wide Bore Capillary Columns. (Continued)

Analyte	Retention Time (minutes)			MDL[d]
	Column 1[a]	Column 2[b]	Column 2′ [c]	(µg/L)
Styrene	17.31	14.60	22.40	0.04
Bromoform	17.93	14.88	22.77	0.12
Isopropylbenzene	18.06	15.46	23.30	0.15
1,1,2,2-Tetrachloroethane	18.72	16.35	24.07	0.04
Bromobenzene	18.95	15.86	24.00	0.03
1,2,3-Trichloropropane	19.02	15.86	24.00	0.32
n-Propylbenzene	19.06	16.41	24.33	0.04
2-Chlorotoluene	19.34	16.42	24.53	0.04
1,3,5-Trimethylbenzene	19.47	16.90	24.83	0.05
4-Chlorotoluene	19.50	16.72	24.77	0.06
tert-Butylbenzene	20.28	17.57	26.60	0.14
1,2,4-Trimethylbenzene	20.34	17.70	31.50	0.13
sec-Butylbenzene	20.79	18.09	26.13	0.13
p-Isopropyltoluene (4-Isopropyl toluene)	21.20	18.52	26.50	0.12
1,3-Dichlorobenzene	21.22	18.14	26.37	0.12
1,4-Dichlorobenzene	21.55	18.39	26.60	0.03
n-Butylbenzene	22.22	19.49	27.32	0.11
1,2-Dichlorobenzene	22.52	19.17	27.43	0.03
1,2-Dibromo-3-chloropropane	24.53	21.08	—	0.26
1,2,4-Trichlorobenzene	26.55	23.08	31.50	0.04
Hexachlorobutadiene	26.99	23.68	32.07	0.11
Naphthalene	27.17	23.52	32.20	0.04
1,2,3-Trichlorobenzene	27.78	24.18	32.97	0.03
Internal Standards/Surrogates				
1,4-Difluorobenzene	13.26			
Chlorobenzene-d_5	23.10			
1,4-Dichlorobenzene-d_4	31.16			
4-Bromofluorobenzene	27.83	15.71	23.63	
1,2-Dichlorobenzene-d_4	32.30	19.08	27.25	
Dichloroethane-d_4	12.08			
Dibromofluoromethane	—			
Toluene-d_8	18.27			
Pentafluorobenzene	—			
Fluorobenzene	13.00	6.27	14.06	

(a) Column 1 - 60 meter by 0.75 mm ID VOCOL capillary. Hold at 10 °C for 8 minutes, then program to 180 °C at 4 °C/min.

(b) Column 2 - 30 meter by 0.53 mm ID DB-624 wide-bore capillary using cryogenic oven. Hold at 10 °C for 5 minutes, then program to 160 °C at 6 °C/min.

(c) Column 2′ - 30 meter by 0.53 ID DB-624 wide-bore capillary, cooling GC oven to ambient temperatures. Hold at 10 °C for 6 minutes, program to 70 °C at 10 °C/min, program to 120 °C at 5 °C/min, then program to 180 °C at 8 °C/min.

(d) MDL based on a 25 mL sample volume.

Method 8260B – *Table 2*

Chromatographic Retention Times and Method Detection Limits (MDL) for Volatile Organic Compounds on Narrow Bore Capillary Columns.

Analyte	Retention Time (min.)	
	Column 3[a]	MDL[b] (μg/L)
Dichlorodifluoromethane	0.88	0.11
Chloromethane	0.97	0.05
Vinyl chloride	1.04	0.04
Bromomethane	1.29	0.06
Chloroethane	1.45	0.02
Trichlorofluoromethane	1.77	0.07
1,1-Dichloroethene	2.33	0.05
Methylene chloride	2.66	0.09
trans-1,2-Dichloroethene	3.54	0.03
1,1-Dichloroethane	4.03	0.03
cis-1,2-Dichloroethene	5.07	0.06
2,2-Dichloropropane	5.31	0.08
Chloroform	5.55	0.04
Bromochloromethane	5.63	0.09
1,1,1-Trichloroethane	6.76	0.04
1,2-Dichloroethane	7.00	0.02
1,1-Dichloropropene	7.16	0.12
Carbon tetrachloride	7.41	0.02
Benzene	7.41	0.03
1,2-Dichloropropane	8.94	0.02
Trichloroethene	9.02	0.02
Dibromomethane	9.09	0.01
Bromodichloromethane	9.34	0.03
Toluene	11.51	0.08
1,1,2-Trichloroethane	11.99	0.08
1,3-Dichloropropane	12.48	0.08
Dibromochloromethane	12.80	0.07
Tetrachloroethene	13.20	0.05
1,2-Dibromoethane	13.60	0.10
Chlorobenzene	14.33	0.03
1,1,1,2-Tetrachloroethane	14.73	0.07
Ethyl benzene	14.73	0.03
p-Xylene	15.30	0.06
m-Xylene	15.30	0.03
Bromoform	15.70	0.20
o-Xylene	15.78	0.06
Styrene	15.78	0.27
1,1,2,2-Tetrachloroethane	15.78	0.20

(a) Column 3 - 30 meter by 0.32 mm ID DB5 Capillary with 1 μm film thickness.

(b) MDL based on a 25 mL sample volume.

Method 8260B – *Table 2*

Chromatographic Retention Times and Method Detection Limits (MDL) for Volatile Organic Compounds on Narrow Bore Capillary Columns. (Continued)

Analyte	Retention Time (min.)	
	Column 3[a]	MDL[b] (µg/L)
1,2,3-Trichloropropane	16.26	0.09
Isopropylbenzene	16.42	0.10
Bromobenzene	16.42	0.11
2-Chlorotoluene	16.74	0.08
n-Propylbenzene	16.82	0.10
4-Chlorotoluene	16.82	0.06
1,3,5-Trimethylbenzene	16.99	0.06
tert-Butylbenzene	17.31	0.33
1,2,4-Trimethylbenzene	17.31	0.09
sec-Butylbenzene	17.47	0.12
1,3-Dichlorobenzene	17.47	0.05
p-Isopropyltoluene (4-Isopropyl toluene)	17.63	0.26
1,4-Dichlorobenzene	17.63	0.04
1,2-Dichlorobenzene	17.79	0.05
n-Butylbenzene	17.95	0.10
1,2-Dibromo-3-chloropropane	18.03	0.50
1,2,4-Trichlorobenzene	18.84	0.20
Naphthalene	19.09	0.10
Hexachlorobutadiene	19.24	0.10
1,2,3-Trichlorobenzene	19.24	0.14

(a)   Column 3 - 30 meter by 0.32 mm ID DB5 Capillary with 1 µm film thickness.

(b)   MDL based on a 25 mL sample volume.

Method 8260B – *Table 3*

Estimated Quantitation Limits for Volatile Analytes[a]

	Estimated Quantitation Limits		
	Ground water µg/L		Low Soil/Sediment[b] µg/kg
Volume of water purged	5 mL	25 mL	
All analytes in Table 1	5	1	5

Other Matrices	Factor[c]
Water miscible liquid waste	50
High-concentration soil and sludge	125
Non-water miscible waste	500

(a) Estimated Quantitation Limit (EQL) - The lowest concentration that can be reliably achieved within specified limits of precision and accuracy during routine laboratory operating conditions. The EQL is generally 5 to 10 times the MDL. However, it may be nominally chosen within these guidelines to simplify data reporting. For many analytes the EQL analyte concentration is selected for the lowest non-zero standard in the calibration curve. Sample EQLs are highly matrix-dependent. The EQLs listed herein are provided for guidance and may not always be achievable. See the following information for further guidance on matrix-dependent EQLs.

(b) EQLs listed for soil/sediment are based on wet weight. Normally data is reported on a dry weight basis; therefore, EQLs will be higher, based on the percent dry weight in each sample.

(c) EQL = [EQL for low soil sediment (table 3)] X [Factor]. For non-aqueous samples, the factor is on a wet-weight basis.

Method 8260B – *Table 4*

BFB Mass - Intensity Specifications (4-Bromofluorobenzene)

Mass	Intensity Required (relative abundance)
50	15 to 40% of mass 95
75	30 to 60% of mass 95
95	base peak, 100% relative abundance
96	5 to 9% of mass 95
173	less than 2% of mass 174
174	greater than 50% of mass 95
175	5 to 9% of mass 174
176	greater than 95% but less than 101% of mass 174
177	5 to 9% of mass 176

Method 8260B – *Table 5*

Characteristic Masses (M/Z) for Purgeable Organic Compounds

Analyte	Primary Characteristic Ion	Secondary Characteristic Ion(s)
Acetone	58	43
Acetonitrile	41	41, 40, 39
Acrolein	56	55, 58
Acrylonitrile	53	52, 51
Allyl alcohol	57	57, 58, 39
Allyl chloride	76	76, 41, 39, 78
Benzene	78	—
Benzyl chloride	91	91, 126, 65, 128
Bromoacetone	136	43, 136, 138, 93, 95
Bromobenzene	156	77, 158
Bromochloromethane	128	49, 130
Bromodichloromethane	83	85, 127
Bromoform	173	175, 254
Bromomethane	94	96
iso-Butanol	74	43
n-Butanol	56	41
2-Butanone	72	43, 72
n-Butylbenzene	91	92, 134
sec-Butylbenzene	105	134
tert-Butylbenzene	119	91, 134
Carbon disulfide	76	78
Carbon tetrachloride	117	119
Chloral hydrate	82	44, 84, 86, 111
Chloroacetonitrile	48	75
Chlorobenzene	112	77, 114
1-Chlorobutane	56	49
Chlorodibromomethane	129	208, 206
Chloroethane	64 (49*)	66 (51*)
2-Chloroethanol	49	49, 44, 43, 51, 80
bis-(2-Chloroethyl) sulfide	109	111, 158, 160
2-Chloroethyl vinyl ether	63	65, 106
Chloroform	83	85
Chloromethane	50 (49*)	52 (51*)
Chloroprene	53	53, 88, 90, 51
3-Chloropropionitrile	54	54, 49, 89, 91
2-Chlorotoluene	91	126
4-Chlorotoluene	91	126
1,2-Dibromo-3-chloropropane	75	155, 157
Dibromochloromethane	129	127
1,2-Dibromoethane	107	109,188

* Characteristic ion for an ion trap mass spectrometer (to be used when ion-molecule reactions are observed)

Method 8260B – *Table 5*

Characteristic Masses (M/Z) for Purgeable Organic Compounds (Continued)

Analyte	Primary Characteristic Ion	Secondary Characteristic Ion(s)
Dibromomethane	93	95, 174
1,2-Dichlorobenzene	146	111, 148
1,2-Dichlorobenzene d$_4$	152	115, 150
1,3-Dichlorobenzene	146	111, 148
1,4-Dichlorobenzene	146	111, 148
cis-1,4-Dichloro-2-butene	75	75, 53, 77, 124, 89
trans-1,4-Dichloro-2-butene	53	88, 75
Dichlorodifluoromethane	85	87
1,1-Dichloroethane	63	65, 83
1,2-Dichloroethane	62	98
1,1-Dichloroethene	96	61, 63
cis-1,2-Dichloroethene	96	61, 98
trans-1,2-Dichloroethene	96	61, 98
1,2-Dichloropropane	63	112
1,3-Dichloropropane	76	78
2,2-Dichloropropane	77	97
1,3-Dichloro-2-propanol	79	79, 43, 81, 49
1,1-Dichloropropene	75	110, 77
cis-1,3-Dichloropropene	75	77, 39
trans-1,3-Dichloropropene	75	77, 39
1,2,3,4-Diepoxybutane	55	55, 57, 56
Diethyl ether	74	45, 59
1,4-Dioxane	88	88, 58, 43, 57
Epichlorohydrin	57	57, 49, 62, 51
Ethanol	31	45, 27, 46
Ethyl acetate	88	43, 45, 61
Ethylbenzene	91	106
Ethylene oxide	44	44, 43, 42
Ethyl methacrylate	69	69, 41, 99, 86, 114
Hexachlorobutadiene	225	223, 227
Hexachloroethane	201	166, 199, 203
2-Hexanone	43	58, 57, 100
2-Hydroxypropionitrile	44	44, 43, 42, 53
Iodomethane	142	127, 141
Isobutyl alcohol	43	43, 41, 42, 74
Isopropylbenzene	105	120
p-Isopropyltoluene	119	134, 91
Malononitrile	66	66, 39, 65, 38
Methyacrylonitrile	41	41, 67, 39, 52, 66

Characteristic Masses (M/Z) for Purgeable Organic Compounds (Continued)

Analyte	Primary Characteristic Ion	Secondary Characteristic Ion(s)
Methyl acrylate	55	85
Methyl-t-butyl ether	73	57
Methylene chloride	84	86, 49
Methyl ethyl ketone	72	43
Methyl iodide	142	142, 127, 141
Methyl methacrylate	69	69, 41, 100, 39
4-Methyl-2-pentanone	100	43, 58, 85
Naphthalene	128	—
Nitrobenzene	123	51, 77
2-Nitropropane	46	—
2-Picoline	93	93, 66, 92, 78
Pentachloroethane	167	167, 130, 132, 165, 169
Propargyl alcohol	55	55, 39, 38, 53
ß-Propiolactone	42	42, 43, 44
Propionitrile (ethyl cyanide)	54	54, 52, 55, 40
n-Propylamine	59	59, 41, 39
n-Propylbenzene	91	120
Pyridine	79	52
Styrene	104	78
1,2,3-Trichlorobenzene	180	182, 145
1,2,4-Trichlorobenzene	180	182, 145
1,1,1,2-Tetrachloroethane	131	133, 119
1,1,2,2-Tetrachloroethane	83	131, 85
Tetrachloroethene	164	129, 131, 166
Toluene	92	91
1,1,1-Trichloroethane	97	99, 61
1,1,2-Trichloroethane	83	97, 85
Trichloroethene	95	97, 139, 132
Trichlorofluoromethane	151	101, 153
1,2,3-Trichloropropane	75	77
1,2,4-Trimethylbenzene	105	120
1,3,5-Trimethylbenzene	105	120
Vinyl acetate	43	86
Vinyl chloride	62	64
o-Xylene	106	91
m-Xylene	106	91
p-Xylene	106	91

Method 8260A – *Table 5*

Characteristic Masses (M/Z) for Purgeable Organic Compounds (Continued)

Analyte	Primary Characteristic Ion	Secondary Characteristic Ion(s)
Internal Standards/Surrogates		
1,4-Difluorobenzene	114	
Chlorobenzene-d$_5$	117	
1,4-Dichlrobenzene-d$_4$	152	115, 150
4-Bromofluorobenzene	95	174, 176
Dibromofluoromethane	113	
Dichloroethane-d$_4$	102	
Toluene-d$_8$	98	
Pentafluorobenzene	168	
Fluorobenzene	96	77

Method 8260B – *Table 6*

Single Laboratory Accuracy and Precision Data for Volatile Organic Compounds in Water Determined with a Wide Bore Capillary Column

Analyte	Conc. Range, μg/L	Number of Samples	Recovery,[a] %	Standard Deviation of Recovery[b]	Percent Rel. Std. Dev.
Benzene	0.1 - 10	31	97	6.5	5.7
Bromobenzene	0.1 - 10	30	100	5.5	5.5
Bromochloromethane	0.5 - 10	24	90	5.7	6.4
Bromodichloromethane	0.1 - 10	30	95	5.7	6.1
Bromoform	0.5 - 10	18	101	6.4	6.3
Bromomethane	0.5 - 10	18	95	7.8	8.2
n-Butylbenzene	0.5 - 10	18	100	7.6	7.6
sec-Butylbenzene	0.5 - 10	16	100	7.6	7.6
tert-Butylbenzene	0.5 - 10	18	102	7.4	7.3
Carbon tetrachloride	0.5 - 10	24	84	7.4	8.8
Chlorobenzene	0.1 - 10	31	98	5.8	5.9
Chloroethane	0.5 - 10	24	89	8.0	9.0
Chloroform	0.5 - 10	24	90	5.5	6.1
Chloromethane	0.5 - 10	23	93	8.3	8.9
2-Chlorotoluene	0.1 - 10	31	90	5.6	6.2
4-Chlorotoluene	0.1 - 10	31	99	8.2	8.3
1,2-Dibromo-3-chloropropane	0.5 - 10	24	83	16.6	19.9
Dibromochloromethane	0.1 - 10	31	92	6.5	7.0
1,2-Dibromoethane	0.5 - 10	24	102	4.0	3.9
Dibromomethane	0.5 - 10	24	100	5.6	5.6
1,2-Dichlorobenzene	0.1 - 10	31	93	5.8	6.2
1,3-Dichlorobenzene	0.5 - 10	24	99	6.8	6.9
1,4-Dichlorobenzene	0.2 - 20	31	103	6.6	6.4
Dichlorodifluoromethane	0.5 - 10	18	90	6.9	7.7
1,1-Dichloroethane	0.5 - 10	24	96	5.1	5.3
1,2-Dichloroethane	0.1 - 10	31	95	5.1	5.4
1,1-Dichloroethene	0.1 - 10	34	94	6.3	6.7
cis-1,2-Dichloroethene	0.5 - 10	18	101	6.7	6.7
trans-1,2-Dichloroethene	0.1 - 10	30	93	5.2	5.6
1,2-Dichloropropane	0.1 - 10	30	97	5.9	6.1
1,3-Dichloropropane	0.1 - 10	31	96	5.7	6.0
2,2-Dichloropropane	0.5 - 10	12	86	14.6	16.9
1,1-Dichloropropene	0.5 - 10	18	98	8.7	8.9
Ethyl benzene	0.1 - 10	31	99	8.4	8.6
Hexachlorobutadiene	0.5 - 10	18	100	6.8	6.8
Isopropylbenzene	0.5 - 10	16	101	7.7	7.6
p-Isopropyltoluene (4-Isopropyl toluene)	0.1 - 10	23	99	6.7	6.7
Methylene chloride	0.1 - 10	30	95	5.0	5.3
Naphthalene	0.1 - 100	31	104	8.6	8.2
n-Propylbenzene	0.1 - 10	31	100	5.8	5.8

(a) Recoveries were calculated using internal standard method. Internal standard was fluorobenzene.

(b) Standard deviation was calculated by pooling data from three concentrations.

Method 8260B – *Table 6*

Single Laboratory Accuracy and Precision Data for Volatile Organic Compounds in Water Determined with a Wide Bore Capillary Column (Continued)

Analyte	Conc. Range, µg/L	Number of Samples	Recovery,[a] %	Standard Deviation of Recovery[b]	Percent Rel. Std. Dev.
Styrene	0.1 - 100	39	102	7.3	7.2
1,1,1,2-Tetrachloroethane	0.5 - 10	24	90	6.1	6.8
1,1,2,2-Tetrachloroethane	0.1 - 10	30	91	5.7	6.3
Tetrachloroethene	0.5 - 10	24	89	6.0	6.8
Toluene	0.5 - 10	18	102	8.1	8.0
1,2,3-Trichlorobenzene	0.5 - 10	18	109	9.4	8.6
1,2,4-Trichlorobenzene	0.5 - 10	18	108	9.0	8.3
1,1,1-Trichloroethane	0.5 - 10	18	98	7.9	8.1
1,1,2-Trichloroethane	0.5 - 10	18	104	7.6	7.3
Trichloroethene	0.5 - 10	24	90	6.5	7.3
Trichlorofluoromethane	0.5 - 10	24	89	7.2	8.1
1,2,3-Trichloropropane	0.5 - 10	16	108	15.6	14.4
1,2,4-Trimethylbenzene	0.5 - 10	18	99	8.0	8.1
1,3,5-Trimethylbenzene	0.5 - 10	23	92	6.8	7.4
Vinyl chloride	0.5 - 10	18	98	6.5	6.7
o-Xylene	0.1 - 31	18	103	7.4	7.2
m-Xylene	0.1 - 10	31	97	6.3	6.5
p-Xylene	0.5 - 10	18	104	8.0	7.7

(a) Recoveries were calculated using internal standard method. Internal standard was fluorobenzene.

(b) Standard deviation was calculated by pooling data from three concentrations.

Method 8260B – *Table 7*

Single Laboratory Accuracy and Precision Data for Volatile Organic Compounds in Water Determined with a Narrow Bore Capillary Column

Analyte	Conc. Range, µg/L	Number of Samples	Recovery,(a) %	Standard Deviation of Recovery	Percent Rel. Std. Dev.
Benzene	0.1	7	99	6.2	6.3
Bromobenzene	0.5	7	97	7.4	7.6
Bromochloromethane	0.5	7	97	5.8	6.0
Bromodichloromethane	0.1	7	100	4.6	4.6
Bromoform	0.5	7	101	5.4	5.3
Bromomethane	0.5	7	99	7.1	7.2
n-Butylbenzene	0.5	7	94	6.0	6.4
sec-Butylbenzene	0.5	7	110	7.1	6.5
tert-Butylbenzene	0.5	7	110	2.5	2.3
Carbon tetrachloride	0.1	7	108	6.8	6.3
Chlorobenzene	0.1	7	91	5.8	6.4
Chloroethane	0.1	7	100	5.8	5.8
Chloroform	0.1	7	105	3.2	3.0
Chloromethane	0.5	7	101	4.7	4.7
2-Chlorotoluene	0.5	7	99	4.6	4.6
4-Chlorotoluene	0.5	7	96	7.0	7.3
1,2-Dibromo-3-chloropropane	0.5	7	92	10.0	10.9
Dibromochloromethane	0.1	7	99	5.6	5.7
1,2-Dibromoethane	0.5	7	97	5.6	5.8
Dibromomethane	0.5	7	93	5.6	6.0
1,2-Dichlorobenzene	0.1	7	97	3.5	3.6
1,3-Dichlorobenzene	0.1	7	101	6.0	5.9
1,4-Dichlorobenzene	0.1	7	106	6.5	6.1
Dichlorodifluoromethane	0.1	7	99	8.8	8.9
1,1-Dichloroethane	0.5	7	98	6.2	6.3
1,2-Dichloroethane	0.1	7	100	6.3	6.3
1,1-Dichloroethene	0.1	7	95	9.0	9.5
cis-1,2-Dichloroethene	0.1	7	100	3.7	3.7
trans-1,2-Dichloroethene	0.1	7	98	7.2	7.3
1,2-Dichloropropane	0.5	7	96	6.0	6.3
1,3-Dichloropropane	0.5	7	99	5.8	5.9
2,2-Dichloropropane	0.5	7	99	4.9	4.9
1,1-Dichloropropene	0.5	7	102	7.4	7.3
Ethyl benzene	0.5	7	99	5.2	5.3
Hexachlorobutadiene	0.5	7	100	6.7	6.7
Isopropylbenzene	0.5	7	102	6.4	6.3
p-Isopropyltoluene (4-Isopropyl toluene)	0.5	7	113	13.0	11.5
Methylene chloride	0.5	7	97	13.0	13.4
Naphthalene	0.5	7	98	7.2	7.3
n-Propylbenzene	0.5	7	99	6.6	6.7

(a) Recoveries were calculated using internal standard method. Internal standard was fluorobenzene.

Method 8260B – *Table 7*

Single Laboratory Accuracy and Precision Data for Volatile Organic Compounds in Water Determined with a Narrow Bore Capillary Column (Continued)

Analyte	Conc. Range, µg/L	Number of Samples	Recovery,[a] %	Standard Deviation of Recovery	Percent Rel. Std. Dev.
Styrene	0.5	7	96	19.0	19.8
1,1,1,2-Tetrachloroethane	0.5	7	100	4.7	4.7
1,1,2,2-Tetrachloroethane	0.5	7	100	12.0	12.0
Tetrachloroethene	0.1	7	96	5.0	5.2
Toluene	0.5	7	100	5.9	5.9
1,2,3-Trichlorobenzene	0.5	7	102	8.9	8.7
1,2,4-Trichlorobenzene	0.5	7	91	16.0	17.6
1,1,1-Trichloroethane	0.5	7	100	4.0	4.0
1,1,2-Trichloroethane	0.5	7	102	4.9	4.8
Trichloroethene	0.1	7	104	2.0	1.9
Trichlorofluoromethane	0.1	7	97	4.6	4.7
1,2,3-Trichloropropane	0.5	7	96	6.5	6.8
1,2,4-Trimethylbenzene	0.5	7	96	6.5	6.8
1,3,5-Trimethylbenzene	0.5	7	101	4.2	4.2
Vinyl chloride	0.1	7	104	0.2	0.2
o-Xylene	0.5	7	106	7.5	7.1
m-Xylene	0.5	7	106	4.6	4.3
p-Xylene	0.5	7	97	6.1	6.3

(a) Recoveries were calculated using internal standard method. Internal standard was fluorobenzene.

Method 8260B – *Table 8*

Surrogate Spike Recovery Limits for Water and Soil/Sediment Samples

Surrogate Compound	Low/High Water	Low/High Soil/Sediment
4-Bromofluorobenzene [a]	86 -115	74 - 121
Dibromofluoromethane [a]	86 - 118	80 - 120
Toluene-d$_8$ [a]	88 - 110	81 - 117
Dichloroethane-d$_4$ [a]	80 - 120	80 - 120

(a) Single laboratory data for guidance only.

Characteristic Ions for Semivolatile Compounds

Compound	Retention Time (min)	Primary Ion	Secondary Ion(s)
2-Picoline	3.75[a]	93	66,92
Aniline	5.68	93	66,65
Phenol	5.77	94	65,66
Bis(2-chloroethyl) ether	5.82	93	63,95
2-Chlorophenol	5.97	128	64,130
1,3-Dichlorobenzene	6.27	146	148,111
1,4-Dichlorobenzene-d_4 (IS)	6.35	152	150,115
1,4-Dichlorobenzene	6.40	146	148,111
Benzyl alcohol	6.78	108	79,77
1,2-Dichlorobenzene	6.85	146	148,111
N-Nitrosomethylethylamine	6.97	88	42,43,56
Bis(2-chloroisopropyl) ether	7.22	45	77,121
Ethyl carbamate	7.27	62	44,45,74
Thiophenol (Benzenethiol)	7.42	110	66,109,84
Methyl methanesulfonate	7.48	80	79,65,95
N-Nitrosodi-n-propylamine	7.55	70	42,101,130
Hexachloroethane	7.65	117	201,199
Maleic anhydride	7.65	54	98,53,44
Nitrobenzene	7.87	77	123,65
Isophorone	8.53	82	95,138
N-Nitrosodiethylamine	8.70	102	42,57,44,56
2-Nitrophenol	8.75	139	109,65
2,4-Dimethylphenol	9.03	122	107,121
p-Benzoquinone	9.13	108	54,82,80
Bis(2-chloroethoxy)methane	9.23	93	95,123
Benzoic acid	9.38	122	105,77
2,4-Dichlorophenol	9.48	162	164,98
Trimethyl phosphate	9.53	110	79,95,109,140
Ethyl methanesulfonate	9.62	79	109,97,45,65
1,2,4-Trichlorobenzene	9.67	180	182,145
Naphthalene-d_8 (IS)	9.75	136	68
Naphthalene	9.82	128	129,127
Hexachlorobutadiene	10.43	225	223,227
Tetraethyl pyrophosphate	11.07	99	155,127,81,109
Diethyl sulfate	11.37	139	45,59,99,111,125
4-Chloro-3-methylphenol	11.68	107	144,142
2-Methylnaphthalene	11.87	142	141
2-Methylphenol	12.40	107	108,77,79,90
Hexachloropropene	12.45	213	211,215,117,106,141
Hexachlorocyclopentadiene	12.60	237	235,272
N-Nitrosopyrrolidine	12.65	100	41,42,68,69
Acetophenone	12.67	105	71,51,120

IS   Internal standard
surr   surrogate
(a)   Estimated retention times
(b)   Substitute for the non-specific mixture, tricresyl phosphate

Method 8270C - Table 1

Characteristic Ions for Semivolatile Compounds (Continued)

Compound	Retention Time (min)	Primary Ion	Secondary Ion(s)
4-Methylphenol	12.82	107	108,77,79,90
2,4,6-Trichlorophenol	12.85	196	198,200
o-Toluidine	12.87	106	107,77,51,79
3-Methylphenol	12.93	107	108,77,79,90
2-Chloronaphthalene	13.30	162	127,164
N-Nitrosopiperidine	13.55	114	42,55,56,41
1,4-Phenylenediamine	13.62	108	80,53,54,52
1-Chloronaphthalene	13.65[a]	162	127,164
2-Nitroaniline	13.75	65	92,138
5-Chloro-2-methylaniline	14.28	106	141,140,77,89
Dimethyl phthalate	14.48	163	194,164
Acenaphthylene	14.57	152	151,153
2,6-Dinitrotoluene	14.62	165	63,89
Phthalic anhydride	14.62	104	76,50,148
o-Anisidine	15.00	108	80,123,52
3-Nitroaniline	15.02	138	108,92
Acenaphthene-d$_{10}$ (IS)	15.05	164	162,160
Acenaphthene	15.13	154	153,152
2,4-Dinitrophenol	15.35	184	63,154
2,6-Dinitrophenol	15.47	162	164,126,98,63
4-Chloroaniline	15.50	127	129,65,92
Isosafrole	15.60	162	131,101,77,51
Dibenzofuran	15.63	168	139
2,4-Diaminotoluene	15.78	121	122,94,77,104
2,4-Dinitrotoluene	15.80	165	63,89
4-Nitrophenol	15.80	139	109,65
2-Naphthylamine	16.00[a]	143	115,116
1,4-Naphthoquinone	16.23	158	104,102,76,50,130
p-Cresidine	16.45	122	94,137,77,93
Dichlorovos	16.48	109	185,79,145
Diethyl phthalate	16.70	149	177,150
Fluorene	16.70	166	165,167
2,4,5-Trimethylaniline	16.70	120	135,134,91,77
N-Nitrosodi-n-butylamine	16.73	84	57,41,116,158
4-Chlorophenyl phenyl ether	16.78	204	206,141
Hydroquinone	16.93	110	81,53,55
4,6-Dinitro-2-methylphenol	17.05	198	51,105
Resorcinol	17.13	110	81,82,53,69
N-Nitrosodiphenylamine	17.17	169	168,167
Safrole	17.23	162	104,77,103,135
Hexamethyl phosphoramide	17.33	135	44,179,92,42

IS   Internal standard
surr   surrogate
(a)   Estimated retention times
(b)   Substitute for the non-specific mixture, tricresyl phosphate

Method 8270C - Table 1

Characteristic Ions for Semivolatile Compounds (Continued)

Compound	Retention Time (min)	Primary Ion	Secondary Ion(s)
3-(Chloromethyl)pyridine hydrochloride	17.50	92	127,129,65,39
Diphenylamine	17.54[a]	169	168,167
1,2,4,5-Tetrachlorobenzene	17.97	216	214,179,108,143,218
1-Naphthylamine	18.20	143	115,89,63
1-Acetyl-2-thiourea	18.22	118	43,42,76
4-Bromophenyl phenyl ether	18.27	248	250,141
Toluene diisocyanate	18.42	174	145,173,146,132,91
2,4,5-Trichlorophenol	18.47	196	198,97,132,99
Hexachlorobenzene	18.65	284	142,249
Nicotine	18.70	84	133,161,162
Pentachlorophenol	19.25	266	264,268
5-Nitro-o-toluidine	19.27	152	77,79,106,94
Thionazime	19.35	107	96,97,143,79,68
4-Nitroaniline	19.37	138	65,108,92,80,39
Phenanthrene-d_{10} (IS)	19.55	188	94,80
Phenanthrene	19.62	178	179,176
Anthracene	19.77	178	176,179
1,4-Dinitrobenzene	19.83	168	75,50,76,92,122
Mevinphos	19.90	127	192,109,67,164
Naled	20.03	109	145,147,301,79,189
1,3-Dinitrobenzene	20.18	168	76,50,75,92,122
Diallate (cis or trans)	20.57	86	234,43,70
1,2-Dinitrobenzene	20.58	168	50,62,74
Diallate (trans or cis)	20.78	86	234,43,70
Pentachlorobenzene	21.35	250	252,108,248,215,254
5-Nitro-o-anisidine	21.50	168	79,52,138,153,77
Pentachloronitrobenzene	21.72	237	142,214,249,295,265
4-Nitroquinoline-1-oxide	21.73	174	101,128,75,116
Di-n-butyl phthalate	21.78	149	150,104
2,3,4,6-Tetrachlorophenol	21.88	232	131,230,166,234,168
Dihydrosaffrole	22.42	135	64,77
Demeton-O	22.72	88	89,60,61,115,171
Fluoranthene	23.33	202	101,203
1,3,5-Trinitrobenzene	23.68	75	74,213,120,91,63
Dicrotophos	23.82	127	67,72,109,193,237
Benzidine	23.87	184	92,185
Trifluralin	23.88	306	43,264,41,290
Bromozynil	23.90	277	279,88,275,168
Pyrene	24.02	202	200,203
Monocrotophos	24.08	127	192,67,97,109
Phorate	24.10	75	121,97,93,260

IS   Internal standard
surr   surrogate
(a)   Estimated retention times
(b)   Substitute for the non-specific mixture, tricresyl phosphate

Method 8270C - Table 1

Characteristic Ions for Semivolatile Compounds (Continued)

Compound	Retention Time (min)	Primary Ion	Secondary Ion(s)
Sulfallate	24.23	188	88,72,60,44
Demeton-S	24.30	88	60,81,89,114,115
Phenacetin	24.33	108	180,179,109,137,80
Dimethoate	24.70	87	93,125,143,229
Phenobarbital	24.70	204	117,232,146,161
Carbofuran	24.90	164	149,131,122
Octamethyl pyrophosphoramide	24.95	135	44,199,286,153,243
4-Aminobiphenyl	25.08	169	168,170,115
Dioxathion	25.25	97	125,270,153
Terbufos	25.35	231	57,97,153,103
a,a-Dimethylphenylamine	25.43	58	91,65,134,42
Pronamide	25.48	173	175,145,109,147
Aminoazobenzene	25.72	197	92,120,65,77
Dichlone	25.77	191	163,226,228,135,193
Dinoseb	25.83	211	163,147,117,240
Disulfoton	25.83	88	97,89,142,186
Fluchloralin	25.88	306	63,326,328,264,65
Mexacarbate	26.02	165	150,134,164,222
4,4'-Oxydianiline	26.08	200	108,171,80,65
Butyl benzyl phthalate	26.43	149	91,206
4-Nitrobiphenyl	26.55	199	152,141,169,151
Phosphamidon	26.85	127	264,72,109,138
2-Cyclohexyl-4,6-Dinitrophenol	26.87	231	185,41,193,266
Methyl parathion	27.03	109	125,263,79,93
Carbaryl	27.17	144	115,116,201
Dimethylaminoazobenzene	27.50	225	120,77,105,148,42
Propylthiouracil	27.68	170	142,114,83
Benz(a)anthracene	27.83	228	229,226
Chrysene-d$_{12}$ (IS)	27.88	240	120,236
3,3'-Dichlorobenzidine	27.88	252	254,126
Chrysene	27.97	228	226,229
Malathion	28.08	173	125,127,93,158
Kepone	28.18	272	274,237,178,143,270
Fenthion	28.37	278	125,109,169,153
Parathion	28.40	109	97,291,139,155
Anilazine	28.47	239	241,143,178,89
Bis(2-ethylhexyl) phthalate	28.47	149	167,279
3,3'-Dimethylbenzidine	28.55	212	106,196,180
Carbophenothion	28.58	157	97,121,342,159,199
5-Nitroacenaphthene	28.73	199	152,169,141,115
Methapyrilene	28.77	97	50,191,71
Isodrin	28.95	193	66,195,263,265,147
Captan	29.47	79	149,77,119,117
Chlorfenvinphos	29.53	267	269,323,325,295
Crotoxyphos	29.73	127	105,193,166

IS   Internal standard
surr surrogate
(a)   Estimated retention times
(b)   Substitute for the non-specific mixture, tricresyl phosphate

Method 8270C - Table 1

Characteristic Ions for Semivolatile Compounds (Continued)

Compound	Retention Time (min)	Primary Ion	Secondary Ion(s)
Phosmet	30.03	160	77,93,317,76
EPN	30.11	157	169,185,141,323
Tetrachlorvinphos	30.27	329	109,331,79,333
Di-n-octyl phthalate	30.48	149	167,43
2-Aminoanthraquinone	30.63	223	167,195
Barban	30.83	222	51,87,224,257,153
Aramite	30.92	185	191,319,334,197,321
Benzo(b)fluoranthene	31.45	252	253,125
Nitrofen	31.48	283	285,202,139,253
Benzo(k)fluoranthene	31.55	252	253,125
Chlorbenzilate	31.77	251	139,253,111,141
Fensulfothion	31.87	293	97,308,125,292
Ethion	32.08	231	97,153,125,121
Diethylstilbestrol	32.15	268	145,107,239,121,159
Famphur	32.67	218	125,93,109,217
Tri-p-tolyl phosphate[b]	32.75	368	367,107,165,198
Benzo(a)pyrene	32.80	252	253,125
Perylene-d$_{12}$ (IS)	33.05	264	260,265
7,12-Dimethylbenz(a)anthracene	33.25	256	241,239,120
5,5-Diphenylhydantion	33.40	180	104,252,223,209
Captafol	33.47	79	77,80,107
Dinocap	33.47	69	41,39
Methoxychlor	33.55	227	228,152,114,274,212
2-Acetylaminofluorene	33.58	181	180,223,152
4,4′-Methylenebis(2-chloroaniline)	34.38	231	266,268,140,195
3,3′-Dimethoxybenzidine	34.47	244	201,229
3-Methylcholanthrene	35.07	268	252,253,126,134,113
Phosalone	35.23	182	184,367,121,379
Azinphos-methyl	35.25	160	132,93,104,105
Leptophos	35.28	171	377,375,77,155,379
Mirex	35.43	272	237,274,270,239,235
Tris(2,3-dibromoprophl) phosphate	35.68	201	137,119,217,219,199
Dibenz(a,j)acridine	36.40	279	280,277,250
Mestranol	36.48	277	310,174,147,242
Coumaphos	37.08	362	226,210,364,97,109
Indena(1,2,3-cd)pyrene	39.52	276	138,227
Dibenz(a,h)anthracene	39.82	278	139,279
Benzo(g,h,i)perylene	41.43	276	138,277
1,2:4,5-Dibenzopyrene	41.60	302	151,150,300
Strychnine	45.15	334	334,335,333

IS   Internal standard
surr surrogate
(a)   Estimated retention times
(b)   Substitute for the non-specific mixture, tricresyl phosphate

Method 8270C - Table 1

Characteristic Ions for Semivolatile Compounds (Continued)

Compound	Retention Time (min)	Primary Ion	Secondary Ion(s)
Piperonyl sulfoxide	46.43	162	135,105,77
Hexachlorophene	47.98	196	198,209,211,406,408
Aldrin	—	66	263,220
Aroclor 1016	—	222	260,292
Aroclor 1221	—	190	224,260
Aroclor 1232	—	190	224,260
Aroclor 1242	—	222	256,292
Aroclor 1248	—	292	362,326
Aroclor 1254	—	292	362,326
Aroclor 1260	—	360	362,394
a-BHC	—	183	181,109
b-BHC	—	181	183,109
s-BHC	—	183	181,109
y-BHC (Lindane)	—	183	181,109
4,4'-DDD	—	235	237,165
4,4'-DDE	—	246	248,176
4,4'-DDT	—	235	237,165
Dieldrin	—	79	263,279
1,2-Diphenylhydrazine	—	77	105,182
Endosulfan I	—	195	339,341
Endosulfan II	—	337	339,341
Endosulfan sulfate	—	272	387,422,
Endrin	—	263	82,81
Endrin aldehyde	—	67	345,250
Endrin ketone	—	317	67,319
2-Fluorobiphenyl (surr)	—	172	171
2-Flourophenol (surr)	—	112	64
Heptachlor	—	100	272,274
Heptachlor epoxide	—	353	355,351
Nitrobenzene-d_5 (surr)	—	82	128,54
N-Nitrosodimethylamine	—	42	74,44
Phenol-d_6 (surr)	—	99	42,71
Terphenyl-d_{14} (surr)	—	244	122,21,
2.4.6-Tribromophenol (surr)	—	330	332,141
Toxaphene	—	159	231,233

IS   Internal standard
surr   surrogate
(a)   Estimated retention times
(b)   Substitute for the non-specific mixture, tricresyl phosphate

Method 8270C – *Table 2*

Estimated Quantitation Limits (EQL) for Semivolatile Organics[a]

Semivolatiles	CAS Number	Estimated Quantitation Limits[b]	
		Ground Water µg/L	Low Soil/ Sediment[1] µg/kg
Acenapthene	83-32-9	10	660
Acenaphthylene	208-96-8	10	660
Acetophenone	98-86-2	10	ND
2-Acetylaminofluorene	53-96-3	20	ND
1-Acetyl-2-thiourea	591-08-2	1000	ND
2-Aminoanthraquinone	117-79-3	20	ND
Aminoazobenzene	60-09-3	10	ND
4-Aminobiphenyl	92-67-1	20	ND
Anilazine	101-05-3	100	ND
o-Anisidine	90-04-0	10	ND
Anthracene	120-12-7	10	660
Aramite	140-57-8	20	ND
Azinphos-methyl	86-50-0	100	ND
Barban	101-27-9	200	ND
Benzo(a)anthracene	56-55-3	10	660
Benzo(b)fluoranthene	205-99-2	10	660
Benzo(k)fluoranthene	207-08-9	10	660
Benzoic acid	65-85-0	50	3300
Benzo(g,h,i)perylene	191-24-2	10	660
Benzo(a)pyrene	50-32-8	10	660
p-Benzoquinone	106-51-4	10	ND
Benzyl alcohol	100-51-6	20	1300
bis(2-Chloroethoxy) methane	111-91-1	10	660
bis(2-Chloroethyl) ether	111-44-4	10	660
bis(2-Chloroisopropyl) ether	108-60-1	10	660
4-Bromophenyl phenyl ether	101-55-3	10	660
Bromoxynil	1689-84-5	10	ND
Butyl benzyl phthalate	85-68-7	10	660
Captafol	2425-06-1	20	ND
Captan	133-06-2	50	ND
Carbaryl	63-25-2	10	ND
Carbofuran	1563-66-2	10	ND
Carbophenothion	786-19-6	10	ND
Chlorfenvinphos	470-90-6	20	ND
4-Chloroaniline	106-47-8	20	1300
Chlorobenzilate	510-15-6	10	ND
5-Chloro-2-methylaniline	95-79-4	10	ND
4-Chloro-3-methylphenol	59-50-7	20	1300
3-(Chloromethyl) pyridine hydrochloride	6959-48-4	100	ND

(a) EQLs listed for soil/sediment are based on wet weight. Normally data is reported on a dry weight basis, therefore, EQLs will be higher based on the % dry weight of each sample. This is based on a 30-g sample and gel permeation chromatography cleanup.

(b) Sample EQLs are highly matrix-dependent. The EQLs listed herein are provided for guidance and may not always be achievable.

ND = Not determined

Appendix A

Estimated Quantitation Limits (EQL) for Semivolatile Organics[a] (Continued)

Semivolatiles	CAS Number	Estimated Quantitation Limits[b]	
		Ground Water µg/L	Low Soil/Sediment[1] µg/kg
2-Chloronaphthalene	91-58-7	10	660
2-Chlorophenol	95-57-8	10	660
4-Chlorophenyl phenyl ether	7005-72-3	10	660
Chrysene	218-01-9	10	660
Coumaphos	56-72-4	40	ND
p-Cresidine	120-71-8	10	ND
Crotoxyphos	7700-17-6	20	ND
2-Cyclohexyl-4,6-dinitrophenol	131-89-5	100	ND
Demeton-o	298-03-3	10	ND
Demeton-s	126-75-0	10	ND
Diallate (*cis* or *trans*)	2303-16-4	10	ND
Diallate (*trans* or *cis*)	2303-16-4	10	ND
2,4-Diaminotoluene	95-80-7	20	ND
Dibenz(a,j)acridine	224-42-0	10	ND
Dibenz(a,h)anthracene	53-70-3	10	660
Dibenzofuran	132-64-9	10	660
Dibenzo(a,e)pyrene	192-65-4	10	ND
Di-n-butylphthalate	84-74-2	10	ND
Dichlone	117-80-6	NA	ND
1,2-Dichlorobenzene	95-50-1	10	660
1,3-Dichlorobenzene	541-73-1	10	660
1,4-Dichlorobenzene	106-46-7	10	660
3,3′-Dichlorobenzidine	91-94-1	20	1300
2,4-Dichlorophenol	120-83-2	10	660
2,6-Dichlorophenol	84-65-0	10	ND
Dichlorovos	62-73-7	10	ND
Dicrotophos	141-66-2	10	ND
Diethylphthalate	84-66-2	10	660
Diethylstilbesterol	56-53-1	20	ND
Diethyl sulfate	64-67-5	100	ND
Dimethoate	60-51-5	20	ND
3,3′-Dimethoxybenzidine	119-90-4	100	ND
Dimethylaminoazobenzene	60-11-7	10	ND
7,12-Dimethylbenz(a)anthracene	57-97-6	10	ND
3,3′-Dimethylbenzidine	119-93-7	10	ND
α,α-Dimethylphenethylamine	122-09-8	ND	ND
2,4-Dimethylphenol	105-67-9	10	660
Dimethyl phthalate	99-65-0	10	660
1,2-Dinitrobenzene	131-11-3	40	ND

(a) EQLs listed for soil/sediment are based on wet weight. Normally data is reported on a dry weight basis, therefore, EQLs will be higher based on the % dry weight of each sample. This is based on a 30-g sample and gel permeation chromatography cleanup.

(b) Sample EQLs are highly matrix-dependent. The EQLs listed herein are provided for guidance and may not always be achievable.

ND = Not determined
NA = Not applicable

Method 8270C – *Table 2*

Estimated Quantitation Limits (EQL) for Semivolatile Organics[a] (Continued)

Semivolatiles	CAS Number	Estimated Quantitation Limits[b]	
		Ground Water µg/L	Low Soil/ Sediment[1] µg/kg
1,3-Dinitrobenzene	528-29-0	20	ND
1,4-Dinitrobenzene	100-25-4	40	ND
4,6-Dinitro-2-methylphenol	534-52-1	50	3300
2,4-Dinitrophenol	51-28-5	50	3300
2,4-Dinitrotoluene	121-14-2	10	660
2,6-Dinitrotoluene	606-20-2	10	660
Dinocap	39300-45-3	100	ND
Dinoseb	88-85-7	20	ND
5,5-Diphenylhydantoin	57-41-0	20	ND
Di-n-octyl phthalate	117-84-0	10	660
Disulfoton	298-04-4	10	ND
EPN	2104-64-5	10	ND
Ethion	563-12-2	10	ND
Ethyl carbamate	51-79-6	50	ND
bis(2-Ethylhexyl)phthalate	117-81-7	10	660
Ethyl methanesulfonate	62-50-0	20	ND
Famphur	52-85-7	20	ND
Fensulfothion	115-90-2	40	ND
Fenthion	55-38-9	10	ND
Fluchloralin	33245-39-5	20	ND
Fluoranthene	206-44-0	10	660
Fluorene	86-73-7	10	660
Hexachlorobenzene	118-74-1	10	660
Hexachlorobutadiene	87-68-3	10	660
Hexachlorocyclopentadiene	77-47-4	10	660
Hexachloroethane	67-72-1	10	660
Hexachlorophene	70-30-4	50	ND
Hexachloropropene	1888-71-7	10	ND
Hexamethyl phosphoramide	680-31-9	20	ND
Hydroquinone	123-31-9	ND	ND
Indeno(1,2,3-cd)pyrene	193-39-5	10	660
Isodrin	465-73-6	20	ND
Isophorone	78-59-1	10	660
Isosafrole	120-58-1	10	ND
Kepone	143-50-0	20	ND
Leptophos	21609-90-5	10	ND
Malathion	121-75-5	50	ND
Maleic anhydride	108-31-6	NA	ND

(a) EQLs listed for soil/sediment are based on wet weight. Normally data is reported on a dry weight basis, therefore, EQLs will be higher based on the % dry weight of each sample. This is based on a 30-g sample and gel permeation chromatography cleanup.

(b) Sample EQLs are highly matrix-dependent. The EQLs listed herein are provided for guidance and may not always be achievable.

ND = Not determined

NA = Not applicable

Method 8270C – *Table 2*

Estimated Quantitation Limits (EQL) for Semivolatile Organics[a] (Continued)

Semivolatiles	CAS Number	Estimated Quantitation Limits[b]	
		Ground Water µg/L	Low Soil/ Sediment[1] µg/kg
Mestranol	72-33-3	20	ND
Methapyrilene	91-80-5	100	ND
Methoxychlor	72-43-5	10	ND
3-Methylcholanthrene	56-49-5	10	ND
4,4′-Methylenebis(2-chloroaniline)	101-14-4	NA	ND
Methylmethanesulfonate	66-27-3	10	ND
2-Methylnaphthalene	91-57-6	10	660
Methyl parathion	298-00-0	10	ND
2-Methylphenol	95-48-7	10	660
3-Methylphenol	108-39-4	10	ND
4-Methylphenol	106-44-5	10	660
Mevinphos	7786-34-7	10	ND
Mexacarbate	315-18-4	20	ND
Mirex	2385-85-5	10	ND
Monocrotophos	6923-22-4	40	ND
Naled	300-76-5	20	ND
Naphthalene	91-20-3	10	660
1,4-Naphthoquinone	130-15-4	10	ND
1-Naphthylamine	134-32-7	10	ND
2-Naphthylamine	91-59-8	10	ND
Nicotine	54-11-5	20	ND
5-Nitroacenaphthene	602-87-9	10	ND
2-Nitroaniline	88-74-4	50	3300
3-Nitroaniline	99-09-2	50	3300
4-Nitroaniline	100-01-6	20	ND
5-Nitro-o-anisidine	99-59-2	10	ND
Nitrobenzene	98-95-3	10	660
4-Nitrobiphenyl	92-93-3	10	ND
Nitrofen	1836-75-5	20	ND
2-Nitrophenol	88-75-5	10	660
4-Nitrophenol	100-02-7	50	3300
5-Nitro-o-toluidine	99-55-8	10	ND
4-Nitroquinoline-1-oxide	56-57-5	40	ND
N-Nitrosodibutylamine	924-16-3	10	ND
N-Nitrosodiethylamine	55-18-5	20	ND
N-Nitrosodiphenylamine	86-30-6	10	660
N-Nitroso-di-n-propylamine	621-64-7	10	660
N-Nitrosopiperidine	100-75-4	20	ND
N-Nitrosopyrrolidine	930-55-2	40	ND

(a) EQLs listed for soil/sediment are based on wet weight. Normally data is reported on a dry weight basis, therefore, EQLs will be higher based on the % dry weight of each sample. This is based on a 30-g sample and gel permeation chromatography cleanup.

(b) Sample EQLs are highly matrix-dependent. The EQLs listed herein are provided for guidance and may not always be achievable.

ND = Not determined
NA = Not applicable

Method 8270C – *Table 2*

Estimated Quantitation Limits (EQL) for Semivolatile Organics[a] (Continued)

Semivolatiles	CAS Number	Estimated Quantitation Limits[b]	
		Ground Water µg/L	Low Soil/ Sediment[1] µg/kg
Octamethyl pyrophosphoramide	152-16-9	200	ND
4,4'-Oxydianiline	101-80-4	20	ND
Parathion	56-38-2	10	ND
Pentachlorobenzene	608-93-5	10	ND
Pentachloronitrobenzene	82-68-8	20	ND
Pentachlorophenol	87-86-5	50	3300
Phenacetin	62-44-2	20	ND
Phenanthrene	85-01-8	10	660
Phenobarbital	50-06-6	10	ND
Phenol	108-95-2	10	660
1,4-Phenylenediamine	106-50-3	10	ND
Phorate	298-02-2	10	ND
Phosalone	2310-17-0	100	ND
Phosmet	732-11-6	40	ND
Phosphamidon	13171-21-6	100	ND
Phthalic anhydride	85-44-9	100	ND
2-Picoline	109-06-8	ND	ND
Piperonyl sulfoxide	120-62-7	100	ND
Pronamide	23950-58-5	10	ND
Propylthiouracil	51-52-5	100	ND
Pyrene	129-00-0	10	660
Pyridine	110-86-1	ND	ND
Resorcinol	108-46-3	100	ND
Safrole	94-59-7	10	ND
Strychnine	57-24-9	40	ND
Sulfallate	95-06-7	10	ND
Terbufos	13071-79-9	20	ND
1,2,4,5-Tetrachlorobenzene	95-94-3	10	ND
2,3,4,6-Tetrachlorophenol	58-90-2	10	ND
Tetrachlorvinphos	961-11-5	20	ND
Tetraethyl pyrophosphate	107-49-3	40	ND
Thionazine	297-97-2	20	ND
Thiophenol (Benzenethiol)	108-98-5	20	ND
Toluene diisocyanate	584-84-9	100	ND
o-Toluidine	95-53-4	10	ND
1,2,4-Trichlorobenzene	120-82-1	10	660
2,4,5-Trichlorophenol	95-95-4	10	660

(a) EQLs listed for soil/sediment are based on wet weight. Normally data is reported on a dry weight basis, therefore, EQLs will be higher based on the % dry weight of each sample. This is based on a 30-g sample and gel permeation chromatography cleanup.

(b) Sample EQLs are highly matrix-dependent. The EQLs listed herein are provided for guidance and may not always be achievable.

ND = Not determined

NA = Not applicable

Method 8270C – *Table 2*

Estimated Quantitation Limits (EQL) for Semivolatile Organics[a] (Continued)

Semivolatiles	CAS Number	Estimated Quantitation Limits[b]	
		Ground Water µg/L	Low Soil/Sediment[1] µg/kg
2,4,6-Trichlorophenol	88-06-2	10	660
Trifluralin	1582-09-8	10	ND
2,4,5-Trimethylaniline	137-17-7	10	ND
Trimethyl phosphate	512-56-1	10	ND
1,3,5-Trinitrobenzene	99-35-4	10	ND
Tris(2,3-dibromopropyl) phosphate	126-72-7	200	ND
Tri-p-tolyl phosphate(h)	78-32-0	10	ND
0,0,0-Triethylphosphorothioate	126-68-1	NT	ND

(a) EQLs listed for soil/sediment are based on wet weight. Normally data is reported on a dry weight basis, therefore, EQLs will be higher based on the % dry weight of each sample. This is based on a 30-g sample and gel permeation chromatography cleanup.

(b) Sample EQLs are highly matrix-dependent. The EQLs listed herein are provided for guidance and may not always be achievable.

ND = Not determined
NA = Not applicable
NT = Not tested

Other Matrice	Factor[1]
Medium-level soil and sludges by sonicator	7.5
Non-water miscible waste	75

(1) EQL - [EQL for Low Soil/Sediment (Table2)] X [Factor]

Method 8270C – *Table 3*

DFTPP KEY Ions and Ion Abundance Criteria[a]

Mass	Ion Abundance Criteria
51	30-60% of mass 198
68	<2% of mass 69
70	<2% of mass 69
127	40-60% of mass 198
197	<1% of mass 198
198	Base peak, 100% relative abundance
199	5-9% of mass 198
275	10-30% of mass 198
365	>1% of mass 198
441	Present but less than mass 443
442	>40% of mass 198
443	17-23% of mass 442

(a) J.W. Eichelberger, L.E. Harris, and W.L. Budde. Reference Compound to Calibrate Ion Abundance Measurement in Gas Chromatograpy-Mass Spectrometry, Analytical Chemistry, 47, 995 (1975).

Method 8270C – *Table 6*

QC Acceptance Criteria[(a)]

Parameter	Test conc. (μg/L)	Limit for s (μg/L)	Range for $\bar{x}$ (μg/L)	Range for P, P_s (%)
Acenaphthene	100	27.6	60.1-132.3	47-145
Acenaphthylene	100	40.2	53.5-126.0	33-145
Aldrin	100	39.0	7.2-152.2	D-166
Anthracene	100	32.0	43.4-118.0	27-133
Benzo(a)anthracene	100	27.6	41.8-133.0	33-143
Benzo(b)fluoranthene	100	38.8	42.0-140.4	24-159
Benzo(k)fluoranthene	100	32.3	25.2-145.7	11-162
Benzo(a)pyrene	100	39.0	31.7-148.0	17-163
Benzo(g,h,i)perylene	100	58.9	D-195.0	D-219
Benzyl butyl phthalate	100	23.4	D-139.9	D-152
β-BHC	100	31.5	41.5-130.6	24-149
δ-BHC	100	21.6	D-100.0	D-110
bis(2-Chloroethyl)ether	100	55.0	42.9-126.0	12-158
bis(2-Chloroethoxy)methane	100	34.5	49.2-164.7	33-184
bis(2-Chloroisopropyl)ether	100	46.3	62.8-138.6	36-166
bis(2-Ethylhexyl)phthalate	100	41.1	28.9-136.8	8-158
4-Bromophenyl phenyl ether	100	23.0	64.9-114.4	53-127
2-Chloronaphthalene	100	13.0	64.5-113.5	60-118
4-Chlorophenyl phenyl ether	100	33.4	38.4-144.7	25-158
Chrysene	100	48.3	44.1-139.9	17-168
4,4'-DDD	100	31.0	D-134.5	D-145
4,4'-DDE	100	32.0	19.2-119.7	4-136
4,4'-DDT	100	61.6	D-170.6	D-203
Dibenzo(a,h)anthracene	100	70.0	D-199.7	D-227
Di-n-butyl phthalate	100	16.7	8.4-111.0	1-118
1,2-Dichlorobenzene	100	30.9	48.6-112.0	32-129
1,3-Dichlorobenzene	100	41.7	16.7-153.9	D-127
1,4-Dichlorobenzene	100	32.1	37.3-105.7	20-124
3,3'-Dichlorobenzidine	100	71.4	8.2-212.5	D-262
Dieldrin	100	30.7	44.3-119.3	29-136
Diethyl phthalate	100	26.5	D-100.0	D-114
Dimethyl phthalate	100	23.2	D-100.0	D-112

s   Standard deviation of four recovery measurements, in μg/L
$\bar{x}$ Average recovery for four recovery measurements, in μg/L
P, P_s Percent recovery measured
D   Detected; result must be greater than zero.
(a)   Criteria from 40 CFR Part 136 for Method 625. These criteria are based directly on the method performance data in Table 7. Where necessary, the limits for recovery have been broadened to assure applicability of the limits to concentrations below those used to develop Table 7.

Method 8270C – *Table 6*

QC Acceptance Criteria[a] (Continued)

Parameter	Test conc. (μg/L)	Limit for s (μg/L)	Range for x̄ (μg/L)	Range for P, Ps (%)
2,4-Dinitrotoluene	100	21.8	47.5-126.9	39-139
2,6-Dinitrotoluene	100	29.6	68.1-136.7	50-158
Di-n-octylphthalate	100	31.4	18.6-131.8	4-146
Endosulfan sulfate	100	16.7	D-103.5	D-107
Endrin aldehyde	100	32.5	D-188.8	D-209
Fluoranthene	100	32.8	42.9-121.3	26-137
Fluorene	100	20.7	71.6-108.4	59-121
Heptachlor	100	37.2	D-172.2	D-192
Heptachlor epoxide	100	54.7	70.9-109.4	26-155
Hexachlorobenzene	100	24.9	7.8-141.5	D-152
Hexachlorobutadiene	100	26.3	37.8-102.2	24-116
Hexachloroethane	100	24.5	55.2-100.0	40-113
Indeno(1,2,3-cd)pyrene	100	44.6	D-150.9	D-171
Isophorone	100	63.3	46.6-180.2	21-196
Naphthalene	100	30.1	35.6-119.6	21-133
Nitrobenzene	100	39.3	54.3-157.6	35-180
N-Nitrosodi-n-propylamine	100	55.4	13.6-197.9	D-230
PCB-1260	100	54.2	19.3-121.0	D-164
Phenanthrene	100	20.6	65.2-108.7	54-120
Pyrene	100	25.2	69.6-100.0	52-115
1,2,4-Trichlorobenzene	100	28.1	57.3-129.2	44-142
4-Chloro-3-methylphenol	100	37.2	40.8-127.9	22-147
2-Chlorophenol	100	28.7	36.2-120.4	23-134
2,4-Chlorophenol	100	26.4	52.5-121.7	39-135
2,4-Dimethylphenol	100	26.1	41.8-109.0	32-119
2,4-Dinitrophenol	100	49.8	D-172.9	D-191
2-Methyl-4,6-dinitrophenol	100	93.2	53.0-100.0	D-181
2-Nitrophenol	100	35.2	45.0-166.7	29-182
4-Nitrophenol	100	47.2	13.0-106.5	D-132
Pentachlorophenol	100	48.9	38.1-151.8	14-176
Phenol	100	22.6	16.6-100.0	5-112
2,4,6-Trichlorophenol	100	31.7	52.4-129.2	37-144

s   Standard deviation of four recovery measurements, in μg/L
x̄   Average recovery for four recovery measurements, in μg/L
P, Ps Percent recovery measured
D   Detected; result must be greater than zero.
[a] Criteria from 40 CFR Part 136 for Method 625. These criteria are based directly on the method performance data in Table 7. Where necessary, the limits for recovery have been broadened to assure applicability of the limits to concentrations below those used to develop Table 7.

Appendix B

All organic analytes mentioned in any of the test methods covered in the method comparison sections in this reference are listed in alphabetical order followed by their CAS number. If the analyte is included in one of the listed test methods it is marked with a "•" or one of the following letters in the appropriate column. "S" for Surrogate compounds, "I" for Internal Standard compounds, or "B" if the compound is both a Surrogate and an Internal Standard.

Organic Analyte Cross-Reference to Methods

Key:
S = Surrogate
I = Internal Standard
B = Both Surrogate and Internal Standard

Analyte	CAS Number	SW846 - 8151A	EPA 500 - 515.1	EPA 500 - 515.2	Std. Methods 6640	SW846 - 8081A	SW846 - 8082	EPA500 - 508	EPA600 - 608	Std. Methods 6630	CLP-PEST- Organic	SW846 - 8270C	EPA500 - 525.2	EPA600 - 625	Std. Methods 6410	CLP - SVOA
		Herbicides (p. 2)				Pesticides & PCBs (p. 8)						Semivolatile Organic (p. 20)				
Acenaphthene	83-32-9											•		•	•	•
Acenaphthene d10	15067-26-2											I	I			I
Acenaphthylene	208-96-8											•	•	•	•	•
Acetone	67-64-1															
Acetonitrile	75-05-8															
Acetophenone	98-86-2											•				
2-Acetylaminofluorene	53-96-3											•				
1-Acetyl-2-thiourea	591-08-2											•				
Acifluorfen	50594-66-6	•	•	•												
Acrolein	107-02-8															
Acrylonitrile	107-13-1															
Alachlor	15972-60-8					•	•						•			
Aldrin	309-00-2					•		•	•	•	•	•		•	•	
Allyl alcohol	107-18-6															
Allyl chloride	107-05-1															
Ametryn	834-12-8												•			
2-Aminoanthraquinone	117-79-3											•				
Aminoazobenzene	60-09-3											•				
4-Aminobiphenyl	92-67-1											•				
3-Amino-9-ethylcarbazole	132-32-1											•				
Anilazine	101-05-3											•				
Aniline	62-53-3											•				
Aniline d5	4165-61-1													B	B	
o-Anisidine	90-04-0											•				
Anthracene	120-12-7											•	•	•	•	•
Anthracene d10	1719-06-8													B	B	
Aramite	140-57-8											•				
Aroclor-1016	12674-11-2						•	•	•	•	•	•	•	•	•	
Aroclor-1221	11104-28-2						•	•	•	•	•	•	•	•	•	
Aroclor-1232	11141-16-5						•	•	•	•		•	•	•	•	
Aroclor-1242	53469-21-9						•	•	•	•	•	•	•	•	•	
Aroclor-1248	12672-29-6						•	•	•	•	•	•	•	•	•	
Aroclor-1254	11097-69-1						•	•	•	•	•	•	•	•	•	
Aroclor-1260	11096-82-5						•	•	•	•	•	•	•	•	•	
Atraton	1610-17-9												•			
Atrazine	1912-24-9												•			
Azabenzene	110-86-1															
Azinphos methyl	86-50-0											•				
Barban	101-27-9											•				

Organic Analyte Cross-Reference to Methods

Volatile Organic by GC/MS (p. 36)							Volatile Organic by GC (p. 48)									Analyte
SW846 - 8260B	EPA500 - 524.2	EPA600 - 624	Std. Methods 6210B	Std. Methods 6210C	Std. Methods 6210D	CLP - VOA	EPA600 - 602	Std. Methods 6220B	Std. Methods 6220C	SW-846 - 8021B	EPA500 - 502.2	EPA600 - 601	Std. Methods 6230B	Std. Methods 6230C	Std. Methods 6230D	
																Acenaphthene
																Acenaphthene d10
																Acenaphthylene
•	•					•										Acetone
•																Acetonitrile
																Acetophenone
																2-Acetylaminofluorene
																1-Acetyl-2-thiourea
																Acifluorfen
•																Acrolein
•	•															Acrylonitrile
																Alachlor
																Aldrin
•																Allyl alcohol
•	•									•						Allyl chloride
																Ametryn
																2-Aminoanthraquinone
																Aminoazobenzene
																4-Aminobiphenyl
																3-Amino-9-ethylcarbazole
																Anilazine
																Aniline
																Aniline d5
																o-Anisidine
																Anthracene
																Anthracene d10
																Aramite
																Aroclor-1016
																Aroclor-1221
																Aroclor-1232
																Aroclor-1242
																Aroclor-1248
																Aroclor-1254
																Aroclor-1260
																Atraton
																Atrazine
																Azabenzene
																Azinphos methyl
																Barban

Organic Analyte Cross-Reference to Methods

Key:
S = Surrogate
 I = Internal Standard
B = Both Surrogate and Internal
    Standard

Analyte	CAS Number	Herbicides (p. 2) SW846 - 8151A	EPA 500 - 515.1	EPA 500 - 515.2	Std. Methods 6640	Pesticides & PCBs (p. 8) SW846 - 8081A	SW846 - 8082	EPA500 - 508	EPA600 - 608	Std. Methods 6630	CLP-PEST - Organic	Semivolatile Organic (p. 20) SW846 - 8270C	EPA500 - 525.2	EPA600 - 625	Std. Methods 6410	CLP - SVOA
Bentazon	25057-89-0	•	•	•	•											
Benzene	71-43-2															
Benzene d6	1076-43-3															
Benzidine	92-87-5											•		•	•	
Benzo(a)anthracene	56-55-3											•	•	•	•	•
Benzo(a)anthracene d12	1718-53-2													B	B	
Benzo(b)fluoranthene	205-99-2											•	•	•	•	•
Benzo(k)fluoranthene	207-08-9											•	•	•	•	•
Benzoic acid	65-85-0											•				
Benzo(g,h,i)perylene	191-24-2											•	•	•	•	•
Benzo(a)pyrene	50-32-8											•	•	•	•	•
p-Benzoquinone	106-51-4											•				
Benzyl alcohol	100-51-6											•				
Benzyl butyl phthalate	85-68-7													•	•	
Benzyl chloride	100-44-7															
alpha-BHC	319-84-6					•		•	•	•	•	•		•	•	
beta-BHC	319-85-7					•		•	•	•	•	•		•	•	
delta-BHC	319-86-8					•		•	•	•	•	•		•	•	
gamma-BHC (Lindane)	58-89-9					•		•	•	•	•	•	•	•	•	
Bromacil	314-40-9												•			
Bromoacetone	598-31-2															
Bromobenzene	108-86-1															
2-Bromochlorobenzene	694-80-4															
4-Bromochlorobenzene	106-39-8															
2-Bromo-1-chloropropane	3017-95-6															
Bromochloromethane	74-97-5															
Bromochloromethane	74-97-5															
Bromodichloromethane	75-27-4															
4-Bromofluorobenzene	460-00-4															
Bromoform	75-25-2															
Bromomethane	74-83-9															
4-Bromophenyl-phenylether	101-55-3											•		•	•	•
Bromoxynil	1689-84-5											•				
Butachlor	23184-66-9												•			
n-Butanol	71-36-3															
2-Butanone	78-93-3															
t-Butyl alcohol	75-65-0															
Butylate	2008-41-5												•			
n-Butylbenzene	104-51-8															

Organic Analyte Cross-Reference to Methods

Volatile Organic by GC/MS (p. 36)							Volatile Organic by GC (p. 48)									
SW846 - 8260B	EPA500 - 524.2	EPA600 - 624	Std. Methods 6210B	Std. Methods 6210C	Std. Methods 6210D	CLP - VOA	EPA600 - 602	Std. Methods 6220B	Std. Methods 6220C	SW-846 - 8021B	EPA500 - 502.2	EPA600 - 601	Std. Methods 6230B	Std. Methods 6230C	Std. Methods 6230D	Analyte
																Bentazon
•	•	•	•	•	•	•	•	•	•	•	•				•	Benzene
		B	B	B	B								I	I	I	Benzene d6
																Benzidine
																Benzo(a)anthracene
																Benzo(a)anthracene d12
																Benzo(b)fluoranthene
																Benzo(k)fluoranthene
																Benzoic acid
																Benzo(g,h,i)perylene
																Benzo(a)pyrene
																p-Benzoquinone
																Benzyl alcohol
																Benzyl butyl phthalate
•											•					Benzyl chloride
																alpha-BHC
																beta-BHC
																delta-BHC
																gamma-BHC (Lindane)
																Bromacil
•											•					Bromoacetone
•	•			•	•					•	•	•		•	•	Bromobenzene
									S							2-Bromochlorobenzene
																4-Bromochlorobenzene
		B	B	B	B					I	I	B	B	B	B	2-Bromo-1-chloropropane
I		B	B	B	B	I						B	B	B	B	Bromochloromethane
•	•		•	•	•						•	•		•	•	Bromochloromethane
•	•	•	•	•	•	•					•	•	•	•	•	Bromodichloromethane
S	S	B	B	B	B	S							I	I	I	4-Bromofluorobenzene
•	•	•	•	•	•	•					•	•	•	•	•	Bromoform
•	•	•	•	•	•	•					•	•	•	•	•	Bromomethane
																4-Bromophenyl-phenylether
																Bromoxynil
																Butachlor
•																n-Butanol
•	•				•											2-Butanone
•																t-Butyl alcohol
																Butylate
•	•				•					•	•	•			•	n-Butylbenzene

Appendix B

Organic Analyte Cross-Reference to Methods

Key:
S = Surrogate
I = Internal Standard
B = Both Surrogate and Internal Standard

Analyte	CAS Number	SW846 - 8151A	EPA 500 - 515.1	EPA 500 - 515.2	Std. Methods 6640	SW846 - 8081A	SW846 - 8082	EPA500 - 508	EPA600 - 608	Std. Methods 6630	CLP-PEST - Organic	SW846 - 8270C	EPA500 - 525.2	EPA600 - 625	Std. Methods 6410	CLP - SVOA
		Herbicides (p. 2)				Pesticides & PCBs (p. 8)						Semivolatile Organic (p. 20)				
sec-Butylbenzene	135-98-8															
tert-Butylbenzene	98-06-6															
Butylbenzylphthalate	85-68-7											•	•	•	•	•
2-sec-Butyl-4,6-Dinitrophenol	2425-06-1											•				
Caffeine 15N2	58-08-2												S			
Captafol	2425-06-1					•						•				
Captan	133-06-2								•			•				
Carbaryl	63-25-2											•				
Carbazole	86-74-8															•
Carbofuran	1563-66-2											•				
Carbon disulfide	75-15-0															
Carbon tetrachloride	56-23-5															
Carbophenothion	786-19-6											•				
Chloral Hydrate	302-17-0															
Chloramben	133-90-4	•	•			•										
Chlordane	57-74-9					•		•	•	•		•		•	•	
alpha-Chlordane	5103-71-9					•		•			•		•			
gamma-Chlordane	5103-74-2					•		•			•		•			
trans-Nonachlor-Chlordane	39765-80-5												•			
Chlorfenvinphos	470-90-6											•				
Chlorneb	2675-77-6					•		•					•			
Chloroacetonitrile	107-14-2															
4-Chloroaniline	106-47-8											•				•
Chlorobenzene	108-90-7															
Chlorobenzene d5	3114-55-4															
Chlorobenzilate	510-15-6					•		•				•	•			
2-Chlorobiphenyl	2051-60-7						•						•			
1-Chlorobutane	109-69-3															
2-Chloro-1,3-butadiene	126-99-8															
Chloroethane	75-00-3															
2-Chloroethanol	107-07-3															
Chloroethene	75-01-4															
bis(2-Chloroethoxy) methane	111-91-1											•		•	•	•
bis(2-Chloroethyl)ether	111-44-4											•		•	•	•
bis-(2-chloroethyl) sulfide	505-60-2															
2-Chloroethylvinyl ether	110-75-8															
1-Chloro-2-fluorobenzene	348-51-6															
Chloroform	67-66-3															
1-Chlorohexane	544-10-5															

Organic Analyte Cross-Reference to Methods

SW846 - 8260B	EPA500 - 524.2	EPA600 - 624	Std. Methods 6210B	Std. Methods 6210C	Std. Methods 6210D	CLP - VOA	EPA600 - 602	Std. Methods 6220B	Std. Methods 6220C	SW-846 - 8021B	EPA500 - 502.2	EPA600 - 601	Std. Methods 6230B	Std. Methods 6230C	Std. Methods 6230D	Analyte
•	•			•	•			•	•	•					•	sec-Butylbenzene
•	•			•	•			•	•	•					•	tert-Butylbenzene
																Butylbenzylphthalate
																2-sec-Butyl-4,6-Dinitrophenol
																Caffeine 15N2
																Captafol
																Captan
																Carbaryl
																Carbazole
																Carbofuran
•	•					•										Carbon disulfide
•	•	•	•	•	•	•	•		•	•	•	•	•	•	•	Carbon tetrachloride
																Carbophenothion
•																Chloral Hydrate
																Chloramben
																Chlordane
																alpha-Chlordane
																gamma-Chlordane
																trans-Nonachlor-Chlordane
																Chlorfenvinphos
																Chlorneb
•	•															Chloroacetonitrile
																4-Chloroaniline
•	•	•	•	•	•	•	•	•	•	•	•	•	•	•	•	Chlorobenzene
I						I										Chlorobenzene d5
																Chlorobenzilate
																2-Chlorobiphenyl
•	•															1-Chlorobutane
•																2-Chloro-1,3-butadiene
•	•	•	•	•	•	•				•	•	•	•	•	•	Chloroethane
•											•					2-Chloroethanol
•	•	•	•	•	•	•				•	•	•	•	•	•	Chloroethene
																bis(2-Chloroethoxy) methane
																bis(2-Chloroethyl)ether
•																bis-(2-chloroethyl) sulfide
•			•	•							•	•				2-Chloroethylvinyl ether
											I					1-Chloro-2-fluorobenzene
•	•	•	•	•	•	•	•		•	•	•	•	•	•	•	Chloroform
•																1-Chlorohexane

Organic Analyte Cross-Reference to Methods

Key:
S = Surrogate
I = Internal Standard
B = Both Surrogate and Internal Standard

Analyte	CAS Number	SW846-8151A	EPA 500-515.1	EPA 500-515.2	Std. Methods 6640	SW846-8081A	SW846-8082	EPA500-508	EPA600-608	Std. Methods 6630	CLP-PEST-Organic	SW846-8270C	EPA500-525.2	EPA600-625	Std. Methods 6410	CLP-SVOA
		Herbicides (p. 2)				**Pesticides & PCBs (p. 8)**						**Semivolatile Organic (p. 20)**				
bis(2-Chloroisopropyl)ether	108-60-1											•		•	•	•
Chloromethane	74-87-3															
Chloromethyl Methyl Ether	107-30-2															
5-Chloro-2-methylaniline	95-79-4											•				
4-Chloro-3-methylphenol	59-50-7											•		•	•	•
3-(Chloromethyl) pyridine hydrochloride	6959-48-4											•				
1-Chloronaphthalene	90-13-1											•				
2-Chloronaphthalene	91-58-7											•		•	•	•
2-Chlorophenol	95-57-8											•		•	•	•
2-Chlorophenol d4	95-57-8															S
4-Chloro-1,2-phenylenediamine	95-83-0											•				
4-Chloro-1,3-phenylenediamine	5131-60-2											•				
4-Chlorophenyl-phenylether	7005-72-3											•		•	•	•
Chloroprene	126-99-8															
3-Chloropropionitrile	542-76-7															
3-Chloropropene	107-05-1															
Chlorpropham	101-21-3												•			
Chlorpyrifos	2921-88-2												•			
Chloropropylate	99516-95-7					•										
Chlorothalonil	1897-45-6					•	•						•			
alpha-Chlorotoluene	100-44-7															
ortho-Chlorotoluene	95-49-8															
para-Chlorotoluene	106-43-4															
2-Chlorotoluene	95-49-8															
4-Chlorotoluene	106-43-4															
Chrysene	218-01-9											•	•	•	•	•
Chrysene d12	1719-03-5											I	I			I
Coumaphos	56-72-4											•				
p-Cresidine	120-71-8											•				
m-Cresol	108-39-4											•				
o-Cresol	95-48-7											•				•
p-Cresol	106-44-5											•				•
Crotonaldehyde	4170-30-3															
Crotoxyphos	7700-17-6											•				
Cumene	98-82-8															
Cyanazine	21725-46-2												•			
Cycloate	1134-23-2												•			
2-Cyclohexyl-4,6-dinitrophenol	131-89-5											•				
2,4-D	94-75-7	•	•	•	•											

Appendix B

Organic Analyte Cross-Reference to Methods

Volatile Organic by GC/MS (p. 36)							Volatile Organic by GC (p. 48)									Analyte
SW846 - 8260B	EPA500 - 524.2	EPA600 - 624	Std. Methods 6210B	Std. Methods 6210C	Std. Methods 6210D	CLP - VOA	EPA600 - 602	Std. Methods 6220B	Std. Methods 6220C	SW-846 - 8021B	EPA500 - 502.2	EPA600 - 601	Std. Methods 6230B	Std. Methods 6230C	Std. Methods 6230D	
																bis(2-Chloroisopropyl)ether
•	•	•	•	•	•	•			•	•	•	•	•	•	•	Chloromethane
																Chloromethyl Methyl Ether
																5-Chloro-2-methylaniline
																4-Chloro-3-methylphenol
																3-(Chloromethyl) pyridine hydrochloride
																1-Chloronaphthalene
																2-Chloronaphthalene
																2-Chlorophenol
																2-Chlorophenol d4
																4-Chloro-1,2-phenylenediamine
																4-Chloro-1,3-phenylenediamine
																4-Chlorophenyl-phenylether
•										•						Chloroprene
•																3-Chloropropionitrile
•																3-Chloropropene
																Chlorpropham
																Chlorpyrifos
																Chloropropylate
																Chlorothalonil
•																alpha-Chlorotoluene
•	•			•	•			•	•	•				•	•	ortho-Chlorotoluene
•	•			•	•			•	•	•				•	•	para-Chlorotoluene
•	•			•	•			•	•	•				•	•	2-Chlorotoluene
•	•			•	•			•	•	•				•	•	4-Chlorotoluene
																Chrysene
																Chrysene d12
																Coumaphos
																p-Cresidine
																m-Cresol
																o-Cresol
																p-Cresol
•																Crotonaldehyde
																Crotoxyphos
•	•			•	•		•	•	•						•	Cumene
																Cyanazine
																Cycloate
																2-Cyclohexyl-4,6-dinitrophenol
																2,4-D

Organic Analyte Cross-Reference to Methods

Key:
S = Surrogate
I = Internal Standard
B = Both Surrogate and Internal Standard

Analyte	CAS Number	Herbicides (p. 2) SW846 - 8151A	EPA 500 - 515.1	EPA 500 - 515.2	Std. Methods 6640	Pesticides & PCBs (p. 8) SW846 - 8081A	SW846 - 8082	EPA500 - 508	EPA600 - 608	Std. Methods 6630	CLP-PEST - Organic	Semivolatile Organic (p. 20) SW846 - 8270C	EPA500 - 525.2	EPA600 - 625	Std. Methods 6410	CLP - SVOA
Dacthal	1861-32-1				•								•			
Dalapon	75-99-0	•	•		•											
2,4-DB	94-82-6	•	•	•												
DBCP	96-12-8					•										
DCPA	1861-32-1					•		•								
DCPA acid metabolites	N/A	•	•													
4,4'-DDD	72-54-8					•		•	•	•	•	•	•	•	•	
4,4'-DDE	72-55-9					•		•	•	•	•	•	•	•	•	
4,4'-DDT	50-29-3					•		•	•	•	•	•	•	•	•	
Decachlorobiphenyl	2051-27-3					S	S				S					
Decafluorobiphenyl	434-90-2												B	B		
Demeton-o	298-03-3											•				
Demeton-s	126-75-0											•				
Diallate (trans or cis)	2303-16-4					•						•				
2,4-Diaminotoluene	95-80-7											•				
Diazinon	333-41-5												•			
Dibenz(a,h)anthracene	53-70-3											•	•	•	•	•
Dibenz(a,j)acridine	224-42-0											•				
Dibenzofuran	132-64-9											•				•
Dibenzo(a,e)pyrene	192-65-4											•				
4,4'-Dibromobiphenyl	92-86-4												B	B		
Dibromochloromethane	124-48-1															
1,2-Dibromo-3-chloropropane	96-12-8											•				
Dibromoflouromethane	1868-53-7															
4,4'-Dibromooctafluorobiphenyl	10386-84-2	I	I	I									B	B		
1,2-Dibromoethane	106-93-4															
Dibromomethane	74-95-3															
Di-n-butylphthalate	84-74-2											•	•	•	•	•
Dicamba	1918-00-9	•	•	•	•											
Dichlone	117-80-6					•						•				
Dichloran	N/A								•							
cis-1,4-Dichloro-2-butene	1476-11-5															
trans-1,4-Dichloro-2-butene	110-57-6															
1,4-Dichloro-2-butene	764-41-0															
1,2-Dichlorobenzene d4	2199-69-1															S
1,2-Dichlorobenzene	95-50-1											•		•	•	•
1,3-Dichlorobenzene	541-73-1											•		•	•	•
1,4-Dichlorobenzene	106-46-7											•		•	•	•
1,4-Dichlorobenzene d4	3855-82-1											I				I

Organic Analyte Cross-Reference to Methods

SW846 - 8260B	EPA500 - 524.2	EPA600 - 624	Std. Methods 6210B	Std. Methods 6210C	Std. Methods 6210D	CLP - VOA	EPA600 - 602	Std. Methods 6220B	Std. Methods 6220C	SW-846 - 8021B	EPA500 - 502.2	EPA600 - 601	Std. Methods 6230B	Std. Methods 6230C	Std. Methods 6230D	Analyte
																Dacthal
																Dalapon
																2,4-DB
																DBCP
																DCPA
																DCPA acid metabolites
																4,4'-DDD
																4,4'-DDE
																4,4'-DDT
																Decachlorobiphenyl
																Decafluorobiphenyl
																Demeton-o
																Demeton-s
																Diallate (trans or cis)
																2,4-Diaminotoluene
																Diazinon
																Dibenz(a,h)anthracene
																Dibenz(a,j)acridine
																Dibenzofuran
																Dibenzo(a,e)pyrene
																4,4'-Dibromobiphenyl
•	•	•	•	•	•	•				•	•	•	•	•	•	Dibromochloromethane
•	•		•	•						•	•				•	1,2-Dibromo-3-chloropropane
S																Dibromoflouromethane
																4,4'-Dibromooctafluorobiphenyl
•	•		•	•						•	•			•	•	1,2-Dibromoethane
•	•		•	•						•	•			•	•	Dibromomethane
																Di-n-butylphthalate
																Dicamba
																Dichlone
																Dichloran
•																cis-1,4-Dichloro-2-butene
•	•															trans-1,4-Dichloro-2-butene
																1,4-Dichloro-2-butene
S	S															1,2-Dichlorobenzene d4
•	•	•	•	•	•		•	•	•	•	•	•	•	•	•	1,2-Dichlorobenzene
•	•	•	•	•	•		•	•	•	•	•	•	•	•	•	1,3-Dichlorobenzene
•	•	•	•	•	•		•	•	•	•	•	•	•	•	•	1,4-Dichlorobenzene
I																1,4-Dichlorobenzene d4

Organic Analyte Cross-Reference to Methods

Key:
S = Surrogate
I = Internal Standard
B = Both Surrogate and Internal Standard

Analyte	CAS Number	SW846 - 8151A	EPA 500 - 515.1	EPA 500 - 515.2	Std. Methods 6640	SW846 - 8081A	SW846 - 8082	EPA500 - 508	EPA600 - 608	Std. Methods 6630	CLP-PEST - Organic	SW846 - 8270C	EPA500 - 525.2	EPA600 - 625	Std. Methods 6410	CLP - SVOA
		Herbicides (p. 2)				Pesticides & PCBs (p. 8)						Semivolatile Organic (p. 20)				
3,3'-Dichlorobenzidine	91-94-1											•		•	•	•
3,5-Dichlorobenzoic acid	51-36-5	•	•	•												
2,3-Dichlorobiphenyl	16605-91-7												•			
4,4'-Dichlorobiphenyl	2050-68-2					S										
1,4-Dichlorobutane	110-56-5															
Dichlorodifluoromethane	75-71-8															
1,1-Dichloroethane	75-34-3															
1,2-Dichloroethane	107-06-2															
1,2-Dichloroethane d4	17060-07-0															
1,1-Dichloroethene	75-35-4															
cis-1,2-Dichloroethene	156-59-4															
trans-1,2-Dichloroethene	156-60-5															
1,2-Dichloroethene (total)	540-59-0															
1,1-Dichloroethylene	75-35-4															
Dichloromethane	75-09-2															
2,4-Dichlorophenol	120-83-2											•		•	•	•
2,6-Dichlorophenol	87-65-0											•				
2,4-Dichlorophenylacetic acid		S	S	S												
Dichloroprop	120-36-5	•	•	•												
1,2-Dichloropropane	78-87-5															
1,3-Dichloropropane	142-28-9															
2,2-Dichloropropane	590-20-7															
1,3-Dichloro-2-propanol	96-23-1															
1,1-Dichloropropanone	513-88-2															
1,1-Dichloropropene	563-58-6															
cis-1,3-Dichloropropene	10061-01-5															
trans-1,3-Dichloropropene	10061-02-6															
Dichlorovos	62-73-7											•	•			
Dicofol	115-32-2					•										
Dicrotophos	141-66-2											•				
Dieldrin	60-57-1					•		•	•	•	•	•	•	•	•	
1,2,3,4-Diepoxybutane	298-18-0															
Diethyl ether	60-29-7															
Diethyl sulfate	64-67-5											•				
Di(2-ethylhexyl)adipate	103-23-1												•			
Diethylphthalate	84-66-2											•	•	•	•	•
Diethylstilbesterol	56-53-1											•				
1,4-Difluorobenzene	540-36-3															
2,2'-Difluorobiphenyl	388-82-9													B	B	

Organic Analyte Cross-Reference to Methods

Column groups: **Volatile Organic by GC/MS (p. 36)** = SW846-8260B, EPA500-524.2, EPA600-624, Std. Methods 6210B, Std. Methods 6210C, Std. Methods 6210D, CLP-VOA. **Volatile Organic by GC (p. 48)** = EPA600-602, Std. Methods 6220B, Std. Methods 6220C, SW-846-8021B, EPA500-502.2, EPA600-601, Std. Methods 6230B, Std. Methods 6230C, Std. Methods 6230D.

SW846 - 8260B	EPA500 - 524.2	EPA600 - 624	Std. Methods 6210B	Std. Methods 6210C	Std. Methods 6210D	CLP - VOA	EPA600 - 602	Std. Methods 6220B	Std. Methods 6220C	SW-846 - 8021B	EPA500 - 502.2	EPA600 - 601	Std. Methods 6230B	Std. Methods 6230C	Std. Methods 6230D	Analyte
																3,3'-Dichlorobenzidine
																3,5-Dichlorobenzoic acid
																2,3-Dichlorobiphenyl
																4,4'-Dichlorobiphenyl
		B	B	B	B					S		B	B	B	B	1,4-Dichlorobutane
•	•			•	•					•	•	•	•	•	•	Dichlorodifluoromethane
•	•	•	•	•	•	•				•	•	•	•	•	•	1,1-Dichloroethane
•	•	•	•	•	•	•				•	•	•	•	•	•	1,2-Dichloroethane
S		B	B	B	B	S				I						1,2-Dichloroethane d4
•	•	•	•	•	•	•				•	•	•	•	•	•	1,1-Dichloroethene
•	•			•	•					•	•			•	•	cis-1,2-Dichloroethene
•	•	•	•	•	•					•	•	•	•	•	•	trans-1,2-Dichloroethene
						•										1,2-Dichloroethene (total)
•	•	•	•	•	•	•				•	•	•	•	•	•	1,1-Dichloroethylene
•	•	•	•	•	•	•				•	•	•	•	•	•	Dichloromethane
										•	•			•	•	2,4-Dichlorophenol
																2,6-Dichlorophenol
																2,4-Dichlorophenylacetic acid
																Dichloroprop
•	•	•	•	•	•	•				•	•	•	•	•	•	1,2-Dichloropropane
•	•			•	•					•	•			•	•	1,3-Dichloropropane
•	•			•	•					•	•			•	•	2,2-Dichloropropane
•										•						1,3-Dichloro-2-propanol
	•															1,1-Dichloropropanone
•	•			•	•					•	•			•	•	1,1-Dichloropropene
•	•	•	•			•				•	•	•	•			cis-1,3-Dichloropropene
•	•	•	•			•				•	•	•	•			trans-1,3-Dichloropropene
																Dichlorovos
																Dicofol
																Dicrotophos
																Dieldrin
•																1,2,3,4-Diepoxybutane
•	•															Diethyl ether
																Diethyl sulfate
																Di(2-ethylhexyl)adipate
																Diethylphthalate
																Diethylstilbesterol
I		B	B	B	B	I				I						1,4-Difluorobenzene
																2,2'-Difluorobiphenyl

Organic Analyte Cross-Reference to Methods

Key:
S = Surrogate
I = Internal Standard
B = Both Surrogate and Internal Standard

Analyte	CAS Number	SW846 - 8151A	EPA 500 - 515.1	EPA 500 - 515.2	Std. Methods 6640	SW846 - 8081A	SW846 - 8082	EPA500 - 508	EPA600 - 608	Std. Methods 6630	CLP-PEST - Organic	SW846 - 8270C	EPA500 - 525.2	EPA600 - 625	Std. Methods 6410	CLP - SVOA
		Herbicides (p. 2)				Pesticides & PCBs (p. 8)						Semivolatile Organic (p. 20)				
Dihydrosaffrole	56312-13-1											•				
Dimethoate	60-51-5											•				
3,3'-Dimethoxybenzidine	119-90-4											•				
p-Dimethylaminoazobenzene	60-11-7											•				
7,12-Dimethylbenz[A]anthracene	57-97-6											•				
1,2-Dimethylbenzene	95-47-6															
1,3-Dimethylbenzene	108-38-3															
1,4-Dimethylbenzene	106-42-3															
3,3'-Dimethylbenzidine	119-93-7											•				
2,4-Dimethylphenol	105-67-9											•		•	•	•
alpha, alpha-Dimethylphenethylamine	122-09-8											•				
Dimethylphthalate	131-11-3											•	•	•	•	•
Di-n-butylphthalate	84-74-2											•	•	•	•	•
1,2-Dinitrobenzene	528-29-0											•				
1,3-Dinitrobenzene	99-65-0											•				
1,4-Dinitrobenzene	100-25-4											•				
4,6-Dinitro-2-methylphenol	121-14-2											•				•
2,4-Dinitrophenol	51-28-5													•	•	•
2,4-Dinitrotoluene	121-14-2											•	•	•	•	•
2,6-Dinitrotoluene	606-20-2											•	•	•	•	•
Dinocap	6119-92-2											•				
Dinoseb	88-85-7	•	•	•	•							•				
Di-n-octylphthalate	117-84-0											•		•	•	•
1,4 Dioxane	123-91-1															
Dioxathion	78-34-2											•				
Diphenamid	957-51-7												•			
Diphenylamine	122-39-4											•				
5,5-Diphenylhydantoin	57-41-0											•				
1,2 Diphenylhydrazine	122-66-7											•				
Disulfoton	298-04-4											•	•			
Disulfoton sulfoxide	2497-07-6												•			
Disulfoton sulfone	2497-06-5												•			
Endosulfan I	959-98-8					•		•	•	•	•	•	•	•	•	
Endosulfan II	33213-65-9					•		•	•	•	•	•	•	•	•	
Endosulfan sulfate	1031-07-8					•		•	•	•	•	•		•	•	
Endrin	72-20-8					•		•	•	•	•	•		•	•	
Endrin aldehyde	7421-93-4					•		•	•	•	•	•		•	•	
Endrin ketone	53494-70-5					•					•	•				
Epichlorohydrin	106-89-8															

Organic Analyte Cross-Reference to Methods

Volatile Organic by GC/MS (p. 36)							Volatile Organic by GC (p. 48)									Analyte
SW846 - 8260B	EPA500 -524.2	EPA600 - 624	Std. Methods 6210B	Std. Methods 6210C	Std. Methods 6210D	CLP - VOA	EPA600 - 602	Std. Methods 6220B	Std. Methods 6220C	SW-846 - 8021B	EPA500 - 502.2	EPA600 - 601	Std. Methods 6230B	Std. Methods 6230C	Std. Methods 6230D	
																Dihydrosaffrole
																Dimethoate
																3,3'-Dimethoxybenzidine
																p-Dimethylaminoazobenzene
																7,12-Dimethylbenz[A]anthracene
•	•		•	•	•			•	•	•					•	1,2-Dimethylbenzene
•	•		•	•	•			•	•	•					•	1,3-Dimethylbenzene
•	•		•	•	•			•	•	•					•	1,4-Dimethylbenzene
																3,3'-Dimethylbenzidine
																2,4-Dimethylphenol
																alpha, alpha-Dimethylphenethylamine
																Dimethylphthalate
																Di-n-butylphthalate
																1,2-Dinitrobenzene
																1,3-Dinitrobenzene
																1,4-Dinitrobenzene
																4,6-Dinitro-2-methylphenol
																2,4-Dinitrophenol
																2,4-Dinitrotoluene
																2,6-Dinitrotoluene
																Dinocap
																Dinoseb
																Di-n-octylphthalate
•																1,4 Dioxane
																Dioxathion
																Diphenamid
																Diphenylamine
																5,5-Diphenylhydantoin
																1,2 Diphenylhydrazine
																Disulfoton
																Disulfoton sulfoxide
																Disulfoton sulfone
																Endosulfan I
																Endosulfan II
																Endosulfan sulfate
																Endrin
																Endrin aldehyde
																Endrin ketone
•										•						Epichlorohydrin

Organic Analyte Cross-Reference to Methods

Key:
S = Surrogate
I = Internal Standard
B = Both Surrogate and Internal Standard

Analyte	CAS Number	SW846 - 8151A	EPA 500 - 515.1	EPA 500 - 515.2	Std. Methods 6640	SW846 - 8081A	SW846 - 8082	EPA500 - 508	EPA600 - 608	Std. Methods 6630	CLP-PEST - Organic	SW846 - 8270C	EPA500 - 525.2	EPA600 - 625	Std. Methods 6410	CLP - SVOA
		Herbicides (p. 2)				Pesticides & PCBs (p. 8)						Semivolatile Organic (p. 20)				
EPN	2104-64-5											•				
EPTC	759-94-4												•			
Ethanenitrile	75-05-8															
Ethanol	64-17-5															
Ethenylbenzene	100-42-5															
Ethion	563-12-2											•				
Ethoprop	13194-48-4												•			
Ethyl Acetate	141-78-6															
Ethyl Alcohol	64-17-5															
Ethyl benzene	100-41-4															
Ethylbenzene d5	20302-26-5															
Ethylbenzene d10	25837-05-2															
Ethyl carbamate	51-79-6											•				
Ethylene dibromide	106-93-4															
Ethylene oxide	75-21-8															
bis(2-Ethylhexyl)phthalate	117-81-7											•	•	•	•	•
Ethyl methacrylate	97-63-2															
Ethyl methanesulfonate	62-50-0											•				
Ethyl Parathion	56-38-2											•				
Etridiazole	2593-15-9					•		•					•			
Famphur	52-85-7											•				
Fensulfothion	115-90-2											•				
Fenamiphos	22224-92-6												•			
Fenarimol	60168-88-9												•			
Fenthion	55-38-9											•				
Fluchloralin	33245-39-5											•				
Fluoranthene	206-44-0											•		•	•	•
Fluorene	86-73-7											•	•	•	•	•
4-Fluoroaniline	371-40-4												B	B		
Fluorobenzene	462-06-6															
2-Fluorobiphenyl	321-60-8											S				S
1-Fluoronaphthalene	321-38-0												B	B		
2-Fluoronaphthalene	323-09-1												B	B		
2-Fluorophenol	367-12-4											S		B	B	S
Fluridone	59756-60-4												•			
Hallowax-1000	58718-67-5						•									
Halowax-1001	58718-67-5						•									
Halowax-1013	12616-35-2						•									
Halowax-1014	12616-36-3						•									

Appendix B

Organic Analyte Cross-Reference to Methods

SW846 - 8260B	EPA500 - 524.2	EPA600 - 624	Std. Methods 6210B	Std. Methods 6210C	Std. Methods 6210D	CLP - VOA	EPA600 - 602	Std. Methods 6220B	Std. Methods 6220C	SW-846 - 8021B	EPA500 - 502.2	EPA600 - 601	Std. Methods 6230B	Std. Methods 6230C	Std. Methods 6230D	Analyte
																EPN
																EPTC
•																Ethanenitrile
•																Ethanol
•	•		•	•	•			•	•	•					•	Ethenylbenzene
																Ethion
																Ethoprop
•																Ethyl Acetate
•																Ethyl Alcohol
•	•	•	•	•	•	•	•	•	•	•	•	•			•	Ethyl benzene
			B	B	B	B							I			Ethylbenzene d5
			B	B	B								I			Ethylbenzene d10
																Ethyl carbamate
•	•		•	•						•	•			•	•	Ethylene dibromide
•																Ethylene oxide
																bis(2-Ethylhexyl)phthalate
•	•															Ethyl methacrylate
																Ethyl methanesulfonate
																Ethyl Parathion
																Etridiazole
																Famphur
																Fensulfothion
																Fenamiphos
																Fenarimol
																Fenthion
																Fluchloralin
																Fluoranthene
																Fluorene
																4-Fluoroaniline
I	I		B	B	B	B				I	I		I	I	I	Fluorobenzene
																2-Fluorobiphenyl
																1-Fluoronaphthalene
																2-Fluoronaphthalene
																2-Fluorophenol
																Fluridone
																Hallowax-1000
																Halowax-1001
																Halowax-1013
																Halowax-1014

Appendix B

Organic Analyte Cross-Reference to Methods

Key:
S = Surrogate
I = Internal Standard
B = Both Surrogate and Internal Standard

Analyte	CAS Number	SW846 - 8151A	EPA 500 - 515.1	EPA 500 - 515.2	Std. Methods 6640	SW846 - 8081A	SW846 - 8082	EPA500 - 508	EPA600 - 608	Std. Methods 6630	CLP-PEST - Organic	SW846 - 8270C	EPA500 - 525.2	EPA600 - 625	Std. Methods 6410	CLP - SVOA
		Herbicides (p. 2)				Pesticides & PCBs (p. 8)						Semivolatile Organic (p. 20)				
Halowax-1051	2234-13-1					•										
Halowax-1099	39450-05-0					•										
Heptachlor	76-44-8					•		•	•	•	•	•	•	•	•	
Heptachlor epoxide	1024-57-3					•		•	•	•	•	•	•	•	•	
2,2',3,3',4,4',5'-Heptachlorobiphenyl	35065-30-6						•									
2,2',3,3',4,4',6-Heptachlorobiphenyl	52663-71-5												•			
2,2',3,4,4',5,5'-Heptachlorobiphenyl	35065-29-3						•									
2,2',3,4,4',5',6-Heptachlorobiphenyl	52663-69-1						•									
2,2',3,4',5,5',6-Heptachlorobiphenyl	52663-68-0						•									
Hexachlorobenzene	118-74-1					•		•				•	•	•	•	•
2,2',3,4,4',5'-Hexachlorobiphenyl	35065-28-2						•									
2,2',3,4,5,5'-Hexachlorobiphenyl	52712-04-6						•									
2,2',3,5,5',6-Hexachlorobiphenyl	52663-63-5						•									
2,2',4,4',5,5'-Hexachlorobiphenyl	35065-27-1						•									
2,2',4,4',5,6'-Hexachlorobiphenyl	60145-22-4												•			
Hexachlorobutadiene	87-68-3											•		•	•	•
Hexachlorocyclohexane, alpha	319-84-6												•			
Hexachlorocyclohexane, beta	319-85-7												•			
Hexachlorocyclohexane, delta	319-86-8												•			
Hexachlorocyclopentadiene	77-47-4					•						•	•	•	•	•
Hexachloroethane	67-72-1											•		•	•	•
Hexachlorophene	70-30-4											•				
Hexachloropropene	1888-71-7											•				
Hexamethyl phosphoramide	680-31-9											•				
2-Hexanone	591-78-6															
Hexazinone	51235-04-2												•			
Hydroquinone	123-31-9											•				
5-Hydroxydicamba	7600-50-2	•	•	•												
2-Hydroxypropionitrile	78-97-7															
Indeno(1,2,3-cd)pyrene	193-39-5											•	•	•	•	•
Iodomethane	74-88-4															
Isobutanol	78-83-1															
Isobutyl Alcohol	78-83-1															
Isodrin	465-73-6					•						•				
Isophorone	78-59-1											•	•	•	•	•
Isopropylbenzene	98-82-8															
4-Isopropyltoluene	99-87-6															
p-Isopropyltoluene	99-87-6															
Isosafrole	120-58-1											•				

Appendix B

Organic Analyte Cross-Reference to Methods

Volatile Organic by GC/MS (p. 36)							Volatile Organic by GC (p. 48)									Analyte
SW846 - 8260B	EPA500 - 524.2	EPA600 - 624	Std. Methods 6210B	Std. Methods 6210C	Std. Methods 6210D	CLP - VOA	EPA600 - 602	Std. Methods 6220B	Std. Methods 6220C	SW-846 - 8021B	EPA500 - 502.2	EPA600 - 601	Std. Methods 6230B	Std. Methods 6230C	Std. Methods 6230D	
																Halowax-1051
																Halowax-1099
																Heptachlor
																Heptachlor epoxide
																2,2',3,3',4,4',5'-Heptachlorobiphenyl
																2,2',3,3',4,4',6-Heptachlorobiphenyl
																2,2',3,4,4',5,5'-Heptachlorobiphenyl
																2,2',3,4,4',5',6-Heptachlorobiphenyl
																2,2',3,4',5,5',6-Heptachlorobiphenyl
																Hexachlorobenzene
																2,2',3,4,4',5'-Hexachlorobiphenyl
																2,2',3,4,5,5'-Hexachlorobiphenyl
																2,2',3,5,5',6-Hexachlorobiphenyl
																2,2',4,4',5,5'-Hexachlorobiphenyl
																2,2',4,4',5,6'-Hexachlorobiphenyl
•	•		•	•				•	•	•					•	Hexachlorobutadiene
																Hexachlorocyclohexane, alpha
																Hexachlorocyclohexane, beta
																Hexachlorocyclohexane, delta
																Hexachlorocyclopentadiene
•	•															Hexachloroethane
																Hexachlorophene
																Hexachloropropene
																Hexamethyl phosphoramide
•	•				•											2-Hexanone
																Hexazinone
																Hydroquinone
																5-Hydroxydicamba
•																2-Hydroxypropionitrile
																Indeno(1,2,3-cd)pyrene
	•															Iodomethane
•																Isobutanol
•																Isobutyl Alcohol
																Isodrin
																Isophorone
•	•		•	•				•	•	•					•	Isopropylbenzene
•	•				•				•	•	•					•	4-Isopropyltoluene
•	•				•				•	•	•					•	p-Isopropyltoluene
																Isosafrole

Appendix B

Organic Analyte Cross-Reference to Methods

Key:
S = Surrogate
I = Internal Standard
B = Both Surrogate and Internal Standard

Analyte	CAS Number	SW846 - 8151A	EPA 500 - 515.1	EPA 500 - 515.2	Std. Methods 6640	SW846 - 8081A	SW846 - 8082	EPA500 - 508	EPA600 - 608	Std. Methods 6630	CLP-PEST- Organic	SW846 - 8270C	EPA500 - 525.2	EPA600 - 625	Std. Methods 6410	CLP - SVOA
		Herbicides (p. 2)				Pesticides & PCBs (p. 8)						Semivolatile Organic (p. 20)				
Kepone	143-50-0											•				
Leptophos	21609-90-5											•				
Lindane	58-89-9												•			
Malathion	121-75-5										•	•				
Maleic anhydride	108-31-6											•				
Malononitrile	109-77-3															
MCPA	94-74-6	•														
MCPP	93-65-2	•														
Merphos	150-50-5												•			
Mestranol	72-33-3											•				
Methacrylonitrile	126-98-7															
Methanol	67-56-1															
Methapyrilene	91-80-5											•				
Methoxychlor	72-43-5					•		•		•	•	•	•			
Methylacrylate	18358-13-9															
Methylbenzene	108-88-3															
Methyl Butyl Ketone	591-78-6															
3-Methylcholanthrene	56-49-5											•				
2-Methyl-4,6-dinitrophenol	534-52-1											•		•	•	
Methyl ethyl ketone	78-93-3															
4,4'-Methylenebis(2-chloroaniline)	101-14-4											•				
4,4'-Methylenebis(N,N-dimethylaniline)	101-61-1											•				
Methylene chloride	75-09-2															
Methyl iodide	74-88-4															
Methyl isobutyl ketone	108-10-1															
Methyl methacrylate	80-62-6															
Methylmethanesulfonate	66-27-3											•				
2-Methylnaphthalene	91-57-6											•				•
2-Methyl-5-Nitroaniline	99-55-8											•				
Methyl paraoxon	950-35-6												•			
Methyl parathion	298-00-0									•		•				
4-Methyl-2-pentanone	108-10-1															
2-Methylphenol	95-48-7											•				•
3-Methylphenol	108-39-4											•				
4-Methylphenol	106-44-5											•				•
2-Methyl-2-Propene Nitrile	126-98-7															
2-Methylpyridine	109-06-8											•				
Methyl-t-butyl ether	1634-04-4															
Metolachlor	51218-45-2												•			

Organic Analyte Cross-Reference to Methods

Volatile Organic by GC/MS (p. 36)							Volatile Organic by GC (p. 48)									Analyte
SW846 - 8260B	EPA500 - 524.2	EPA600 - 624	Std. Methods 6210B	Std. Methods 6210C	Std. Methods 6210D	CLP - VOA	EPA600 - 602	Std. Methods 6220B	Std. Methods 6220C	SW-846 - 8021B	EPA500 - 502.2	EPA600 - 601	Std. Methods 6230B	Std. Methods 6230C	Std. Methods 6230D	
																Kepone
																Leptophos
																Lindane
																Malathion
																Maleic anhydride
•																Malononitrile
																MCPA
																MCPP
																Merphos
																Mestranol
•	•															Methacrylonitrile
•																Methanol
																Methapyrilene
																Methoxychlor
•	•															Methylacrylate
•	•	•	•	•	•	•	•	•	•	•	•				•	Methylbenzene
						•										Methyl Butyl Ketone
																3-Methylcholanthrene
																2-Methyl-4,6-dinitrophenol
•	•					•										Methyl ethyl ketone
																4,4'-Methylenebis(2-chloroaniline)
																4,4'-Methylenebis(N,N-dimethylaniline)
•	•	•	•	•	•	•			•	•	•	•	•	•	•	Methylene chloride
	•															Methyl iodide
•	•					•										Methyl isobutyl ketone
•	•															Methyl methacrylate
																Methylmethanesulfonate
																2-Methylnaphthalene
																2-Methyl-5-Nitroaniline
																Methyl paraoxon
																Methyl parathion
•	•					•										4-Methyl-2-pentanone
																2-Methylphenol
																3-Methylphenol
																4-Methylphenol
•	•															2-Methyl-2-Propene Nitrile
																2-Methylpyridine
•	•															Methyl-t-butyl ether
																Metolachlor

Appendix B

Organic Analyte Cross-Reference to Methods

Key:
S = Surrogate
I = Internal Standard
B = Both Surrogate and Internal Standard

Analyte	CAS Number	SW846 - 8151A	EPA 500 - 515.1	EPA 500 - 515.2	Std. Methods 6640	SW846 - 8081A	SW846 - 8082	EPA500 - 508	EPA600 - 608	Std. Methods 6630	CLP-PEST - Organic	SW846 - 8270C	EPA500 - 525.2	EPA600 - 625	Std. Methods 6410	CLP - SVOA
		Herbicides (p. 2)				Pesticides & PCBs (p. 8)						Semivolatile Organic (p. 20)				
Metribuzin	21087-64-9												•			
Mevinphos	7786-34-7											•				
Mexacarbate	315-18-4											•				
MGK 264	113-48-4												•			
Mirex	2385-85-5					•				•		•				
Molinate	2212-67-1												•			
Monocrotophos	6923-22-4											•				
Naled	300-76-5											•				
Naphthalene	91-20-3											•		•	•	•
Naphthalene d8	1146-65-2											I		B	B	I
Naphthene	91-20-3											•		•	•	•
1,4-Naphthoquinone	130-15-4											•				
1-Naphthylamine	134-32-7											•				
2-Naphthylamine	91-59-8											•				
Napropamide	15299-99-7												•			
Nicotine	54-11-5											•				
5-Nitroacenaphthene	602-87-9											•				
2-Nitroaniline	88-74-4											•				•
3-Nitroaniline	99-09-2											•				•
4-Nitroaniline	100-01-6											•				•
5-Nitro-o-anisidine	99-59-2											•				
Nitrobenzene	98-95-3											•		•	•	•
Nitrobenzene d5	4165-60-0											S		B	B	S
4-Nitrobiphenyl	92-93-3											•				
Nitrofen	1836-75-5					•						•				
2-Nitrophenol	88-75-5											•		•	•	•
4-Nitrophenol	100-02-7	•	•									•		•	•	•
2-Nitropropane	79-46-9															
Nitroquinoline-1-oxide	56-57-5											•				
N-Nitrosodibutylamine	924-16-3											•				
N-Nitrosodiethylamine	55-18-5											•				
N-Nitrosodimethylamine	62-75-9											•		•	•	
N-Nitroso-di-n-butylamine	924-16-3											•				
N-Nitrosodiphenylamine	86-30-6											•		•	•	•
N-Nitroso-di-n-propylamine	621-64-7											•		•	•	•
N-Nitrosomethylethylamine	10595-95-6											•				
N-Nitrosomorpholine	59-89-2											•				
N-Nitrosopiperidine	100-75-4											•				
N-Nitrosopyrrolidine	930-55-2											•				

Organic Analyte Cross-Reference to Methods

SW846 - 8260B	EPA500 - 524.2	EPA600 - 624	Std. Methods 6210B	Std. Methods 6210C	Std. Methods 6210D	CLP - VOA	EPA600 - 602	Std. Methods 6220B	Std. Methods 6220C	SW-846 - 8021B	EPA500 - 502.2	EPA600 - 601	Std. Methods 6230B	Std. Methods 6230C	Std. Methods 6230D	Analyte
																Metribuzin
																Mevinphos
																Mexacarbate
																MGK 264
																Mirex
																Molinate
																Monocrotophos
																Naled
•	•				•				•	•	•				•	Naphthalene
																Naphthalene d8
•	•				•				•	•	•				•	Naphthene
																1,4-Naphthoquinone
																1-Naphthylamine
																2-Naphthylamine
																Napropamide
																Nicotine
																5-Nitroacenaphthene
																2-Nitroaniline
																3-Nitroaniline
																4-Nitroaniline
																5-Nitro-o-anisidine
•	•															Nitrobenzene
																Nitrobenzene d5
																4-Nitrobiphenyl
																Nitrofen
																2-Nitrophenol
																4-Nitrophenol
•	•															2-Nitropropane
																Nitroquinoline-1-oxide
																N-Nitrosodibutylamine
																N-Nitrosodiethylamine
																N-Nitrosodimethylamine
•																N-Nitroso-di-n-butylamine
																N-Nitrosodiphenylamine
																N-Nitroso-di-n-propylamine
																N-Nitrosomethylethylamine
																N-Nitrosomorpholine
																N-Nitrosopiperidine
																N-Nitrosopyrrolidine

Organic Analyte Cross-Reference to Methods

Key:
S = Surrogate
I = Internal Standard
B = Both Surrogate and Internal Standard

Analyte	CAS Number	SW846 - 8151A	EPA 500 - 515.1	EPA 500 - 515.2	Std. Methods 6640	SW846 - 8081A	SW846 - 8082	EPA500 - 508	EPA600 - 608	Std. Methods 6630	CLP-PEST - Organic	SW846 - 8270C	EPA500 - 525.2	EPA600 - 625	Std. Methods 6410	CLP - SVOA
		Herbicides (p. 2)				Pesticides & PCBs (p. 8)						Semivolatile Organic (p. 20)				
5-Nitro-o-toluidine	99-55-8											•				
trans-Nonachlor	39765-80-5					•	•									
2,2',3,3',4,4',5,5',6-Nonachlorobiphenyl	40186-72-9						•									
Norflurazon	27314-13-2												•			
2,2',3,3',4,5',6,6'-Octachlorobiphenyl	40186-71-8												•			
Octamethyl pyrophosphoramide	152-16-9											•				
4,4'-Oxydianiline	101-80-4											•				
Paraldehyde	123-63-7															
Parathion	56-38-2								•							
Pebulate	1114-71-2												•			
Pentachlorobenzene	608-93-5											•				
2,2',3,4,5'-Pentachlorobiphenyl	38380-02-8						•									
2,2',3',4,6-Pentachlorobiphenyl	60233-25-2												•			
2,2',4,5,5'-Pentachlorobiphenyl	37680-73-2						•									
2,3,3',4',6-Pentachlorobiphenyl	38380-03-9						•									
Pentachloroethane	76-01-7															
Pentachloronitrobenzene (PCNB)	82-68-8					•		•		•		•				
Pentachloronitrobenzene (PCNB)	82-68-8					I		I								
Pentachlorophenol	87-86-5	•	•	•	•							•	•	•	•	•
Pentafluorobenzene	363-72-4															
2,3,4,5,6-Pentafluorobiphenyl												B	B			
Pentafluorophenol	771-61-9											B	B			
2-Pentanone	107-87-9															
2-Perfluoromethyl phenol												B	B			
cis-Permethrin	52645-53-1							•					•			
trans-Permethrin	51877-74-8					•		•					•			
Perthane	72-56-0					•										
Perylene d12	1520-96-3											I	S			I
Phenacetin	62-44-2											•				
Phenanthrene	85-01-8											•	•	•	•	•
Phenanthrene d10	1517-22-2											I	I	B	B	I
Phenobarbital	50-06-6											•				
Phenol	108-95-2											•		•	•	•
Phenol d6 d5 (see note)	4165-62-2											S		B	B	S
1,4-Phenylenediamine	106-50-3											•				
Phorate	298-02-2											•				
Phosalone	2310-17-0											•				
Phosmet	732-11-6											•				
Phosphamidon	13171-21-6											•				

Organic Analyte Cross-Reference to Methods

Volatile Organic by GC/MS (p. 36)							Volatile Organic by GC (p. 48)									
SW846 - 8260B	EPA500 - 524.2	EPA600 - 624	Std. Methods 6210B	Std. Methods 6210C	Std. Methods 6210D	CLP - VOA	EPA600 - 602	Std. Methods 6220B	Std. Methods 6220C	SW-846 - 8021B	EPA500 - 502.2	EPA600 - 601	Std. Methods 6230B	Std. Methods 6230C	Std. Methods 6230D	Analyte
																5-Nitro-o-toluidine
																trans-Nonachlor
																2,2',3,3',4,4',5,5',6-Nonachlorobiphenyl
																Norflurazon
																2,2',3,3',4,5',6,6'-Octachlorobiphenyl
																Octamethyl pyrophosphoramide
																4,4'-Oxydianiline
•																Paraldehyde
																Parathion
																Pebulate
																Pentachlorobenzene
																2,2',3,4,5'-Pentachlorobiphenyl
																2,2',3',4,6-Pentachlorobiphenyl
																2,2',4,5,5'-Pentachlorobiphenyl
																2,3,3',4',6-Pentachlorobiphenyl
•	•															Pentachloroethane
																Pentachloronitrobenzene (PCNB)
																Pentachloronitrobenzene (PCNB)
																Pentachlorophenol
B			B	B	B	B							I			Pentafluorobenzene
																2,3,4,5,6-Pentafluorobiphenyl
																Pentafluorophenol
•																2-Pentanone
																2-Perfluoromethyl phenol
																cis-Permethrin
																trans-Permethrin
																Perthane
																Perylene d12
																Phenacetin
																Phenanthrene
																Phenanthrene d10
																Phenobarbital
																Phenol
																Phenol d6 d5 (see note)
																1,4-Phenylenediamine
																Phorate
																Phosalone
																Phosmet
																Phosphamidon

Note for Phenol d6: This method calls for a Phenol d6 standard. Phenol d6 raw material is used to manufacture this standard. However, Phenolic Hydrogen exchange within the ampul converts Phenol d6 to Phenol d5. The analysis will detect Phenol d5, not Phenol d6.

Organic Analyte Cross-Reference to Methods

Key:
S = Surrogate
I = Internal Standard
B = Both Surrogate and Internal Standard

Analyte	CAS Number	SW846-8151A	EPA500-515.1	EPA500-515.2	Std. Methods 6640	SW846-8081A	SW846-8082	EPA500-508	EPA600-608	Std. Methods 6630	CLP-PEST-Organic	SW846-8270C	EPA500-525.2	EPA600-625	Std. Methods 6410	CLP-SVOA
		Herbicides (p. 2)				Pesticides & PCBs (p. 8)						Semivolatile Organic (p. 20)				
Phthalic anhydride	85-44-9											•				
Picloram	1918-02-1	•	•	•	•											
2-Picoline	109-06-8											•				
Piperonyl sulfoxide	120-62-7											•				
Prometon	1610-18-0												•			
Prometryn	7287-19-6												•			
Pronamide	23950-58-5											•	•			
Propachlor	1918-16-7					•		•				•				
Propane Dinitrile	109-77-3															
Propanenitrile	107-12-0															
Propazine	139-40-2												•			
1-Propanol	71-23-8															
2-Propanol	67-63-0															
2-Propanone	67-64-1															
Propargyl alcohol	107-19-7															
2-Propenal	107-02-8															
2-Propenenitrile	107-13-1															
2-Propenol	107-18-6															
beta-Propiolactone	57-57-8															
Propionitrile	107-12-0															
n-Propylamine	107-10-8															
n-Propylbenzene	103-65-1															
2-Propyn-1-ol	107-19-7															
Propylthiouracil	51-52-5											•				
Pyrene	129-00-0											•	•	•	•	•
Pyrene d10	1718-52-1												S			
Pyridine	110-86-1											•				
Pyridine d5	7291-22-7														B	B
Resorcinol	108-46-3											•				
Safrole	94-59-7											•				
Silvex (2,4,5-TP)	93-72-1	•	•	•	•											
Simazine	122-34-9												•			
Simetryn	1014-70-6												•			
Stirofos	22248-79-9												•			
Strobane	8001-50-1					•				•						
Strychnine	57-24-9											•				
Styrene	100-42-5															
Sulfallate	95-06-7											•				
2,4,5-T	93-76-5	•	•	•	•											

Appendix B

Organic Analyte Cross-Reference to Methods

Volatile Organic by GC/MS (p. 36)							Volatile Organic by GC (p. 48)									
SW846 - 8260B	EPA500 - 524.2	EPA600 - 624	Std. Methods 6210B	Std. Methods 6210C	Std. Methods 6210D	CLP - VOA	EPA600 - 602	Std. Methods 6220B	Std. Methods 6220C	SW-846 - 8021B	EPA500 - 502.2	EPA600 - 601	Std. Methods 6230B	Std. Methods 6230C	Std. Methods 6230D	Analyte
																Phthalic anhydride
																Picloram
•																2-Picoline
																Piperonyl sulfoxide
																Prometon
																Prometryn
																Pronamide
																Propachlor
•																Propane Dinitrile
•	•															Propanenitrile
																Propazine
•																1-Propanol
•																2-Propanol
•	•					•										2-Propanone
•																Propargyl alcohol
•																2-Propenal
•	•															2-Propenenitrile
•																2-Propenol
•																beta-Propiolactone
•	•															Propionitrile
•																n-Propylamine
•	•			•	•				•	•	•				•	n-Propylbenzene
•																2-Propyn-1-ol
																Propylthiouracil
																Pyrene
																Pyrene d10
•																Pyridine
																Pyridine d5
																Resorcinol
																Safrole
																Silvex (2,4,5-TP)
																Simazine
																Simetryn
																Stirofos
																Strobane
																Strychnine
•	•			•	•	•			•	•	•				•	Styrene
																Sulfallate
																2,4,5-T

Organic Analyte Cross-Reference to Methods

Key:
S = Surrogate
I = Internal Standard
B = Both Surrogate and Internal Standard

Analyte	CAS Number	SW846 - 8151A	EPA 500 - 515.1	EPA 500 - 515.2	Std. Methods 6640	SW846 - 8081A	SW846 - 8082	EPA500 - 508	EPA600 - 608	Std. Methods 6630	CLP-PEST - Organic	SW846 - 8270C	EPA500 - 525.2	EPA600 - 625	Std. Methods 6410	CLP - SVOA
		Herbicides (p. 2)				Pesticides & PCBs (p. 8)						Semivolatile Organic (p. 20)				
2,4,5-TP (Silvex)	93-72-1	•	•	•	•											
Tebuthiuron	34014-18-1												•			
Terbacil	5902-51-2												•			
Terbufos	13071-79-9											•	•			
Terbutryn	886-50-0												•			
Terphenyl d14	1718-51-0											S				S
1,2,4,5-Tetrachlorobenzene	95-94-3											•				
2,2',3,5'-Tetrachlorobiphenyl	41464-39-5						•									
2,2',4,4'-Tetrachlorobiphenyl	2437-79-8												•			
2,2',5,5'-Tetrachlorobiphenyl	35693-99-3						•									
2,3',4,4'-Tetrachlorobiphenyl	32598-10-0						•									
1,1,1,2-Tetrachloroethane	630-20-6															
1,1,2,2-Tetrachloroethane	79-34-5															
Tetrachloroethene	127-18-4															
2,3,4,6-Tetrachlorophenol	58-90-2											•				
Tetrachloromethane	56-23-5															
2,4,5,6-Tetrachloro-m-xylene	877-09-8					S	S				S					
Tetrachlorvinphos	961-11-5											•				
Tetraethyl Dithiopyrophosphate	3689-24-5											•				
Tetraethyl pyrophosphate	107-49-3											•				
2,3,5,6-Tetraflourobenzoic acid	652-18-6				S											
Tetrahydrofuran	109-99-9															
Thionazine	297-97-2											•				
Thiophenol (Benzenethiol)	108-98-5											•				
Toluene	108-88-3															
Toluene d8	2037-26-5															
Toluene diisocyanate	584-84-9											•				
o-Toluidine	95-53-4											•				
Toxaphene	8001-35-2					•		•	•	•	•	•	•	•	•	
2,4,5-TP(Silvex)	93-72-1	•	•	•	•											
Triademefon	43121-43-3												•			
Tribromomethane	75-25-2															
2,4,6-Tribromophenol	118-79-6											S				S
1,2,3-Trichlorobenzene	87-61-6															
1,2,4-Trichlorobenzene	120-82-1											•		•	•	•
2,2',5-Trichlorobiphenyl	37680-65-2						•									
2,4',5-Trichlorobiphenyl	16606-02-3						•									
2,4,5-Trichlorobiphenyl	15862-07-4												•			
1,1,1-Trichloroethane	71-55-6															

Appendix B

Organic Analyte Cross-Reference to Methods

Volatile Organic by GC/MS (p. 36)							Volatile Organic by GC (p. 48)									Analyte
SW846 - 8260B	EPA500 - 524.2	EPA600 - 624	Std. Methods 6210B	Std. Methods 6210C	Std. Methods 6210D	CLP - VOA	EPA600 - 602	Std. Methods 6220B	Std. Methods 6220C	SW-846 - 8021B	EPA500 - 502.2	EPA600 - 601	Std. Methods 6230B	Std. Methods 6230C	Std. Methods 6230D	Analyte
																2,4,5-TP (Silvex)
																Tebuthiuron
																Terbacil
																Terbufos
																Terbutryn
																Terphenyl d14
																1,2,4,5-Tetrachlorobenzene
																2,2',3,5'-Tetrachlorobiphenyl
																2,2',4,4'-Tetrachlorobiphenyl
																2,2',5,5'-Tetrachlorobiphenyl
																2,3',4,4'-Tetrachlorobiphenyl
●	●		●	●				●	●				●	●		1,1,1,2-Tetrachloroethane
●	●	●	●	●	●	●		●	●	●	●	●	●	●	●	1,1,2,2-Tetrachloroethane
●	●	●	●	●	●	●		●	●	●	●	●	●	●	●	Tetrachloroethene
																2,3,4,6-Tetrachlorophenol
●	●	●	●	●	●	●		●	●	●	●	●	●	●	●	Tetrachloromethane
																2,4,5,6-Tetrachloro-m-xylene
																Tetrachlorvinphos
																Tetraethyl Dithiopyrophosphate
																Tetraethyl pyrophosphate
																2,3,5,6-Tetraflourobenzoic acid
	●															Tetrahydrofuran
																Thionazine
																Thiophenol (Benzenethiol)
●	●	●	●	●	●	●	●	●	●	●	●				●	Toluene
S						S										Toluene d8
																Toluene diisocyanate
																o-Toluidine
																Toxaphene
																2,4,5-TP(Silvex)
																Triademefon
●	●	●	●	●	●	●			●	●	●	●	●	●	●	Tribromomethane
																2,4,6-Tribromophenol
●	●				●			●	●	●					●	1,2,3-Trichlorobenzene
●	●				●			●	●	●					●	1,2,4-Trichlorobenzene
																2,2',5-Trichlorobiphenyl
																2,4',5-Trichlorobiphenyl
																2,4,5-Trichlorobiphenyl
●	●	●	●	●	●	●			●	●	●	●	●	●	●	1,1,1-Trichloroethane

Appendix B

Organic Analyte Cross-Reference to Methods

Key:
S = Surrogate
I = Internal Standard
B = Both Surrogate and Internal Standard

Analyte	CAS Number	SW846 - 8151A	EPA 500 - 515.1	EPA 500 - 515.2	Std. Methods 6640	SW846 - 8081A	SW846 - 8082	EPA500 - 508	EPA600 - 608	Std. Methods 6630	CLP-PEST - Organic	SW846 - 8270C	EPA500 - 525.2	EPA600 - 625	Std. Methods 6410	CLP - SVOA
		Herbicides (p. 2)				Pesticides & PCBs (p. 8)						Semivolatile Organic (p. 20)				
1,1,2-Trichloroethane	79-00-5															
Trichloroethene	79-01-6															
Trichloroethylene	79-01-6															
Trichlorofluoromethane	75-69-4															
Trichloromethane	67-66-3															
2,4,5-Trichlorophenol	95-95-4											•				•
2,4,6-Trichlorophenol	88-06-2											•		•	•	•
1,2,3-Trichloropropane	96-18-4				I											
Tricyclazole	41814-78-2												•			
0,0,0-Triethylphosphorothioate	126-68-1											•				
Trifluralin	1582-09-8					•		•			•	•	•			
alpha, alpha, alpha-Trifluorotoluene	98-08-8															
2,4,5-Trimethylaniline	137-17-7											•				
1,2,4-Trimethylbenzene	95-63-6															
1,3,5-Trimethylbenzene	108-67-8															
Trimethyl phosphate	512-56-1											•				
1,3,5-Trinitrobenzene	99-35-4											•				
Tris(2,3-dibromopropyl) phosphate	126-72-7											•				
Tri-p-tolyl phosphate(h)	78-32-0											•				
Vernolate	1929-77-7												•			
Vinyl acetate	108-05-4															
Vinyl chloride	75-01-4															
m-Xylene	108-38-3															
o-Xylene	95-47-6															
p-Xylene	106-42-3															
Xylene (total)	1330-20-7															

Organic Analyte Cross-Reference to Methods

Volatile Organic by GC/MS (p. 36)							Volatile Organic by GC (p. 48)									Analyte
SW846 - 8260B	EPA500 - 524.2	EPA600 - 624	Std. Methods 6210B	Std. Methods 6210C	Std. Methods 6210D	CLP - VOA	EPA600 - 602	Std. Methods 6220B	Std. Methods 6220C	SW-846 - 8021B	EPA500 - 502.2	EPA600 - 601	Std. Methods 6230B	Std. Methods 6230C	Std. Methods 6230D	
•	•	•	•	•	•	•				•	•	•	•	•	•	1,1,2-Trichloroethane
•	•	•	•	•	•	•			•	•	•	•	•	•	•	Trichloroethene
•	•	•	•	•	•	•			•	•	•	•	•	•	•	Trichloroethylene
•	•	•	•	•	•					•	•	•	•	•	•	Trichlorofluoromethane
•	•	•	•	•	•	•				•	•	•	•	•	•	Trichloromethane
																2,4,5-Trichlorophenol
																2,4,6-Trichlorophenol
•	•			•	•						•	•		•	•	1,2,3-Trichloropropane
																Tricyclazole
																0,0,0-Triethylphosphorothioate
																Trifluralin
							B	B	B							alpha, alpha, alpha-Trifluorotoluene
																2,4,5-Trimethylaniline
•	•				•				•	•	•				•	1,2,4-Trimethylbenzene
•	•				•				•	•	•				•	1,3,5-Trimethylbenzene
																Trimethyl phosphate
																1,3,5-Trinitrobenzene
																Tris(2,3-dibromopropyl) phosphate
																Tri-p-tolyl phosphate(h)
																Vernolate
•																Vinyl acetate
•	•	•	•	•	•	•				•	•	•	•	•	•	Vinyl chloride
•	•		•	•	•				•	•	•				•	m-Xylene
•	•		•	•	•				•	•	•				•	o-Xylene
•	•		•	•	•				•	•	•				•	p-Xylene
						•										Xylene (total)

Appendix B

APPENDIX C

Inorganic analytes mentioned in the categories of Trace Metals identified by Flame and Graphite Furnace Atomic Absorption Spectroscopy, Trace Metals identified by Inductively Coupled Plasma (ICP), and Trace Metals identified by Inductively Coupled Plasma/Mass Spectroscopy are listed in alphabetical order. If the analyte is included in one of the listed test methods it is noted with a "•" in the appropriate column.

Determination of Total Trace Metals Cross-Reference

Metal	Symbol	CAS	Flame and Graphite Furnace AA Spect. (p. 74)				ICP (p. 80)				ICP-MS (p. 86)	
			SW-846 7000 Series	EPA 200 Series	Std Methods 3000 Series	CLP Inorganic Method	SW-846 6010B	EPA 200.7	Std. Methods 3120	CLP Inorganic Method	SW-846 6020	EPA 200.8
Aluminum	Al	7429-90-5	•	•	•		•	•	•	•	•	•
Antimony	Sb	7440-36-0	•	•	•	•	•	•	•	•	•	•
Arsenic	As	7440-38-2	•	•	•	•	•	•	•	•	•	•
Barium	Ba	7440-39-3	•	•	•		•	•	•	•	•	•
Beryllium	Be	7440-41-7	•	•	•	•	•	•	•	•	•	•
Bismuth	Bi	7440-69-9			•							
Boron	B	7440-42-8					•	•				
Cadmium	Cd	7440-43-9	•	•	•	•	•	•	•	•	•	•
Calcium	Ca	7440-70-2	•	•	•	•	•	•	•	•		
Cerium	Ce	7440-45-1						•				
Cesium	Cs	7440-46-2			•							
Chromium	Cr	7440-47-3	•	•	•	•	•	•	•	•	•	•
Cobalt	Co	7440-48-4	•	•	•		•	•	•	•	•	•
Copper	Cu	7440-50-8	•	•	•		•	•	•	•	•	•
Gold	Au	7440-57-5		•	•							
Iridium	Ir	7439-88-5		•	•							
Iron	Fe	7439-89-6	•	•	•		•	•	•	•		
Lead	Pb	7439-92-1	•	•	•	•	•	•	•	•	•	•
Lithium	Li	7439-93-2	•		•		•	•	•			
Magnesium	Mg	7439-95-4	•	•	•	•	•	•	•	•		
Manganese	Mn	7439-96-5	•	•	•		•	•	•	•	•	•
Mercury (a)	Hg	7439-97-6	•	•	•							•
Molybdenum	Mo	7439-98-7	•	•	•		•	•	•			•
Nickel	Ni	7440-02-0	•	•	•		•	•	•	•	•	•
Osmium	Os	7440-04-2	•	•	•							
Palladium	Pd	7440-05-3		•	•							
Phosphorus	P	7723-14-0					•	•				
Platinum	Pt	7440-06-4		•	•							
Potassium	K	7440-09-7	•	•	•	•	•	•	•	•		
Rhenium	Re	7440-15-5		•	•							

Inorganic Analyte Cross-Reference to Methods

Determination of Total Trace Metals Cross-Reference (Continued)

Metal	Symbol	CAS	Flame and Graphite Furnace AA Spect. (p. 74)				ICP (p. 80)				ICP-MS (p. 86)	
			SW-846 7000 Series	EPA 200 Series	Std Methods 3000 Series	CLP Inorganic Method	SW-846 6010B	EPA 200.7	Std. Methods 3120	CLP Inorganic Method	SW-846 6020	EPA 200.8
Rhodium	Rh	7440-16-6		•	•							
Ruthenium	Ru	7440-18-8		•	•							
Selenium	Se	7782-49-2	•	•	•	•	•	•	•	•		•
Silicon/Silica (SiO$_2$)	Si	7440-21-3			•			•	•			
Silver	Ag	7440-22-4	•	•	•	•	•	•	•	•	•	•
Sodium	Na	7440-23-5	•	•	•	•	•	•	•	•		
Strontium	Sr	7440-24-6	•		•		•	•	•			
Thallium	Tl	7440-28-0	•	•	•	•	•	•	•	•	•	•
Thorium	Th	7440-29-1			•							•
Tin	Sn	7440-31-5	•					•				
Titanium	Ti	7440-32-6		•	•			•				
Uranium	U	7440-61-1										•
Vanadium	V	7440-62-2	•	•	•		•	•	•	•		•
Zinc	Zn	7440-66-6	•	•	•		•	•	•	•	•	•

(a)Cold vapor technique.

APPENDIX D

The main table contained in this section (starting on page D-3) is a synopsis of methods contained in SW-846 Third Edition and its subsequent updates. This table outlines the methods, updates to existing methods, new methods added during update publications, and methods deleted from SW-846. Also included as an introduction to SW-846 publications is a tabulated history of updates to the Third Edition.

History of SW-846, Third Edition, and Updates

Publication	Date Listed on Methods	Color of Paper	Status of Package
Third Edition	September 1986	White	**Finalized (Promulgated)**
Proposed Update I	December 1987	Green	Obsolete
Final Update I (Accidently Released)	November 1990	White	Obsolete! Never formally finalized
Proposed Update II (Accidently Released)	November 1990	Blue	Obsolete! Never formally proposed
Final Update I	July 1992	White	**Finalized (Promulgated)**
Proposed Update II	November 1992	Yellow	Obsolete
Proposed Update IIA* (Available from EPA by request only)	October 1992	White	Obsolete
Final Update IIA* (Included with Final Update II)	August 1993	White	**Finalized (Promulgated)**
Final Update II	September 1994	White	**Finalized (Promulgated)**
Final Update IIB**	January 1995	White	**Finalized (Promulgated)**
Proposed Update III	January 1995	Pink	Obsolete
Final Update III	December 1996	White	**Finalized (Promulgated)**

* Contains only Method 4010

** Contains only a revised Table of Contents, a revised Chapter Six, and revised Methods 9040B and 9045C.

SW-846 Chapter Text

Status Table for SW-846 Chapter Text and Other Documents

Title	Third Edition (9/86)	Final Update I (7/92)	Final Update II (9/94) IIA (8/93) IIB (1/95)	Final Update III (12/96)	SW-846 Volume Location	Current Final Version
Disclaimer	—	X	—	X	All Vols	Rev 1 (12/96)
Abstract	X	X	X (Up. II)	—	All Vols	Rev 2 (9/94)
Table of Contents	X	X	X (Up. II and IIB)	X	All Vols	Rev 4 (12/96)
Method Index and Conversion Table	X	—	—	—	All Vols	Rev 0 (9/86)
Preface and Overview	X	—	—	X	All Vols	Rev 1 (12/96)
Acknowledgements	X	—	—	—	Vol IA	Rev 0 (9/86)
Chapter One Quality Control	X	X	—	—	All Vols	Rev 1 (7/92)
Chapter Two Choosing the Correct Procedure	X	X	X (Up. II)	X	Vol IA	Rev 3 (12/96)
Chapter Three Inorganic Analytes	X	X	X (Up. II)	X	Vol IA	Rev 3 (12/96)
Chapter Four Organic Analytes	X	—	X (Up. II)	X	Vol IB	Rev 3 (12/96)
Chapter Five Miscellaneous Test Methods	X	—	X (Up. II)	X	Vol IC	Rev 2 (12/96)
Chapter Six Properties	X	—	X (Up. II and IIB)	X	Vol IC	Rev 3 (12/96)
Chapter Seven Characteristics Introduction and Regulatory Definitions	X	X	X (Up. II)	X	Vol IC	Rev 3 (12/96)
Chapter Eight Methods for Determining Characteristics	X	—	X (Up. II)	X	Vol IC	Rev 2 (12/96)
Chapter Nine Sampling Plan	X	—	—	—	Vol II	Rev 0 (9/86)
Chapter Ten Sampling Methods	X	—	—	X	Vol II	Rev 1 (12/96)
Chapter Eleven Ground Water Monitoring	X	—	—	—	Vol II	Rev 0 (9/86)
Chapter Twelve Land Treatment Monitoring	X	—	—	—	Vol II	Rev 0 (9/86)
Chapter Thirteen Incineration	X	—	—	—	Vol II	Rev 0 (9/86)
Appendix Company References	X	—	—	—	All Vols	Rev 0 (9/86)

Appendix D

SW-846 Method Status Table (Compiled May 1997)

Method Number Third Edition (9/86)	Method Number Final Update 1 (7/92)	Method Number Final Update II (9/94) IIA (8/93)	Method Number Final Update III (12/96)	Method Title	SW-846 Volume/ Chapter/ Section	Current Promulgated Method
0010	—	IIB (1/95)	—	Modified Method 5 Sample Train	Vol II Chap 10	0010 Rev 0 (9/86)
—	—	—	0011	Sampling for Selected Aldehyde and Ketone Emissions from Stationary Sources	Vol II Chap 10	0011 Rev 0 (12/97)
0020	—	—	—	Source Assessment Sampling System (SASS)	Vol II Chap 10	0020 Rev 0 (9/86)
—	—	—	0023A (Revision of Method 23, 40CFR Part 60)	Sampling Method for Polychlorinated Dibenzo-p-Dioxins and Polychlorinated Dibenzofuran Emissions from Stationary Sources	Vol II Chap 10	0023A Rev 1 (12/96)
0030	—	—	—	Volatile Organic Sampling Train	Vol II Chap 10	0030 Rev 0 (9/86)
—	—	—	0031	Sampling Method for Volatile Organic Compounds (SMVOC)	Vol II Chap 10	0031 Rev 0 (12/96)
—	—	—	0040	Sampling of Principal Organic Hazardous Constituents from Combustion Sources Using Tedlar® Bags	Vol II Chap 10	0040 Rev 0 (12/96)
—	—	—	0050	Isokinetic HCl/Cl_2 Emission Sampling Train	Vol II Chap 10	0050 Rev 0 (12/96)
—	—	—	0051	Midget Impinger HCl/Cl_2 Emission Sampling Train	Vol II Chap 10	0051 Rev 0 (12/96)
—	—	—	0060	Determination of Metals in Stack Emissions	Vol II Chap 10	0060 Rev 0 (12/96)

NOTE: For this table, the date in parenthesis is the date found at the bottom right-hand corner of the method.

Appendix D

SW-846 Method Status Table (Compiled May 1997) (Continued)

Method Number Third Edition (9/86)	Method Number Final Update 1 (7/92)	Method Number Final Update II (9/94) IIA (8/93) IIB (1/95)	Method Number Final Update III (12/96)	Method Title	SW-846 Volume/ Chapter/ Section	Current Promulgated Method
—	—	—	0061	Determination of Hexavalent Chromium Emissions from Stationary Sources	Vol II Chap 10	0060 Rev 0 (12/96)
—	—	—	0100	Sampling for Formaldehyde and Other Carbonyl Compounds in Indoor Air	Vol II Chap 10	0100 Rev 0 (12/96)
1010	—	—	—	Pensky-Martens Closed-Cup Method for Determining Ignitability	Vol IC Chap 8 Sec 8.1	1010 Rev 0 (9/86)
1020	1020A	—	—	Setaflash Closed-Cup Method for Determining Ignitability	Vol IC Chap 8 Sec 8.1	1020A Rev 1 (7/92)
—	—	—	1030	Ignitability of Solids	Vol IC Chap 6	1030 Rev 0 (12/96)
1110	—	—	—	Corrosivity Toward Steel	Vol IC Chap 8 Sec 8.2	1110 Rev 0 (9/86)
—	—	—	1120	Dermal Corrosion	Vol IC Chap 6	1120 Rev 0 (12/96)
1310	1310A	—	—	Extraction Procedure (EP) Toxicity Test Method and Structural Integrity Test	Vol IC Chap 8 Sec 8.4	1310A Rev 1 (7/92)
—	1311	—	—	Toxicity Characteristic Leaching Procedure	Vol IC Chap 8 Sec 8.4	1311 Rev 0 (7/92)
—	—	1312 (Up. II)	—	Synthetic Precipitation Leaching Procedure	Vol IC Chap 6	1312 Rev 0 (9/94)
1320	—	—	—	Multiple Extraction Procedure	Vol IC Chap 6	1320 Rev 0 (9/86)
1330	1330A	—	—	Extraction Procedure for Oily Wastes	Vol IC Chap 6	1330A Rev 1 (7/92)

SW-846 Method Status Table (Compiled May 1997) (Continued)

Method Number Third Edition (9/86)	Method Number Final Update I (7/92)	Method Number Final Update II (9/94) IIA (8/93) IIB (1/95)	Method Number Final Update III (12/96)	Method Title	SW-846 Volume/ Chapter/ Section	Current Promulgated Method
3005	3005A	—	—	Acid Digestion of Waters for Total Recoverable or Dissolved Metals for Analysis by FLAA or ICP Spectroscopy	Vol IA Chap 3 Sec 3.2	3005A Rev 1 (7/92)
3010	3010A	—	—	Acid Digestion of Aqueous Samples and Extracts for Total Metals for Analysis by FLAA or ICP Spectroscopy	Vol IA Chap 3 Sec 3.2	3010A Rev 1 (7/92)
—	—	3015 (Up. II)	—	Microwave Assisted Acid Digestion of Aqueous Samples and Extracts	Vol IA Chap 3 Sec 3.2	3015 Rev 0 (9/94)
3020	3020A	—	—	Acid Digestion of Aqueous Samples and Extracts for Total Metals for Analysis by GFAA Spectroscopy	Vol IA Chap 3 Sec 3.2	3020A Rev 1 (7/92)
—	—	—	3031	Acid Digestion of Oils for Metals Analysis by Atomic Absorption or ICP Spectrometry.	Vol IA Chap 3 Sec 3.2	3031 Rev 0 (12/96)
3040	—	—	3040A	Dissolution Procedure for Oils, Greases, or Waxes	Vol IA Chap 3 Sec 3.2	3040A Rev 1 (12/96)
3050	3050A	—	3050B	Acid Digestion of Sediments, Sludges, and Soils	Vol IA Chap 3 Sec 3.2	3050B Rev 2 (12/96)
—	—	3051 (Up. II)	—	Microwave Assisted Acid Digestion of Sediments, Sludges, Soils, and Oils	Vol IA Chap 3 Sec 3.2	3051 Rev 0 (9/94)
—	—	—	3052	Microwave Assisted Acid Digestion of Siliceous and Organically Based Matrices	Vol IA Chap 3 Sec 3.2	3052 Rev 0 (12/96)

Appendix D

Method Number Third Edition (9/86)	Method Number Final Update I (7/92)	Method Number Final Update II (9/94) IIA (8/93) IIB (1/95)	Method Number Final Update III (12/96)	Method Title	SW-846 Volume/ Chapter/ Section	Current Promulgated Method
[3060, in the Second Edition]	—	—	3060A	Alkaline Digestion for Hexavalent Chromium	Vol IA Chap 3 Sec 3.2	3060A Rev 1 (12/96)
3500	3500A	—	3500B	Organic Extraction and Sample Preparation	Vol IB Chap 4 Sec 4.2.1	3500B Rev 2 (12/96)
3510	3510A	3510B (Up. II)	3510C	Separatory Funnel Liquid-Liquid Extraction	Vol IB Chap 4 Sec 4.2.1	3510C Rev 3 (12/96)
3520	3520A	3520B (Up. II)	3520C	Continuous Liquid-Liquid Extraction	Vol IB Chap 4 Sec 4.2.1	3520C Rev 3 (12/96)
—	—	—	3535	Solid-Phase Extraction (SPE)	Vol IB Chap 4 Sec 4.2.1	3535 Rev 0 (12/96)
3540	3540A	3540B (Up. II)	3540C	Soxhlet Extraction	Vol IB Chap 4 Sec 4.2.1	3540C Rev 3 (12/96)
—	—	3541 (Up. II)	—	Automated Soxhlet Extraction	Vol IB Chap 4 Sec 4.2.1	3541 Rev 0 (9/94)
—	—	—	3542	Extraction of Semivolatile Analytes Collected Using Method 0010 (Modified Method 5 Sampling Train)	Vol IB Chap 4 Sec 4.2.1	3542 Rev 0 (12/96)
—	—	—	3545	Pressurized Fluid Extraction (PFE)	Vol IB Chap 4 Sec 4.2.1	3545 Rev 0 (12/96)
3550	—	3550A (Up. II)	3550B	Ultrasonic Extraction	Vol IB Chap 4 Sec 4.2.1	3550B Rev 2 (12/96)
—	—	—	3560	Supercritical Fluid Extraction of Total Recoverable Petroleum Hydrocarbons	Vol IB Chap 4 Sec 4.2.1	3560 Rev 0 (12/96)
—	—	—	3561	Supercritical Fluid Extraction of Polynuclear Aromatic Hydrocarbons	Vol IB Chap 4 Sec 4.2.1	3561 Rev 0 (12/96)

Appendix D

SW-846 Method Status Table (Compiled May 1997) (Continued)

Method Number Third Edition (9/86)	Method Number Final Update I (7/92)	Method Number Final Update II (9/94) IIA (8/93) IIB (1/95)	Method Number Final Update III (12/96)	Method Title	SW-846 Volume/ Chapter/ Section	Current Promulgated Method
3580	3580A	—	—	Waste Dilution	Vol IB Chap 4 Sec 4.2.1	3580A Rev 1 (7/92)
—	—	—	3585	Waste Dilution for Volatile Organics	Vol IB Chap 4 Sec 4.2.1	3585 Rev 0 (12/96)
3600	3600A	3600B (Up. II)	3600C	Cleanup	Vol IB Chap 4 Sec 4.2.2	3600C Rev 3 (12/96)
3610	3610A	—	3610B	Alumina Cleanup	Vol IB Chap 4 Sec 4.2.2	3610B Rev 2 (12/96)
3611	3611A	—	3611B	Alumina Column Cleanup and Separation of Petroleum Wastes	Vol IB Chap 4 Sec 4.2.2	3611B Rev 2 (12/96)
3620	3620A	—	3620B	Florisil Cleanup	Vol IB Chap 4 Sec 4.2.2	3620B Rev 2 (12/96)
3630	3630A	3630B (Up. II)	3630C	Silica Gel Cleanup	Vol IB Chap 4 Sec 4.2.2	3630C Rev 3 (12/96)
3640	—	3640A (Up. II)	—	Gel-Permeation Cleanup	Vol IB Chap 4 Sec 4.2.2	3640A Rev 1 (9/94)
3650	3650A	—	3650B	Acid-Base Partition Cleanup	Vol IB Chap 4 Sec 4.2.2	3650B Rev 2 (12/96)
3660	3660A	—	3660B	Sulfur Cleanup	Vol IB Chap 4 Sec 4.2.2	3660B Rev 2 (12/96)
—	—	3665 (Up. II)	3665A	Sulfuric Acid/Permanganate Cleanup	Vol IB Chap 4 Sec 4.2.2	3665A Rev 1 (12/96)
3810	—	—	—	Headspace	Vol IB Chap 4 Sec 4.5	3810 Rev 0 (9/86)
3820	—	—	—	Hexadecane Extraction and Screening of Purgeable Organics	Vol IB Chap 4 Sec 4.5	3820 Rev 0 (9/86)

Appendix D

SW-846 Method Status Table (Compiled May 1997) (Continued)

Method Number Third Edition (9/86)	Method Number Final Update I (7/92)	Method Number Final Update II (9/94) IIA (8/93) IIB (1/95)	Method Number Final Update III (12/96)	Method Title	SW-846 Volume/ Chapter/ Section	Current Promulgated Method
—	—	—	4000	Immunoassay	Vol IB Chap 4 Sec 4.4	4000 Rev 0 (12/96)
—	—	4010 (Up. IIA)	4010A	Screening for Pentachlorophenol by Immunoassay	Vol IB Chap 4 Sec 4.4	4010A Rev 1 (12/96)
—	—	—	4015	Screening for 2,4-Dichlorophenoxyacetic Acid by Immunoassay	Vol IB Chap 4 Sec 4.4	4015 Rev 0 (12/96)
—	—	—	4020	Screening for Polychlorinated Biphenyls by Immunoassay	Vol IB Chap 4 Sec 4.4	4020 Rev 0 (12/96)
—	—	—	4030	Soil Screening for Petroleum Hydrocarbons by Immunoassay	Vol IB Chap 4 Sec 4.4	4030 Rev 0 (12/96)
—	—	—	4035	Soil Screening for Polynuclear Aromatic Hydrocarbons by Immunoassay	Vol IB Chap 4 Sec 4.4	4035 Rev 0 (12/96)
—	—	—	4040	Soil Screening for Toxaphene by Immunoassay	Vol IB Chap 4 Sec 4.4	4040 Rev 0 (12/96)
—	—	—	4041	Soil Screening for Chlordane by Immunoassay	Vol IB Chap 4 Sec 4.4	4041 Rev 0 (12/96)
—	—	—	4042	Soil Screening for DDT by Immunoassay	Vol IB Chap 4 Sec 4.4	4042 Rev 0 (12/96)
—	—	—	4050	TNT Explosives in Soil by Immunoassay	Vol IB Chap 4 Sec 4.4	4050 Rev 0 (12/96)
—	—	—	4051	Hexahydro-1,3,5-trinitro-1,3,5-triazine (RDX) in Soil by Immunoassay	Vol IB Chap 4 Sec 4.4	4051 Rev 0 (12/96)
—	—	—	5000	Sample Preparation for Volatile Organic Compounds	Vol IB Chap 4 Sec 4.2.1	5000 Rev 0 (12/96)

SW-846 Method Status Table (Compiled May 1997) (Continued)

Method Number Third Edition (9/86)	Method Number Final Update I (7/92)	Method Number Final Update II (9/94) IIA (8/93) IIB (1/95)	Method Number Final Update III (12/96)	Method Title	SW-846 Volume/ Chapter/ Section	Current Promulgated Method
—	—	—	5021	Volatile Organic Compounds in Soils and Other Solid Matrices Using Equilibrium Headspace Analysis	Vol IB Chap 4 Sec 4.2.1	5021 Rev 0 (12/96)
5030	5030A	—	5030B	Purge-and-Trap for Aqueous Samples	Vol IB Chap 4 Sec 4.2.1	5030B Rev 2 (12/96)
—	—	—	5031	Volatile, Nonpurgeable, Water-Soluble Compounds by Azetropic Distillation	Vol IB Chap 4 Sec 4.2.1	5031 Rev 0 (12/96)
—	—	—	5032	Volatile Organic Compounds by Vacuum Distillation	Vol IB Chap 4 Sec 4.2.1	5032 Rev 0 (12/96)
—	—	—	5035	Closed-System Purge-and-Trap and Extraction for Volatile Organics in Soil and Waste Samples	Vol IB Chap 4 Sec 4.2.1	5035 Rev 0 (12/96)
5040	—	5040A (Up. II)	Deleted from SW-846	Analysis of Sorbent Cartridges from Volatile Organic Sampling Train (VOST): Gas Chromatography/Mass Spectrometry Technique	Deleted from SW-846	Deleted from SW-846
—	—	5041 (Up. II)	5041A	Analysis for Desorption of Sorbent Cartridges from Volatile Organic Sampling Train (VOST)	Vol IB Chap 4 Sec 4.2.1	5041A Rev 1 (12/96)
—	—	5050 (Up. II)	—	Bomb Preparation Method for Solid Waste	Vol IC Chap 5	5050 Rev 0 (9/94)
6010	6010A	—	6010B	Inductively Coupled Plasma-Atomic Emission Spectrometry	Vol IA Chap 3 Sec 3.3	6010B Rev 2 (12/96)
—	—	6020 (Up. II)	—	Inductively Coupled Plasma - Mass Spectrometry	Vol IA Chap 3 Sec 3.3	6020 Rev 0 (9/94)

SW-846 Method Status Table (Compiled May 1997) (Continued)

Method Number Third Edition (9/86)	Method Number Final Update I (7/92)	Method Number Final Update II (9/94) IIA (8/93) IIB (1/95)	Method Number Final Update III (12/96)	Method Title	SW-846 Volume/ Chapter/ Section	Current Promulgated Method
7000	7000A	—	—	Atomic Absorption Methods	Vol IA Chap 3 Sec 3.3	7000A Rev 1 (7/92)
7020	—	—	—	Aluminum (Atomic Absorption, Direct Aspiration)	Vol IA Chap 3 Sec 3.3	7020 Rev 0 (9/86)
7040	—	—	—	Antimony (Atomic Absorption, Direct Aspiration)	Vol IA Chap 3 Sec 3.3	7040 Rev 0 (9/86)
7041	—	—	—	Antimony (Atomic Absorption, Furnace Technique)	Vol IA Chap 3 Sec 3.3	7041 Rev 0 (9/86)
7060	—	7060A (Up. II)	—	Arsenic (Atomic Absorption, Furnace Technique)	Vol IA Chap 3 Sec 3.3	7060A Rev 1 (9/94)
7061	7061A	—	—	Arsenic (Atomic Absorption, Gaseous Hydride)	Vol IA Chap 3 Sec 3.3	7061A Rev 1 (7/92)
—	—	7062 (Up. II)	—	Antimony and Arsenic (Atomic Absorption Borohydride Reduction)	Vol IA Chap 3 Sec 3.3	7062 Rev 0 (9/94)
—	—	—	7063	Arsenic in Aqueous Samples and Extracts by Anodic Stripping Voltammetry (ASV)	Vol IA Chap 3 Sec 3.3	7063 Rev 0 (12/96)
7080	—	7080A (Up. II)	—	Barium (Atomic Absorption, Direct Aspiration)	Vol IA Chap 3 Sec 3.3	7080A Rev 1 (9/94)
—	7081	—	—	Barium (Atomic Absorption, Furnace Technique)	Vol IA Chap 3 Sec 3.3	7081 Rev 0 (7/92)
7090	—	—	—	Beryllium (Atomic Absorption, Direct Aspiration)	Vol IA Chap 3 Sec 3.3	7090 Rev 0 (9/86)
7091	—	—	—	Beryllium (Atomic Absorption, Furnace Technique)	Vol IA Chap 3 Sec 3.3	7091 Rev 0 (9/86)

Appendix D

SW-846 Method Status Table (Compiled May 1997) (Continued)

Method Number Third Edition (9/86)	Method Number Final Update I (7/92)	Method Number Final Update II (9/94) IIA (8/93) IIB (1/95)	Method Number Final Update III (12/96)	Method Title	SW-846 Volume/ Chapter/ Section	Current Promulgated Method
7130	—	—	—	Cadmium (Atomic Absorption, Direct Aspiration)	Vol IA Chap 3 Sec 3.3	7130 Rev 0 (9/86)
7131	—	7131A (Up. II)	—	Cadmium (Atomic Absorption, Direct Aspiration)	Vol IA Chap 3 Sec 3.3	7131A Rev 1 (9/94)
7140	—	—	—	Calcium (Atomic Absorption, Direct Aspiration)	Vol IA Chap 3 Sec 3.3	7140 Rev 0 (9/86)
7190	—	—	—	Chromium (Atomic Absorption, Direct Aspiration)	Vol IA Chap 3 Sec 3.3	7190 Rev 0 (9/86)
7191	—	—	—	Chromium (Atomic Absorption, Furnace Technique)	Vol IA Chap 3 Sec 3.3	7191 Rev 0 (9/86)
7195	—	—	—	Chromium, Hexavalent (Coprecipitation)	Vol IA Chap 3 Sec 3.3	7195 Rev 0 (9/86)
7196	7196A	—	—	Chromium, Hexavalent (Colorimetric)	Vol IA Chap 3 Sec 3.3	7196A Rev 1 (7/92)
7197	—	—	—	Chromium, Hexavalent (Chelation/Extraction)	Vol IA Chap 3 Sec 3.3	7197 Rev 0 (9/86)
7198	—	—	—	Chromium, Hexavalent (Differential Pulse Polarography)	Vol IA Chap 3 Sec 3.3	7198 Rev 0 (9/86)
—	—	—	7199	Determination of Hexavalent Chromium in Drinking Water, Groundwater and Industrial Wastewater Effluents by Ion Chromatography	Vol IA Chap 3 Sec 3.3	7199 Rev 0 (12/96)
7200	—	—	—	Cobalt (Atomic Absorption, Direct Aspiration)	Vol IA Chap 3 Sec 3.3	7200 Rev 0 (9/86)

SW-846 Method Status Table (Compiled May 1997) (Continued)

Method Number Third Edition (9/86)	Method Number Final Update I (7/92)	Method Number Final Update II (9/94) IIA (8/93) IIB (1/95)	Method Number Final Update III (12/96)	Method Title	SW-846 Volume/ Chapter/ Section	Current Promulgated Method
7201	—	—	—	Cobalt (Atomic Absorption, Furnace Technique)	Vol IA Chap 3 Sec 3.3	7201 Rev 0 (9/86)
7210	—	—	—	Copper (Atomic Absorption, Direct Aspiration)	Vol IA Chap 3 Sec 3.3	7210 Rev 0 (9/86)
—	7211	—	—	Copper (Atomic Absorption, Furnace Technique)	Vol IA Chap 3 Sec 3.3	7211 Rev 0 (7/92)
7380	—	—	—	Iron (Atomic Absorption, Direct Aspiration)	Vol IA Chap 3 Sec 3.3	7380 Rev 0 (9/86)
—	7381	—	—	Iron (Atomic Absorption, Furnace Technique)	Vol IA Chap 3 Sec 3.3	7381 Rev 0 (7/92)
7420	—	—	—	Lead (Atomic Absorption, Direct Aspiration)	Vol IA Chap 3 Sec 3.3	7420 Rev 0 (9/86)
7421	—	—	—	Lead (Atomic Absorption, Furnace Technique)	Vol IA Chap 3 Sec 3.3	7421 Rev 0 (9/86)
—	7430	—	—	Lithium (Atomic Absorption, Direct Aspiration)	Vol IA Chap 3 Sec 3.3	7430 Rev 0 (7/92)
7450	—	—	—	Magnesium (Atomic Absorption, Direct Aspiration)	Vol IA Chap 3 Sec 3.3	7450 Rev 0 (9/86)
7460	—	—	—	Manganese (Atomic Absorption, Direct Aspiration)	Vol IA Chap 3 Sec 3.3	7460 Rev 0 (9/86)
—	7461	—	—	Manganese (Atomic Absorption, Furnace Technique)	Vol IA Chap 3 Sec 3.3	7461 Rev 0 (7/92)
7470	—	7470A (Up. II)	—	Mercury in Liquid Waste (Manual Cold-Vapor Technique)	Vol IA Chap 3 Sec 3.3	7470A Rev 1 (9/94)

SW-846 Method Status Table (Compiled May 1997) (Continued)

Method Number Third Edition (9/86)	Method Number Final Update I (7/92)	Method Number Final Update II (9/94) IIA (8/93) IIB (1/95)	Method Number Final Update III (12/96)	Method Title	SW-846 Volume/ Chapter/ Section	Current Promulgated Method
7471	—	7471A (Up. II)	—	Mercury in Solid or Semisolid Waste (Manual Cold-Vapor Technique)	Vol IA Chap 3 Sec 3.3	7471A Rev 1 (9/94)
—	—	—	7472	Mercury in Aqueous Samples and Extracts by Anodic Stripping Voltammetry (ASV)	Vol IA Chap 3 Sec 3.3	7472 Rev 0 (12/96)
7480	—	—	—	Molybdenum (Atomic Absorption, Direct Aspiration)	Vol IA Chap 3 Sec 3.3	7480 Rev 0 (9/86)
7481	—	—	—	Molybdenum (Atomic Absorption, Furnace Technique)	Vol IA Chap 3 Sec 3.3	7481 Rev 0 (9/86)
7520	—	—	—	Nickel (Atomic Absorption, Direct Aspiration)	Vol IA Chap 3 Sec 3.3	7520 Rev 0 (9/86)
—	—	—	7521	Nickel (Atomic Absorption, Furnace Method)	Vol IA Chap 3 Sec 3.3	7521 Rev 0 (12/96)
7550	—	—	—	Osmium (Atomic Absorption, Direct Aspiration)	Vol IA Chap 3 Sec 3.3	7550 Rev 0 (9/86)
—	—	—	7580	White Phosphorus (P_4) by Solvent Extraction and Gas Chromatography	Vol IA Chap 3 Sec 3.3	7580 Rev 0 (12/96)
7610	—	—	—	Potassium (Atomic Absorption, Direct Aspiration)	Vol IA Chap 3 Sec 3.3	7610 Rev 0 (9/86)
7740	—	—	—	Selenium (Atomic Absorption, Furnace Technique)	Vol IA Chap 3 Sec 3.3	7740 Rev 0 (9/86)
7741	—	7741A (Up. II)	—	Selenium (Atomic Absorption, Gaseous Hydride)	Vol IA Chap 3 Sec 3.3	7741A Rev 1 (9/94)

Appendix D

SW-846 Method Status Table (Compiled May 1997) (Continued)

Method Number Third Edition (9/86)	Method Number Final Update I (7/92)	Method Number Final Update II (9/94) IIA (8/93) IIB (1/95)	Method Number Final Update III (12/96)	Method Title	SW-846 Volume/ Chapter/ Section	Current Promulgated Method
—	—	7742 (Up. II)	—	Selenium (Atomic Absorption, Borohydride Reduction)	Vol IA Chap 3 Sec 3.3	7742 Rev 0 (9/94)
7760	7760A	—	—	Silver (Atomic Absorption, Direct Aspiration)	Vol IA Chap 3 Sec 3.3	7760A Rev 1 (7/92)
—	7761	—	—	Silver (Atomic Absorption, Furnace Technique)	Vol IA Chap 3 Sec 3.3	7761 Rev 0 (7/92)
7770	—	—	—	Sodium (Atomic Absorption, Direct Aspiration)	Vol IA Chap 3 Sec 3.3	7770 Rev 0 (9/86)
—	7780	—	—	Strontium (Atomic Absorption, Direct Aspiration)	Vol IA Chap 3 Sec 3.3	7780 Rev 0 (7/92)
7840	—	—	—	Thallium (Atomic Absorption, Direct Aspiration)	Vol IA Chap 3 Sec 3.3	7840 Rev 0 (9/86)
7841	—	—	—	Thallium (Atomic Absorption, Furnace Technique)	Vol IA Chap 3 Sec 3.3	7841 Rev 0 (9/86)
7870	—	—	—	Tin (Atomic Absorption, Direct Aspiration)	Vol IA Chap 3 Sec 3.3	7870 Rev 0 (9/86)
7910	—	—	—	Vanadium (Atomic Absorption, Direct Aspiration)	Vol IA Chap 3 Sec 3.3	7910 Rev 0 (9/86)
7911	—	—	—	Vanadium (Atomic Absorption, Furnace Technique)	Vol IA Chap 3 Sec 3.3	7911 Rev 0 (9/86)
7950	—	—	—	Zinc (Atomic Absorption, Direct Aspiration)	Vol IA Chap 3 Sec 3.3	7950 Rev 0 (9/86)
—	7951	—	—	Zinc (Atomic Absorption, Furnace Technique)	Vol IA Chap 3 Sec 3.3	7951 Rev 0 (7/92)

Appendix D

SW-846 Method Status Table (Compiled May 1997) (Continued)

Method Number Third Edition (9/86)	Method Number Final Update I (7/92)	Method Number Final Update II (9/94) IIA (8/93) IIB (1/95)	Method Number Final Update III (12/96)	Method Title	SW-846 Volume/ Chapter/ Section	Current Promulgated Method
8000	8000A	—	8000B	Determinative Chromatographic Separations	Vol IB Chap 4 Sec 4.3.1	8000B Rev 2 (12/96)
8010	8010A	8010B (Up. II)	Deleted from SW-846	Halogenated Volatile Organics by Gas Chromatography	Deleted from SW-846	Deleted from SW-846
—	8011	—	—	1,2-Dibromoethane and 1,2-Dibromo-3-chloro-propane by Microextraction and Gas Chromatography	Vol IB Chap 4 Sec 4.3.1	8011 Rev 0 (7/92)
8015	8015A	—	8015B	Nonhalogenated Organics Using GC/FID	Vol IB Chap 4 Sec 4.3.1	8015B Rev 2 (12/96)
8020	—	8020A (Up. II)	Deleted from SW-846	Aromatic Volatile Organics by Gas Chromotography	Deleted from SW-846	Deleted from SW-846
—	8021	8021A (Up. II)	8021B	Aromatic and Halogenated Volatiles by Gas Chromatography Using Photoionization and/or Electrolytic Conductivity Detectors	Vol IB Chap 4 Sec 4.3.1	8021B Rev 2 (12/96)
8030	8030A	—	Deleted from SW-846	Acrolein and Acrylonitrile by Gas Chromatography	Deleted from SW-846	Deleted from SW-846
—	—	8031 (Up. II)	—	Acrylonitrile by Gas Chromatography	Vol IB Chap 4 Sec 4.3.1	8031 Rev 0 (9/94)
—	—	8032 (Up. II)	8032A	Acrylamide by Gas Chromatography	Vol IB Chap 4 Sec 4.3.1	8032A Rev 1 (12/96)
—	—	—	8033	Acetonitrile by Gas Chromatography with Nitrogen-Phosphorus Detection	Vol IB Chap 4 Sec 4.3.1	8033 Rev 0 (12/96)

Appendix D

SW-846 Method Status Table (Compiled May 1997) (Continued)

Method Number Third Edition (9/86)	Method Number Final Update I (7/92)	Method Number Final Update II (9/94) IIA (8/93) IIB (1/95)	Method Number Final Update III (12/96)	Method Title	SW-846 Volume/ Chapter/ Section	Current Promulgated Method
8040	8040A	—	Deleted from SW-846	Phenols by Gas Chromatography	Deleted from SW-846	Deleted from SW-846
—	—	—	8041	Phenols by Gas Chromatography	Vol IB Chap 4 Sec 4.3.1	8041 Rev 0 (12/96)
8060	—	—	Deleted from SW-846	Phthalate Esters	Deleted from SW-846	Deleted from SW-846
—	—	8061 (Up. II)	8061A	Phthalate Esters by Gas Chromatography with Electron Capture Detection (GC/ECD)	Vol IB Chap 4 Sec 4.3.1	8061 Rev 1 (12/96)
—	8070	—	8070A	Nitrosamines by Gas Chromatography	Vol IB Chap 4 Sec 4.3.1	8070A Rev 1 (12/96)
8080	—	8080A (Up. II)	Deleted from SW-846	Organochlorine Pesticides and Polychlorinated Biphenyls by Gas Chromatography	Deleted from SW-846	Deleted from SW-846
—	—	8081 (Up. II)	8081A	Organochlorine Pesticides by Gas Chromatography	Vol IB Chap 4 Sec 4.3.1	8081A Rev 1 (12/96)
—	—	—	8082	Polychlorinated Biphenyls (PCBs) by Gas Chromatography	Vol IB Chap 4 Sec 4.3.1	8082 Rev 0 (12/96)
8090	—	—	Deleted from SW-846	Nitroaromatics and Cyclic Ketones	Deleted from SW-846	Deleted from SW-846
—	—	—	8091	Nitroaromatics and Cyclic Ketones by Gas Chromatography	Vol IB Chap 4 Sec 4.3.1	8091 Rev 0 (12/96)
8100	—	—	—	Polynuclear Aromatic Hydrocarbons	Vol IB Chap 4 Sec 4.3.1	8100 Rev 0 (9/86)

SW-846 Method Status Table (Compiled May 1997) (Continued)

Method Number Third Edition (9/86)	Method Number Final Update I (7/92)	Method Number Final Update II (9/94) IIA (8/93) IIB (1/95)	Method Number Final Update III (12/96)	Method Title	SW-846 Volume/ Chapter/ Section	Current Promulgated Method
—	8110	—	Deleted from SW-846	Haloethers by Gas Chromatography	Deleted from SW-846	Deleted from SW-846
—	—	—	8111	Haloethers by Gas Chromatography	Vol IB Chap 4 Sec 4.3.1	8111 Rev 0 (12/96)
8120	—	8120A (Up. II)	Deleted from SW-846	Chlorinated Hydrocarbons by Gas Chromatography	Deleted from SW-846	Deleted from SW-846
—	—	8121 (Up. II)	—	Chlorinated Hydrocarbons by Gas Chromatography Capillary Column Technique	Vol IB Chap 4 Sec 4.3.1	8121 Rev 0 (9/94)
—	—	—	8131	Aniline and Selected Derivatives by Gas Chromatography	Vol IB Chap 4 Sec 4.3.1	8131 Rev 0 (12/96)
8140	—	—	Deleted from SW-846	Organophosphorus Pesticides	Deleted from SW-846	Deleted from SW-846
—	8141	8141A (Up. II)	—	Organophosphorus Compounds by Gas Chromatography; Capillary Column Technique	Vol IB Chap 4 Sec 4.3.1	8141A Rev 1 (9/94)
8150	8150A	8150B (Up. II)	Deleted from SW-846	Chlorinated Herbicides by Gas Chromatography	Deleted from SW-846	Deleted from SW-846
—	—	8151 (Up. II)	8151A	Chlorinated Herbicides by GC Using Methylation or Pentafluorobenzylation Derivatization	Vol IB Chap 4 Sec 4.3.1	8151A Rev 1 (12/96)
8240	8240A	8240B (Up. II)	Deleted from SW-846	Volatile Organic Compounds by Gas Chromatography/Mass Spectrometry (GC/MS)	Deleted from SW-846	Deleted from SW-846

Appendix D

SW-846 Method Status Table (Compiled May 1997) (Continued)

Method Number Third Edition (9/86)	Method Number Final Update I (7/92)	Method Number Final Update II (9/94) IIA (8/93) IIB (1/95)	Method Number Final Update III (12/96)	Method Title	SW-846 Volume/ Chapter/ Section	Current Promulgated Method
8250	—	8250A (Up. II)	Deleted from SW-846	Semivolatile Organic Compounds by Gas Chromatography/Mass Spectrometry (GC/MS)	Deleted from SW-846	Deleted from SW-846
—	8260	8260A (Up. II)	8260B	Volatile Organic Compounds by Gas Chromatography/Mass Spectrometry (GC/MS)	Vol IB Chap 4 Sec 4.3.2	8260B Rev 2 (12/96)
8270	8270A	8270B (Up. II)	8270C	Semivolatile Organic Compounds by Gas Chromatography/Mass Spectrometry (GC/MS)	Vol IB Chap 4 Sec 4.3.2	8270C Rev 3 (12/96)
—	—	8275 (Up. II)	8275A	Semivolatile Organic Compounds (PAHs and PCBs) in Soils/Sludges and Solid Wastes Using Thermal Extraction/Gas Chromatography/Mass Spectrometry (TE/GC/MS)	Vol IB Chap 4 Sec 4.3.2	8275A Rev 1 (12/96)
8280	—	—	8280A	The Analysis of Polychlorinated Dibenzo-p-Dioxins and Polychlorinated Diben-zofurans by High Reso-lution Gas Chroma-ography/Low Resolution Mass Spectrometry (HRGC/LRMS)	Vol IB Chap 4 Sec 4.3.2	8280A Rev 1 (12/96)
—	—	8290 (Up. II)	—	Polychlorinated Dibenzodioxins (PCDDs) and Polychlorinated Dibenzofurans (PCDFs) by High-Resolution Gas Chromatography/High-Resolution Mass Spectrometry (HRGC/HRMS)	Vol IB Chap 4 Sec 4.3.2	8290 Rev 0 (9/94)
8310	—	—	—	Polynuclear Aromatic Hydrocarbons	Vol IB Chap 4 Sec 4.3.3	8310 Rev 0 (9/86)

SW-846 Method Status Table (Compiled May 1997) (Continued)

Method Number Third Edition (9/86)	Method Number Final Update I (7/92)	Method Number Final Update II (9/94) IIA (8/93) IIB (1/95)	Method Number Final Update III (12/96)	Method Title	SW-846 Volume/ Chapter/ Section	Current Promulgated Method
—	—	8315 (Up. II)	8315A	Determination of Carbonyl Compounds by High Performance Liquid Chromatography (HPLC)	Vol IB Chap 4 Sec 4.3.3	8315A Rev 1 (12/96)
—	—	8316 (Up. II)	—	Acrylamide, Acrylonitrile and Acrolein by High Performance Liquid Chromatography (HPLC)	Vol IB Chap 4 Sec 4.3.3	8316 Rev 0 (9/94)
—	—	8318 (Up. II)	—	N-Methylcarbamates by High Performance Liquid Chromatography (HPLC)	Vol IB Chap 4 Sec 4.3.3	8318 Rev 0 (9/94)
—	—	8321 (Up. II)	8321A	Solvent Extractable Non-volatile Compounds by High Performance Liquid Chromatography-Thermo-spray/ Mass Spectrometry (HPLC/TS/MS) or Ultraviolet (UV) Detection	Vol IB Chap 4 Sec 4.3.3	8321A Rev 1 (12/96)
—	—	—	8325	Solvent Extractable Non-volatile Compounds by High Performance Liquid Chromatography/Particle Beam/ Mass Spectrometry (HPLC/PB/MS)	Vol IB Chap 4 Sec 4.3.3	8325 Rev 0 (12/96)
—	—	8330 (Up. II)	—	Nitroaromatics and Nitramines by High Performance Liquid Chromatography (HPLC)	Vol IB Chap 4 Sec 4.3.3	8330 Rev 0 (9/94)
—	—	8331 (Up. II)	—	Tetrazene by Reverse Phase High Performance Liquid Chromatography (HPLC)	Vol IB Chap 4 Sec 4.3.3	8331 Rev 0 (9/94)
—	—	—	8332	Nitroglycerine by High Performance Liquid Chromatography	Vol IB Chap 4 Sec 4.3.3	8332 Rev 0 (12/96)

Appendix D

SW-846 Method Status Table (Compiled May 1997) (Continued)

Method Number Third Edition (9/86)	Method Number Final Update I (7/92)	Method Number Final Update II (9/94) IIA (8/93) IIB (1/95)	Method Number Final Update III (12/96)	Method Title	SW-846 Volume/ Chapter/ Section	Current Promulgated Method
—	—	8410 (Up. II)	—	Gas Chromatography/ Fourier Transform Infrared (GC/FT-IR) Spectrometry for Semivolatile Organics: Capillary Column	Vol IB Chap 4 Sec 4.3.4	8410 Rev 0 (9/94)
—	—	—	8430	Analysis of Bis(2-chloroethyl) Ether and Hydrolysis Products by Direct Aqueous Injection (GC/FT-IR)	Vol IB Chap 4 Sec 4.3.4	8430 Rev 0 (12/96)
—	—	—	8440	Total Recoverable Petroleum Hydro-carbons by Infrared Spectrophotometry	Vol IB Chap 4 Sec 4.3.4	8440 Rev 0 (12/96)
—	—	—	8515	Colorimetric Screening Method for Trinitro-toluene (TNT) in Soil	Vol IB Chap 4 Sec 4.5	8515 Rev 0 (12/96)
—	—	—	8520	Continuous Measure-ment of Formaldehyde in Ambient Air	Vol IB Chap 4 Sec 4.3.5	8520 Rev 0 (12/96)
9010	9010A	—	9010B	Total and Amenable Cyanide: Distillation	Vol IC Chap 5	9010B Rev 2 (12/96)
9012	—	—	9012A	Total and Amenable Cyanide (Automated Colorimetric, with Off-line Distillation)	Vol IC Chap 5	9012A Rev 1 (12/96)
—	9013	—	—	Cyanide Extraction Procedure for Solids and Oils	Vol IC Chap 5	9013 Rev 0 (7/92)
—	—	—	9014	Titrimetric and Manual Spectrophotometric Determinative Methods for Cyanide	Vol IC Chap 5	9014 Rev 0 (12/96)
9020	9020A	9020B (Up. II)	—	Total Organic Halides (TOX)	Vol IC Chap 5	9020B Rev 2 (9/94)

Appendix D

SW-846 Method Status Table (Compiled May 1997) (Continued)

Method Number Third Edition (9/86)	Method Number Final Update I (7/92)	Method Number Final Update II (9/94) IIA (8/93) IIB (1/95)	Method Number Final Update III (12/96)	Method Title	SW-846 Volume/ Chapter/ Section	Current Promulgated Method
—	9021	—	—	Purgeable Organic Halides (POX)	Vol IC Chap 5	9021 Rev 0 (7/92)
9022	—	—	—	Total Organic Halides (TOX) by Neutron Activation Analysis	Vol IC Chap 5	9022 Rev 0 (9/86)
—	—	—	9023	Extractable Organic Halides (EOX) in Solids	Vol IC Chap 5	9023 Rev 0 (12/96)
9030	9030A	—	9030B	Acid-Soluble and Acid-Insoluble Sulfides: Distillation	Vol IC Chap 5	9030B Rev 2 (12/96)
—	9031	—	—	Extractable Sulfides	Vol IC Chap 5	9031 Rev 0 (7/92)
—	—	—	9034	Titrimetric Procedure for Acid-soluble and Acid-insoluble Sulfides	Vol IC Chap 5	9034 Rev 0 (12/96)
9035	—	—	—	Sulfate (Colorimetric, Automated, Chloranilate)	Vol IC Chap 5	9035 Rev 0 (9/86)
9036	—	—	—	Sulfate (Colorimetric, Automated, Methyl-thymol Blue, AA II)	Vol IC Chap 5	9036 Rev 0 (9/86)
9038	—	—	—	Sulfate (Turbidimetric)	Vol IC Chap 5	9038 Rev 0 (9/86)
9040	—	9040A (Up. II) 9040B (Up. IIB)	—	pH Electrometric Measurement	Vol IC Chap 6	9040B Rev 2 (1/95)
9041	9041A	—	—	pH Paper Method	Vol IC Chap 6	9041A Rev 1 (7/92)

SW-846 Method Status Table (Compiled May 1997) (Continued)

Method Number Third Edition (9/86)	Method Number Final Update I (7/92)	Method Number Final Update II (9/94) IIA (8/93) IIB (1/95)	Method Number Final Update III (12/96)	Method Title	SW-846 Volume/ Chapter/ Section	Current Promulgated Method
9045	9045A	9045B (Up. II) 9045C (Up. II)	—	Soil and Waste pH	Vol IC Chap 6	9045C Rev 3 (1/95)
9050	—	—	9050A	Specific Conductance	Vol IC Chap 6	9050A Rev 1 (12/96)
—	—	9056 (Up. II)	—	Determination of Inorganic Anions by Ion Chromatography	Vol IC Chap 5	9056 Rev 0 (9/94)
—	—	—	9057	Determination of Chloride from HCl/Cl_2 Emission Sampling Train (Methods 0050 and 0051) by Anion Chromatography	Vol IC Chap 5	9057 Rev 0 (12/96)
9060	—	—	—	Total Organic Carbon	Vol IC Chap 5	9060 Rev 0 (9/86)
9065	—	—	—	Phenolics (Spectro-photometric, Manual 4-AAP with Distillation)	Vol IC Chap 5	9065 Rev 0 (9/86)
9066	—	—	—	Phenolics (Colorimetric, Automated 4-AAP with Distillation)	Vol IC Chap 5	9066 Rev 0 (9/86)
9067	—	—	—	Phenolics (Spectrophotometric, MBTH with Distillation)	Vol IC Chap 5	9067 Rev 0 (9/86)
9070	—	—	—	Total Recoverable Oil and Grease (Gravi-metric, Separatory Funnel Extraction)	Vol IC Chap 5	9070 Rev 0 (9/86)
9071	—	9071A (Up. II)	—	Oil and Grease Extraction Method for Sludge and Sediment Samples	Vol IC Chap 5	9071A Rev 1 (9/94)

Method Number Third Edition (9/86)	Method Number Final Update I (7/92)	Method Number Final Update II (9/94) IIA (8/93) IIB (1/95)	Method Number Final Update III (12/96)	Method Title	SW-846 Volume/ Chapter/ Section	Current Promulgated Method
—	—	9075 (Up. II)	—	Test Method for Total Chlorine in New and Used Petroleum Products by X-Ray Fluorescence Spectrometry (XRF)	Vol IC Chap 5	9075 Rev 0 (9/94)
—	—	9076 (Up. II)	—	Test Method for Total Chlorine in New and Used Petroleum Products by Oxidative Combustion and Microcoulometry	Vol IC Chap 5	9076 Rev 0 (9/94)
—	—	9077 (Up. II)	—	Test Methods for Total Chlorine in New and Used Petroleum Products (Field Test Kit Methods)	Vol IC Chap 5	9077 Rev 0 (9/94)
—	—	—	9078	Screening Test Method for Polychlorinated Biphenyls in Soil	Vol IB Chap 4 Sec 4.5	9078 Rev 0 (12/96)
—	—	—	9079	Screening Test Method for Polychlorinated Biphenyls in Transformer Oil	Vol IB Chap 4 Sec 4.5	9079 Rev 0 (12/96)
9080	—	—	—	Cation-Exchange Capacity of Soils (Ammonium Acetate)	Vol IC Chap 6	9080 Rev 0 (9/86)
9081	—	—	—	Cation-Exchange Capacity of Soils (Sodium Acetate)	Vol IC Chap 6	9081 Rev 0 (9/86)
9090	9090A	—	—	Compatibility Test for Wastes and Membrane Liners	Vol IC Chap 6	9090A Rev 1 (7/92)
9095	—	—	9095A	Paint Filter Liquids Test	Vol IC Chap 6	9095A Rev 1 (12/96)
—	—	9096 (Up. II)	—	Liquid Release Test (LRT) Procedure	Vol IC Chap 6	9096 Rev 0 (9/94)

Appendix D

SW-846 Method Status Table (Compiled May 1997) (Continued)

Method Number Third Edition (9/86)	Method Number Final Update I (7/92)	Method Number Final Update II (9/94) IIA (8/93) IIB (1/95)	Method Number Final Update III (12/96)	Method Title	SW-846 Volume/ Chapter/ Section	Current Promulgated Method
9100	—	—	—	Saturated Hydraulic Conductivity, Saturated Leachate Conductivity, and Intrinsic Permeability	Vol IC Chap 6	9100 Rev 0 (9/86)
9131	—	—	—	Total Coliform: Multiple Tube Fermentation Technique	Vol IC Chap 5	9131 Rev 0 (9/86)
9132	—	—	—	Total Coliform: Membrane-Filter Technique	Vol IC Chap 5	9132 Rev 0 (9/86)
9200	—	—	Deleted from SW-846	Nitrate	Deleted from SW-846	Deleted from SW-846
—	—	—	9210	Potentiometric Determination of Nitrate in Aqueous Samples with Ion-Selective Electrode	Vol IC Chap 5	9210 Rev 0 (12/96)
—	—	—	9211	Potentiometric Determination of Bromide in Aqueous Samples with Ion-Selective Electrode	Vol IC Chap 5	9211 Rev 0 (12/96)
—	—	—	9212	Potentiometric Determination of Chloride in Aqueous Samples with Ion-Selective Electrode	Vol IC Chap 5	9212 Rev 0 (12/96)
—	—	—	9213	Potentiometric Determination of Cyanide in Aqueous Samples and Distillates with Ion-Selective Electrode	Vol IC Chap 5	9213 Rev 0 (12/96)
—	—	—	9214	Potentiometric Determination of Fluroide in Aqueous Samples with Ion-Selective Electrode	Vol IC Chap 5	9214 Rev 0 (12/96)

SW-846 Method Status Table (Compiled May 1997) (Continued)

Method Number Third Edition (9/86)	Method Number Final Update I (7/92)	Method Number Final Update II (9/94) IIA (8/93) IIB (1/95)	Method Number Final Update III (12/96)	Method Title	SW-846 Volume/ Chapter/ Section	Current Promulgated Method
—	—	—	9215	Potentiometric Determination of Sulfide in Aqueous Samples and Distillates with Ion-Selective Electrode	Vol IC Chap 5	9215 Rev 0 (12/96)
9250	—	—	—	Chloride (Colorimetric, Automated Ferricyanide AAI)	Vol IC Chap 5	9250 Rev 0 (9/86)
9251	—	—	—	Chloride (Colorimetric, Automated Ferricyanide AAII)	Vol IC Chap 5	9251 Rev 0 (9/86)
9252	—	9252A (Up. II)	Deleted from SW-846	Chloride (Titrimetric, Mercuric Nitrate)	Deleted from SW-846	Deleted from SW-846
	—	9253 (Up. II)	—	Chloride (Titrimetric, Silver Nitrate)	Vol IC Chap 5	9253 Rev 0 (9/94)
9310	—	—	—	Gross Alpha and Gross Beta	Vol IC Chap 6	9310 Rev 0 (9/86)
9315	—	—	—	Alpha-Emitting Radium Isotopes	Vol IC Chap 6	9315 Rev 0 (9/86)
9320	—	—	—	Radium - 228	Vol IC Chap 5	9320 Rev 0 (9/86)
HCN Test Method	HCN Test Method	HCN Test Method (Up. II)	HCN Test Method	Test Method to Determine Hydrogen Cyanide Released from Wastes	Vol IC Chap 7 Sec 7.3.3.2	Guidance Method Only Rev 3 (12/96)
H_2S Test Method	H_2S Test Method	H_2S Test Method (Up. II)	H_2S Test Method	Test Method to Determine Hydrogen Sulfide Released from Wastes	Vol IC Chap 7 Sec 7.3.4.2	Guidance Method Only Rev 3 (12/96)

Appendix D

APPENDIX E

Contained in this section are definitions and explanations of terms and abbreviations used in this book or commonly in use in this field.

AA. Atomic Absorption.

Absorb. To soak up. The incorporation of a liquid into a solid substance, as by capillary, osmotic, solvent, or chemical action.

Absorbance. A measure of the decrease in incident light passing through a sample into the detector. It is defined mathematically as:

$$A = \frac{I \text{ (solvent)}}{I \text{ (solution)}} = \log \frac{Io}{I}$$

Where, I = radiation intensity

Absorption. To absorb. The process of incorporating a substance (liquid or gas) into the body of another substance (solid).

Accuracy. Accuracy means the nearness of a result or the mean ($\bar{x}$) of a set of results to the true value. Accuracy is assessed by means of reference samples and percent recoveries.

Acid. An inorganic or organic compound that 1) reacts with metals to yield hydrogen; 2) reacts with a base to form a salt; 3) dissociates in water to yield hydrogen ions; 4) has a pH of less than 7.0; and 5) neutralizes bases or alkalies. All acids contain hydrogen and turn litmus paper red. They are corrosive to human tissue and are to be handled with care. See Base; pH.

Action Levels. Typically action levels are considered regulatory levels set by federal government agencies such as the EPA, the Food and Drug Administration (FDA), and the U.S. Department of Agriculture. The presence of a contaminant in high enough concentration to warrant action or a response under regulatory guidelines.

Adsorb. To attract and retain gas, liquid, or dissolved substances on the surface of another material.

Adsorbtion. To adsorb. Adhesion of molecules of gas, liquid, or dissolved solids to a surface. Adsorbtion is a surface phenomenon.

Aliquot. A measured portion of a sample taken for analysis.

Alkali. Any compound having highly basic properties; i.e., one that readily ionizes in aqueous solution to yield OH anions, with a pH above 7, and turns litmus paper blue. Examples are oxides and hydroxides of certain metals belonging to group IA of the periodic table (Li, Na, K, Rb, Cs, Fr). Ammonia and amines may also be alkaline. Alkalies are caustic and dissolve human tissue. Treat alkali burns by quickly washing the affected area with large amounts of water for at least 15 min. Common commercial alkalies are sodium carbonate (soda ash), caustic soda and caustic potash, lime, lye, waterglass, regular mortar, Portland cement, and bicarbonate of soda. See Acid; Base; pH.

Alpha Particle. A particle emitted from radioactive decay. The particle is a helium nucleus consisting of two neutrons and two protons.

Ambient. Usual or surrounding conditions of temperatures, humidity, etc.

Analysis Date/Time. The date and military time (24 hour clock) of the introduction of the sample, standard, or blank into the analysis system.

Analyte. The element or ion compound an analyst seeks to determine; the element of interest.

Analytical Batch. The basic unit for analytical quality control is the analytical batch. The analytical batch is defined as samples which are analyzed together with the same method sequence and the same lots of reagents and with the manipulations common to each sample within the same time period or in continuous sequential time periods. Samples in each batch should be of similar composition (e.g. groundwater, sludge, ash, etc.)

Analytical Sample. Any solution or media introduced into an instrument on which an analysis is performed excluding instrument calibration, initial calibration verification, initial calibration blank, continuing calibration verification and continuing calibration blank. Note the following are all defined as analytical samples: undiluted and diluted samples (EPA and non-EPA), predigestion spike samples, duplicate samples, serial dilution samples, analytical spike samples, post-digestion spike samples, interference check samples (ICS), CRDL standard for AA (CRA), CRDL standard for ICP (CRI), laboratory control sample (LCS), preparation blank (PB) and linear range analysis sample (LRS).

Analytical Spike. (Inorganic Analysis) The post-digestion spike. The addition of a known amount of standard after digestion.

Aromatics. A class of cyclic hydorcarbon compounds with the simplest member of the series being benzene. These unsaturated, multi-ringed compounds are found in petroleum products and in emissions produced by coal burning power plants, engines that burn gasoline and diesel fuel, and incinerators used for refuse disposal. Many aromatic hydrocarbons are carcinogenic.

Asbestos. A silicate compound combined with other minerals that form long flexible fibers. Asbestos was used for its fire resistancy. Asbestos was found to cause cancer (asbestosis) and banned from use. Asbestos can exist in many forms. The most common are chrysotile, anthophyllite, amosite, actinolite, tremolite, and crocidolite.

ASTM. American Society for Testing and Materials. An organization that devises consensus standards for materials characterization and use. (1916 Race Street, Philadelphia, PA 19103; [215] 299-5400.)

atm. Atmosphere. A unit of pressure equal to the average pressure the air exerts at sea level. One atm = 1.013×10^5 N/m^2, or 14.7 lb/in.^{2}, or 760 mm Hg. Generally used in connection with high pressures.

Autozero. (Inorganic Analysis) Zeroing the instrument at the proper wavelength. It is equivalent to running a standard blank with the absorbance set at zero.

Average Intensity. (Inorganic Analysis) The average of two different injections (exposures).

Background Correction. (Inorganic Analysis) A technique to compensate for variable back-ground contribution to the instrument signal in the determination of trace elements.

Background Level. In air pollution control, the concentration of air pollutants in a defined area during a fixed period of time prior to initiating or stopping a controlled emission. In toxic substances monitoring, the average presence in the environment, originally referring to naturally occurring phenomena.

Bar Graph Spectrum. (Organic Analysis) A plot of the mass-to-charge ratio (m/e) versus relative intensity of the ion current.

Base. Substances that (usually) liberate OH anions when dissolved in water. Bases react with acids to form salts and water. Bases have a pH > 7, turn litmus paper blue, and may be corrosive to human tissue. A strong base is called alkaline or caustic. Examples are lye and DRANO™. See Acid; Alkali; pH.

Batch. A group of samples prepared at the same time in the same location using the same method.

Bias. Consistent deviation of measured values from the true value, caused by systematic errors in a procedure.

Bioaccumulation. A phenomenon that occurs in living organisms that are exposed to pollution or contamination existing in their environment. Substances are concentrated in organisms by uptake through breathing, ingestion, or physical contact. Certain substances are slowly metabolized or eliminated and therefore accumulate in the organism. At each level of the food chain these bioaccumulated substances can increase, a process known as biomagnification.

Biochemical Oxygen Demand (BOD). A measure of the amount of oxygen consumed in the biological processes that break down organic matter in water. The greater the BOD, the greater the degree of pollution.

Biodegradable. Capable of being broken down, especially into innocuous products, under natural conditions and processes.

Blank. (also see Method Blank) A blank is an artificial sample designed to monitor the introduction of artifacts into the process. For aqueous samples, reagent water is used as a blank matrix; however, a universal blank matrix does not exist for solid samples, but sometimes clean sand is used as a blank matrix. The blank is taken through the appropriate steps of the process. A reagent blank is an aliquot of analyte-free water or solvent analyzed with the analytical batch. Field blanks are aliquots of analyte-free water or solvents brought to the field in sealed containers and transported back to the laboratory with the sample containers. Trip blanks and equipment blanks are two specific types of field blanks. Trip blanks are not opened in the field. They are a check on sample contamination originating from sample transport, shipping and from site conditions. Equipment blanks are opened in the field and the contents are poured appropriately over or through the sample collection device, collected in a sample container, and returned to the laboratory as a sample. Equipment blanks are a check on sampling device cleanliness.

B/N/A. Base, Neutral, Acid Extractable Compounds

Boiling Point, BP. The temperature at which a liquid's vapor pressure equals the surrounding atmospheric pressure so that the liquid rapidly vaporizes. Flammable materials with low BPs generally present special fire hazards [e.g., butane, BP = 31°F (-0.5°C); gasoline, BP = 100°F (38°C)]. For mixtures, a range of temperature is given.

BP. See Boiling Point.

4-Bromofluorobenzene (BFB). (Organic Analysis) Compound chosen to establish mass spectral instrument performance for volatile analyses. Also used as a surrogate for volatile organic analysis.

Buffer. A substance that reduces the change in hydrogen ion concentration (pH) otherwise produced by adding acids or bases to a solution. A pH stabilizer.

By-Product. Material, other than the principal product, that is generated as a consequence of an industrial process.

°C. Degrees Celsius (centigrade). Metric temperature scale on which 0 = water's freezing point and 100 = its boiling point. °F = (°C x 9/5) + 32. °C = (°F - 32) x 5/9. See °F.

Calibration. (Inorganic Analysis) The establishment of an analytical curve based on the absorbance, emission intensity, or other measured characteristic of known standards. The calibration standards must be prepared using the same type of acid and reagents or concentration of acids as used in the sample preparation.

Calibration Blank. (Inorganic Analysis) Usually an organic or aqueous solution that is as free of analyte as possible and prepared with the same volume of chemical reagents used in the preparation of the calibration standards and diluted to the appropriate volume with the same solvent (water or organic) used in the preparation of the calibration standard. The calibration blank is used to give the null reading for the instrument response versus concentration calibration curve. One calibration blank should be analyzed with each analytical batch or every method specified number of samples, whichever is greater.

Calibration Check. Verification of the ratio of instrument response to analyte amount, a calibration check is done by analyzing for analyte standards in an appropriate solvent. Calibration check solutions are made from a stock solution which is different from the stock used to prepare standards.

Calibration Check Standard. Standard used to determine the state of calibration of an instrument between periodic recalibrations.

Calibration Standards. A series of known standard solutions used by the analyst for calibration of the instrument (i.e., preparation of the analytical curve).

Carcinogen. A cancer-causing substance.

CAS Number (CAS Registration Number). An assigned number used to identify a chemical. CAS stands for Chemical Abstracts Service, an organization that indexes information published in *Chemical Abstracts* by the American Chemical Society and that provides index guides by which information about particular substances may be located in the abstracts. Sequentially assigned CAS numbers identify *specific* chemicals, except when followed by an asterisk (*) which signifies a compound (often naturally occurring) of variable composition. The numbers have no chemical significance. The CAS number is a concise, unique means of material identification. (Chemical Abstracts Service, Div. of American Chemical Society, Box 3012, Columbus, OH 43210; [614] 447-3600.)

Caustic. See Alkali.

CCC. Calibration Check Compound.

Appendix E

CERCLA. The Comprehensive Environmental Response, Compensation, and Liability Act. The Superfund Law, Public Law PL 96-510, found at 40 CFR 300. The EPA has jurisdiction. Enacted Dec. 11, 1980, and amended thereafter, CERCLA provides for identification and cleanup of hazardous materials released over the land and into air, waterways, and groundwater. It covers areas affected by newly released materials and older leaking or abandoned dump sites. Report releases of hazardous materials to the National Response Center, (800) 424-8802. CERCLA established the superfund, a trust fund to help pay for cleanup of hazardous materials sites. The EPA has authority to collect cleanup costs from those who release the waste material. Cleanup funds come from fines and penalties, from taxes on chemical/petrochemical feed stocks, and the US Treasury Dept. A separate fund collects taxes on active disposal sites to finance monitoring after they close. CERCLA is a result of the serious problems that arose from release of hazardous materials in Love Canal area near Niagara Falls, NY, in August 1978.

CFR. *Code of Federal Regulations*. A collection of the regulations established by law. Contact the agency that issued the regulation for details, interpretations, etc. Copies are sold by the Superintendent of Documents, Government Printing Office, Washington, DC 20402; (202) 783-3238.

Characteristic. As part of the RCRA Toxicity Characteristic (TC) Rule, wastes are compared against four characteristics in defining if they are hazardous: ignitability, corrosivity, reactivity, and toxicity.

Characterization. A determination of the approximate concentration range of compounds of interest used to choose the appropriate analytical protocol.

Check Sample. A blank which has been spiked with the analyte(s) from an independent source in order to monitor the execution of the analytical method is called a check sample. The level of the spike shall be at the regulatory action level when applicable. Otherwise, the spike shall be at 5 times the estimate of the quantification limit. The matrix used shall be phase matched with the samples and well characterized: for example, reagent grade water is appropriate for an aqueous sample.

Check Standard. A material of known composition that is analyzed concurrently with test samples to evaluate a measurement process. An analytical standard that is analyzed to verify the calibration of the analytical system. One check standard should be analyzed with each analytical batch or every 20 samples, whichever is greater. See Calibration Check Standard; Calibration Check.

Chemical Oxygen Demand (COD). A measure of the oxygen required to oxidize all compounds in water, both organic and inorganic.

CHEMTREC. Chemical Transportation Emergency Center. Established in Washington, DC, by the Chemical Manufacturers Association (CMA) to provide emergency information on materials involved in transportation accidents. Twenty-four-hour number: (800) 424-9300. In Washington, DC, Alaska, and Hawaii call (202) 483-7616.

Chlorinated Hydrocarbons. A broad class of chemical compounds that contain chlorine as part of their molecular structure. Chlorinated pesticides such as DDT, aldrin, dieldrin, and mirex are examples. Other examples are chlorinated solvents such as chloroform, methylene chloride, methyl chloride, vinyl chloride, and 1,2,3-trichloropropane.

Chlorinated Solvent. See Chlorinated Hydrocarbons.

Chlorination. The application of chlorine to drinking water, sewage, or industrial waste to disinfect or oxidize undesirable compounds.

Chlorofluorocarbons (CFCs). A family of inert, nontoxic, and easily liquified chemicals used in refrigeration, air conditioning, packaging and insulation and as solvents and aerosol propellants. Because CFCs are not destroyed in the lower atmosphere, they drift into the upper atmosphere, where their chlorine components destroy ozone.

Clean Air Act (CAA). Originally passed in 1963 and since much amended, all with the goal of reducing air pollution and protecting air quality. It has addressed acid rain, urban smog, automobile emissions, toxic air pollutants, and depletion of the ozone layer.

CLP. Contract Laboratory Program.

COD. Chemical oxygen demand.

Code of Federal Regulations. See CFR.

Coefficient of Variation (CV). The standard deviation as a percent of the arithmetic mean.

Combustible. A term the NFPA, DOT, and others use to classify certain materials with low flash points that ignite easily. Both NFPA and DOT generally define *combustible liquids* as having a flash point of 100°F (38°C) or higher. The NFPA classifies nonliquid materials such as wood and paper as *ordinary combustibles*. OSHA defines *combustible liquid* within the Hazard Communication Law as any liquid with a flash point at or above 100°F (38°C) but below 200°F (93.3°C). See Flammable.

Confidence Coefficient. The probability, %, that a measurement result will lie within the confidence interval or between the confidence limits.

Confidence Interval. Set of possible values within which the true value will lie with a specified level of probability.

Confidence Limit. One of the boundary values defining the confidence interval.

Contaminant. Any physical, chemical, biological, or radiological substance or matter that has an adverse affect on air, water, or soil.

Continuing Calibration. (Organic Analysis) Analytical standard run every 12 hours to verify the calibration of the GC/MS system.

Continuing Calibration. (Inorganic Analysis) Analytical standard run every 10 analytical samples or every 2 hours, whichever is more frequent, to verify the calibration of the analytical system.

Continuous Liquid-Liquid Extraction. Used herein synonymously with the terms continuous extraction, continuous liquid extraction and liquid extraction. This extraction technique involves boiling the extraction solvent in a flask and condensing the solvent above the aqueous sample. The condensed solvent drips through the sample, extracting the compounds of interest from the aqueous phase.

Contract Labs. Laboratories contracted by the EPA to analyze samples taken from wastes, soil, air, and water or to carry out research projects.

Contract Required Detection Limit (CRDL). Minimum level of detection acceptable under the contract Statement of Work.

Control Limits. A range within which specified measurement results must fall to be compliant.

Control limits may be mandatory, requiring corrective action if exceeded, or advisory, requiring that noncompliant data be flagged.

Corrective Action. Under the 1984 amendments to the RCRA, the EPA can require RCRA permittees or applicants for permits to take corrective action for releases of hazardous waste or hazardous constituents from solid waste management units (SWMUs) at a permitted facility. Off-site contamination must also be remediated. Facilities with interim permitting status may also be required to take corrective actions. The corrective action process under RCRA roughly parallels the remedial action process under CERCLA.

Correlation Coefficient. A number (r) which indicates the degree of dependence between two variables (concentration - absorbance). The more dependent they are the closer the value to one. Determined on the basis of the least squares line.

Corrosive. A chemical that causes visible destruction of or irreversible alterations in living tissue by chemical action at the site of contact, or which causes a severe corrosion rate in steel or aluminum. A waste that exhibits a "characteristic of corrosivity (40 CFR 261.22)," as defined by RCRA, may be regulated by EPA as a hazardous waste.

CRQL. Contract Required Quantitation Limit.

Cryogenic. Relating to extremely low temperature as for refrigerated gases.

cu ft, ft³. Cubic foot. Cu ft is more usual.

cu m, m³. Cubic meter. m³ is preferred.

CVAA. Cold Vapor Atomic Absorption.

CWA. Clean Water Act. Public Law PL 92-500. Found at 40 CFR 100-140 and 400-470. Effective November 18, 1972, and amended significantly since then. EPA and Army Corps of Engineers have jurisdiction. CWA regulates the discharge of nontoxic and toxic pollutants into surface waters. Its ultimate goal is to eliminate all discharges into surface waters. Its interim goal is to make surface waters usable for fishing, swimming, etc. EPA sets guidelines, and states issue permits (NPDES, Natural Pollutant Discharge Elimination System permit) specifying the types of control equipment and discharges for each facility.

Appendix E

DDT. The first chlorinated hydrocarbon insecticide. Its chemical name is dichloro-diphenol-trichloroethane. It has a half-life of 15 years and can collect in fatty tissues of certain animals. The EPA banned registration and interstate sale of DDT for virtually all but emergency uses in the United State in 1972 because of the persistence of the substance in the environment and accumulation in the food chain.

Decafluorotriphenylphosphine (DFTPP). (Organic Analysis) Compound chosen to establish mass spectral instrument performance for semivolatile analysis.

Degradability. Ability of materials to break down, by bacterial (biodegradable) or ultraviolet (photo-degradable) action.

Degradation. The process by which a chemical is reduced to a less complex form.

Density. Ratio of weight (mass) to volume of a material, usually in grams per cubic centimeter or pounds per gallon. See Specific Gravity.

Digestion Log. (Inorganic Analysis) An official record of the sample preparation (digestion).

Dioxin. Any of a family of heterocyclic hydro-carbons that occur especially as persistent toxic impurities; they prompt concern because of their presence in commercial products. Tests indicate that they are among the more toxic synthetic chemicals known.

Dissolved Metals. (Inorganic Analysis) Analyte elements which have not been digested prior to analysis and which will pass through a 0.45 μm filter.

Dissolved Oxygen (DO). The oxygen freely available in water. Dissolved oxygen is vital to fish and other aquatic life and for the prevention of odors. Traditionally, the level of dissolved oxygen has been accepted as the single most important indicator of a water body's ability to support desirable aquatic life. Secondary and advanced waste treatment are generally designed to protect DO in waste-receiving waters.

Dissolved Solids. Disintegrated organic and inorganic material contained in water. Excessive amounts make water unfit to drink or use in industrial processes.

DOT. US Dept. of Transportation. Regulates transportation of materials to protect the public as well as fire, law, and other emergency-response personnel. DOT classifications specify the use of appropriate warnings, such as Oxidizing Agent or Flammable Liquid. (400 7th St., SW, Washington, DC 20590.)

DOT Identification Numbers. Four-digit numbers used to identify particular materials for regulation of their transportation. See DOT publi-cations that describe the regulations (49 CFR 172.102). These numbers are called product identification numbers (PINs) under the Canadian Transportation of Dangerous Goods Regulations. Those numbers used internationally may carry a "UN" prefix (e.g., UN 1170, ethyl alcohol), but those used only North America have an "NA" prefix (e.g., NA 9121, ferric sulphate).

Dry Weight. The weight of a sample based on percent solids. The weight after drying in an oven. See Percent Moisture.

Duplicate. A second aliquot of a sample that is treated the same as the original sample in order to determine the precision of the method.

Duplicate Samples. Duplicate samples are two separate samples taken from the same source (i.e., in separate containers and analyzed independently).

EDL. Estimated Detection Limit.

Effluent. Treated or untreated wastewater that flows out of a treatment plant, sewer, or industrial outfall. Generally refers to wastes discharged into surface waters.

EICP. Extracted Ion Current Profile.

Emission. Pollution discharged into the atmos-phere from smokestacks, other vents, and surface areas of commercial or industrial facilities; from residential chimneys; and from motor vehicle, locomotive, and aircraft exhausts.

EMSL. Environmental Monitoring Systems Laboratory.

Environmental Sample. An environmental sample or field sample is a representative sample of any material (aqueous, nonaqueous, or multi-media) collected from any source for which de-termination of composition or contamination is requested or required. Environmental samples are normally classified as follows:

Drinking Water - delivered (treated or untreated) water designated as potable water;

Water/Wastewater - raw source waters for public drinking water supplies, ground waters, municipal influents/effluents, and industrial influents/effluents;

Sludge - municipal sludges and industrial sludges;

Waste - aqueous and nonaqueous liquid wastes, chemical solids, contaminated soils, and industrial liquid and solid wastes.

EPA, (Canada) Environmental Protection Act. Federal legislation, administered by Environment Canada, designed to protect the environment.

EPA, (US) Environmental Protection Agency. A Federal agency with environmental protection regulatory and enforcement authority. Administers the CAA, CWA, RCRA, TSCA, and other Federal environmental laws. (400 M Street, SW, Washington, DC 20460; [202] 382-2090.)

Equipment Blank. Usually an organic or aqueous solution that is as free of analyte as possible and is transported to the site, opened in the field, and poured over or through the sample collection device, collected in a sample container, and returned to the laboratory. This serves as a check on sampling device cleanliness. One equipment blank should be analyzed with each analytical batch or every 20 samples, whichever is greater.

EQL Estimated Quantitation Limit.

Exposure. The amount of radiation or pollutant present in an environment that represents a potential health threat to the living organisms in the environment.

Extractable. (Organic Analysis) A compound that can be partitioned into an organic solvent from the sample matrix and is amenable to gas chromatography. Extractables include semi-volatile (BNA) and pesticide/Aroclor compounds.

Extraction Procedure (EP) Toxicity Test. A series of laboratory tests designed to determine the level of toxicity in solid waste or landfill materials.

°F *or* **F.** Degrees Fahrenheit. See °C.

***Federal Register* (US).** See FR.

Field Blank. Usually an organic or aqueous solution that is as free of analyte as possible and is transferred from one vessel to another at the sampling site and preserved with the appropriate reagents. This serves as a check on reagent and environmental contamination. One field blank should be analyzed with each analytical batch or every 20 samples, whichever is greater.

Field Reagent Blank (FRB). An aliquot of reagent water or other blank matrix that is placed in a sample container in the laboratory and treated as a sample in all respects, including shipment to the sampling site, exposure to the sampling site conditions, storage, preservation, and all analytical procedures. The purpose of the FRB is to determine if method analytes or other interferences are present in the field environment.

Fire Diamond (NFPA Hazard Rating). Per "NFPA 704" publication. Visual system that provides a general idea of the inherent hazards, and their severity, of materials relating to *fire* prevention, exposure and control. Preferred reading order; Health, Flammability, Reactivity, Special.

Position A - Health Hazard (Blue). Degree of hazard; level of short-term protection

0 = Ordinary Combustible Hazards in a Fire
1 = Slightly Hazardous
2 = Hazardous
3 = Extreme Danger
4 = Deadly

Position B - Flammability (Red). Susceptibility to burning

0 = Will Not Burn
1 = Will Ignite if Preheated
2 = Will Ignite if Moderately Heated
3 = Will Ignite at Most Ambient Conditions
4 = Burns Readily at Ambient Conditions

Position C - Reactivity, Instability (Yellow). Energy released if burned, decomposed, or mixed

0 = Stable and Not Reactive with Water
1 = Unstable if Heated
2 = Violent Chemical Change
3 = Shock and Heat May Detonate
4 = May Detonate

Position D - Special Hazard (White).

OX = Oxidizer
W̶ = Use No Water, reacts!

Fire Point. The lowest temperature at which a liquid produces sufficient vapor to flash near its surface and continues to burn. Usually 10 to 30°C higher than the flash point.

Flame Atomic Absorption (FLAA). Atomic absorption which utilizes flame for excitation.

Flammable. Describes any solid, liquid, vapor, or gas that ignites easily and burns rapidly. See Combustible and Inflammable.

Flammable Liquid. A liquid that gives off vapors readily ignitable at room temperature. Defined by the NFPA and DOT as a liquid with a flash point below 100°F (38°C).

Flash Point, FP. Lowest temperature at which a flammable liquid gives off sufficient vapor to form an ignitable mixture with air near its surface or within a vessel. Combustion does not continue. FP is determined by laboratory tests in cups. See Fire Point.

FR. The *Federal Register*. A daily publication that lists and discusses Federal regulations. Available from the Government Printing Office.

10% Frequency. A frequency specification during an analytical sequence allowing for no more than 10 analytical samples between required calibration verification measurements, as specified by the Contract Statement of Work.

FSCC. Fused Silica Capillary Column.

Furans. A class of organic compounds found in the flue gas emitted from the refuse combustion facilities as a result of incomplete or inefficient combustion of carbon compounds. Referred to as polychlorinated dibenzofurans (PCDF), certain isomers of the class have been reported to be toxic.

g. Gram. Metric unit of weight. See kg.

GC/EC. Gas Chromatography/Electron Capture.

GC/MS. Gas Chromatography/Mass Spectrometry.

Graphite Furnace Atomic Absorption (GFAA). Atomic absorption which utilizes a graphite cell for excitation.

Halogenated Solvents. Liquid substances that contain carbon, halogen or carbon hydrogen, and halogen (such as fluorine or chlorine) atoms.

Halons. A family of compounds comtaining bromine used in fighting fires; breakdown of these compounds in the atmosphere depletes stratospheric ozone.

Hazardous Waste. By-products of society that can pose a substantial or potential hazard to human health or the environment when improperly managed. Possesses at least one of four char-

acteristics (ignitability, corrosivity, reactivity, or toxicity) or appears on special EPA lists.

Heavy Metals. Metallic elements with high atomic weights, such as mercury, chromium, cadmium, arsenic, and lead. They can damage living things at low concentrations and tend to accumulate in the food chain.

Holding Time. The elapsed time expressed in days from the date of receipt of the sample by the Contractor until the date of its analysis.

HRGC. High Resolution Gas Chromatography.

HRMS. High Resolution Mass Spectrometry.

Hydrocarbon. Any of a series of chemical compounds that consist entirely of carbon and hydrogen.

Hydrochloric Acid (HCl). An acid gas in the flue gas of refuse combustion. HCl is one of the most significant components of acid gas typically found in the flue gas.

Hydrochlorofluorocarbons (HCFCs). Chlorofluorocarbons that have been chemically altered by the addition of hydrogen and are significantly less damaging to stratospheric ozone than other CFCs.

ICP/MS. Inductively Coupled Plasma/Mass Spectrometry

ICS. Interference Check Standard.

ID. Identification.

IDL. Instrument Detection Limit.

Independent Standard. A contractor-prepared standard solution that is composed of analytes from a different source than those used in the standards for the initial calibration.

Inductively Coupled Plasma (ICP). A technique for the simultaneous or sequential multi-element determination of elements in solution. The basis of the method is the measurement of atomic and ionic emission by an optical spectroscopic technique. Characteristic atomic and ionic line emission spectra are produced by excitation of the sample in a radio frequency inductively coupled plasma.

In-House. At the contractor's facility.

Initial Calibration. Analysis of analytical standards for a series of different specified concentrations; used to define the linearity and dynamic range of the response of the analytical detector or method.

Injection. Introduction of the analytical sample into the instrument excitation system for the purpose of measuring absorbance, emission or concentration of analyte. May also be referred to as exposure.

Instrument Calibration. Analysis of analytical standards for a series of different specified concentrations; used to define the quantitative response, linearity, and dynamic range of the instrument to target analytes.

Instrument Check Standard. A multi-element standard of known concentration prepared by the analyst to monitor and verify instrument performance on a daily basis.

Instrument Detection Limit (IDL). Determined by multiplying by three the standard deviation obtained for the analysis of a standard solution (each analyte in reagent water) at a concentration of 3 times to 5 times IDL on three nonconsecutive days with seven consecutive measurements per day.

Instrument Performance Check (IPC) Solution. A solution of method analytes, used to evaluate the performance of the instrument system with respect to a defined set of method criteria.

Interferents. Substances which affect the analysis for the element of interest.

Internal Standard(s). Compound(s) added to every standard, blank, matrix spike, matrix spike duplicate, sample (for VOAs), sample digestates (for ICP-MS) and sample extract (for semi-volatiles) at a known concentration, prior to analysis. Internal standard(s) are used as the basis for quantitation of the target compounds.

IS. Internal Standard.

Isomers. Chemical compounds with the same molecular weight and atomic composition but differing molecular structure; e.g., *n*-pentane and 2-methylbutane.

kg, kilogram. 1000 grams.

l. Liter. Basic metric unit of volume. One liter of water weighs 1 kg and is equal to 1.057 quarts.

Laboratory Control Sample (LCS). A control sample of known composition. Aqueous and solid laboratory control samples are analyzed using the same sample preparation, reagents, and analytical methods employed for the EPA samples received.

Laboratory Control Standard. A standard, usually certified by an outside agency, used to measure the bias in a procedure. For certain constituents and matrices, use National Institute of Standards and Technology (NIST) Standard Reference Materials when they are available.

Laboratory Fortified Blank (LFB). An aliquot of laboratory reagent water to which known quantities of the method analytes are added in the laboratory the LFB is analyzed exactly like a sample, and its purpose is to determine whether the methodology is in control and whether the laboratory is capable of making accurate and precise measurements.

Laboratory Fortified Sample Matrix (LFM). Same as *Matrix Spike*.

Landfill. (1) Sanitary landfills are land disposal sites for nonhazardous solid wastes at which the waste is spread in layers, compacted to the smallest practical volume, and covered with material applied at the end of each operating day. (2) Secure chemical landfills are disposal sites for hazardous waste. They are selected and designed to minimize the chance of release of hazardous substances into the environment.

Leachate. A liquid that has passed through or emerged from solid waste, collecting contaminants as it trickles through wastes, agricultural pesticides, or fertilizers.

Limit of Quantitation (LOQ). The constituent concentration that produces a signal sufficiently greater than the blank that it can be detected with the specified limits by good laboratories during routine operating conditions. Typically it is the concentration that produces a signal 10σ above the reagent water blank signal.

Linear Range, Linear Dynamic Range. (Inorganic Analysis) The concentration range over which the ICP analytical curve remains linear.

Listed Waste. Wastes listed as hazardous under RCRA but not subjected to the Toxic Characteristics Listing Process because the dangers they possess are considered self-evident.

Appendix E

Lower Limit of Detection (LLD). The constituent concentration in reagent water that produces a signal $2(1.645)\sigma$ above the mean of blank analyses. This sets both Type I and Type II errors at 5%. Other names for this limit are "detection limit" and "limit of detection (LOD)".

LSE. Liquid-Solid Extraction.

m. Meter. Metric unit of length equal to 39.37 in.

m^3 or cu m. Cubic meter; m^3 is preferred.

Material Safety Data Sheet. See MSDS.

Matrix. The predominant material of which the sample to be analyzed is composed. Matrix is not synonymous with phase (liquid or solid).

Matrix Modifier. (Inorganic Analysis) Reagents used in AA to lessen the effects of chemical interferents, viscosity, and surface tension.

Matrix Spike (MS). Aliquot of a sample (water or soil) fortified (spiked) with known quantities of specific compounds and subjected to the entire analytical procedure in order to indicate the appropriateness of the method for the matrix by measuring recovery.

Matrix Spike Duplicate (MSD). A second aliquot of the same matrix as the matrix spike that is spiked in order to determine the precision of the method.

Maximum Contaminant Levels (MCLs). (1) The maximum level of contamination allowed in any surface or ground water under the Safe Drinking Water Act for certain identified pollutants. MCLs are often used as the standard to be achieved for CERCLA and RCRA cleanups. (2) The maximum permissible level of a contaminant in water delivered to any user of a public water system. MCLs are enforceable standards.

MDL. Method Detection Limit

Metabolism. The chemical and physical processes whereby the body functions.

Meter (m). The basic metric measure of length; equivalent to 39.371 in.

Methane. An odorless, colorless flammable gas formed by anaerobic decomposition of organic waste matter or by chemical syntheses. The principal component of natural gas.

Method Blank. An analytical control consisting of all reagents, internal standards and surrogate standards, that is carried through the entire analytical procedure. The method blank is used to define the level of laboratory background and reagent contamination.

Method Detection Limit (MDL). The constituent concentration that, when processed through the complete method, produces a signal with a 99% probability that it is different from the blank. For seven replicates of the sample, the mean must be 3.14 above the blank where σ is the standard deviation of the seven replicates. The MDL will be larger than the LLD because of the few replications and the sample processing steps and may vary with constituent and matrix.

Method of Standard Additions (MSA). (Inorganic Analysis) The addition of 3 increments of a standard solution (spikes) to sample aliquots of the same size. Measurements are made on the original and after each addition. The slope, x-intercept and y-intercept are determined by least-square analysis. The analyte concentration is determined by the absolute value of the x-intercept. Ideally, the spike volume is low relative to the sample volume (approximately 10% of the volume). Standard addition may counteract matrix effects; it will not counteract spectral effects. Also referred to as Standard Addition.

mg. Milligram (1/1000, 10^{-3}, of a gram).

Microgram (μg). One-millionth (10^{-6}) of a gram.

Micrometer (μm). One-millionth (10^{-6}) of a meter; often referred to as a micron.

Millimeter (mm). 1/1000 of a meter.

min. Minute

Miscible. When two liquids or two gases are completely soluble in each other in all proportions. While gases mix with one another in all proportions, the miscibility of liquids depends on the chemical natures.

ml. Milliliter. One thousandth of a liter. A metric unit of capacity, for all practical purposes equal to 1 cubic centimeter. One cubic inch is about 16 ml.

Mole or mol. The quantity of a chemical substance that has a mass in grams numerically equal to the molecular weight. For example, salt (NaCl) has a molecular weight of 58.5 (Na, 23, and chlorine, 35.5). Thus, one mole of NaCl is 58.5 g.

Molecular Weight. The sum of atomic weights of the atoms in a molecule. For example, water (H_2O) has a molecular weight of 18.015, the atomic weights being hydrogen = 2 (1.008) + oxygen = 15.999.

Monitoring Wells. Wells drilled at a hazardous waste management facility or Superfund site to collect groundwater samples for the purpose of physical, chemical, or biological analysis to determine the amounts, types, and distribution of contaminants in the groundwater beneath the site.

MQL. The method quantification limit (MQL) is the minimum concentration of a substance that can be measured and reported.

MSDS. Material safety data sheet. OSHA has established guidelines for the descriptive data that should be concisely provided on a data sheet to serve as the basis for written hazard communication programs. The thrust of the law is to have those who make, distribute, and use hazardous materials responsible for effective communication. See the Hazard Communication Rule, 29 CFR, Part 1910. 1200, as amended, Sec. g. See Schedule I, Sec. 12, of the Canadian Hazardous Products Act.

MW. See Molecular Weight.

m/z. (Organic Analysis) Mass to charge ratio, synonymous with "m/e".

n-. Normal. A chemical name prefix signifying a straight-chain structure; i.e., no branches.

NA, ND. Not applicable, not available; not determined.

Narrative. A descriptive documentation of any problems encountered in processing the samples, along with corrective action taken and problem resolution.

National Pollution Discharge Elimination System (NPDES). (1) A system under which the EPA issues permits (often delegated to the states) according to provisions of the Federal Water Pollution Control Act. There can be no discharge without the NPDES permit. (2) A provision of the Clean Water Act that prohibits discharge of pollutants into waters of the United States unless a special permit is issued by the EPA, a state, or (where delegated) a tribal government.

National Priorities List (NPL). The EPA's list of the most serious uncontrolled or abandoned hazardous waste sites identified for possible long-term remedial action under Superfund. A site must be on the NPL to receive money from the Trust Fund for remedial action. The list is based primarily on the score a site receives when evaluted according to the Hazard Ranking system. The EPA is required to update the NPL at least once each year.

Neutralization. Decreasing the acidity or alkalinity of a substance by adding to it alkaline or acidic materials.

ng. Nanogram. One billionth, 10^{-9}, of a gram.

NIOSH. National Institute of Occupational Safety and Health. The agency of the Public Health Service that tests and certifies respiratory and air-sampling devices. It recommends exposure limits to OSHA for substances, investigates incidents, and researches occupational safety. (NIOSH, 4676 Columbia Parkway, Cincinnati, OH 45226; [513] 533-8328.)

NIST. National Institute for Standards and Technology

NOC. Not otherwise classified.

Occupational Safety and Health Act. See OSH Act.

Occupational Safety and Health Administration. See OSHA.

OSHA. The Occupational Safety and Health Administration. Part of the US Department of Labor. The regulatory and enforcement agency for safety and health in most US industrial sectors. (Documents are available from the OSHA Technical Data Center Docket Office, Room N-3670, 200 Constitution Ave, NW, Washington, DC 20210; [202] 523-7894.)

OSH Act. The Occupational Safety and Health Act of 1970. Effective April 28, 1971. Public Law 91-596. Found at 29 CFR 1910, 1915, 1918, 1926. OSHA jurisdiction. The regulatory vehicle to ensure the safety and health of workers in firms larger than 10 employees. Its goal is to set standards of safety that prevent injury and illness among the workers. Regulating employee exposure and informing employees of the dangers of materials are key factors. This act established the Hazard Communication Rule (29 CFR 1910.1200). See Hazard Communication Rule for details.

Appendix E

PCBs. A group of toxic, persistent chemicals (polychlorinated biphenyls) used in transformers and capacitators for insultating purposes and in gas pipeline systems as a lubricant. Further sale banned by law in 1979.

PE. Performance Evaluation

Percent Moisture. An approximation of the amount of water in a soil/sediment sample made by drying an aliquot of the sample at 105°C. The percent moisture determined in this manner also includes contributions from all compounds that may volatilize at or below 105°C, including water. Percent moisture may be determined from decanted samples and from samples that are not decanted.

Percent Solids. The proportion of solid in a soil sample determined by drying an aliquot of the sample.

Performance Evaluation (PE) Sample. A sample of known composition provided by EPA for contractor analysis. Used by EPA to evaluate contractor performance.

PEST. Pesticides.

Pesticide. Substance or mixture of substances intended for preventing, destroying, repelling, or mitigating any pest. Also, any substance or mixture of substances intended for use as a plant regulator, defoliant, or desiccant. Pesticides can accumulate in the food chain and/or contaminate the environment if misused.

pH. "Hydrogen ion exponent," a measure of hydrogen ion concentration. A scale (0 to 14) representing an aqueous solution's acidity or alkalinity. Low pH values indicate acidity and high values, alkalinity. The scale's mid-point, 7, is neutral. Substances in an aqueous solution ionize to various extents giving different concentrations of H and OH ions. Strong acids have excess H ions and a pH of 1 to 3 (HCl, pH = 1). Strong bases have excess OH ions and a pH of 11 to 13 (NaOH, pH = 12).

Phenols. Organic compounds that are by-products of petroleum refining, tanning, textile, dye, and resin manufacturing. Low concentrations cause taste and odor problems in water, higher concentrations can kill aquatic life and humans.

Plasma Solution. A solution that is used to determine the optimum height above the work coil for viewing the plasma.

ppb. Parts per billion.

ppm. Parts per million. "Parts of vapor or gas per million parts of contaminated air by volume at 25°C and 1 torr pressure" (ACGIH). At 25°C, ppm = (mg/m^3 x 24.45) divided by molecular weight.

ppt. Parts per trillion.

PQL. The practical quantitation limit (PQL) is the lowest level that can be reliably achieved within specified limits of precision and accuracy during routine laboratory operating conditions.

Precision. Precision is the agreement between a set of replicate measurements without assumption of knowledge of the true value. Precision is assessed by means of duplicate/replicate sample analysis.

Preparation Blank (reagent blank, method blank). An analytical control that contains distilled, deionized water and reagents, which is carried through the entire analytical procedure (digested and analyzed). An aqueous method blank is treated with the same reagents as a sample with a water matrix; a solid method blank is treated with the same reagents as a soil sample.

Protocol. Describes the exact procedures to be followed with respect to sample receipt and handling, analytical methods, data reporting and deliverables, and document control. Used synonymously with Statement of Work (SOW).

Purge and Trap (Device). Analytical technique (device) used to isolate volatile (purgeable) organics by stripping the compounds from water or soil by a stream of inert gas, trapping the compounds on an adsorbent such as a porous polymer trap, and thermally desorbing the trapped compounds onto the gas chromatographic column.

QA. Quality Assurance

QAP. Quality Assurance Plan

QC. Quality Control

Quality Assessment. Procedure for determining the quality of laboratory measurements by use of data from internal and external quality control measures.

Quality Assurance. A definitive plan for laboratory operation that specifies the measures used to produce data of known precision and bias.

Quality Control. Set of measures within a sample analysis methodology to assure that the process is in control.

Quality Control Reference Sample. A sample prepared from an independent standard at a concentration other than that used for calibration, but within the calibration range. An independent standard is defined as a standard composed of the analyte(s) of interest from a different source than that used in the preparation of standards for use in the standard curve. A quality control reference sample is intended as an independent check of technique, methodology, and standards and should be run with every analytical batch or every 20 samples, whichever is greater. This is applicable to all organic and inorganic analyses.

Quality Control Sample. A solution obtained from an outside source having known concentration values to be used to verify the calibration standards.

R. Recovery.

Random Error. The deviation in any step in an analytical procedure that can be explained by standard statistical techniques.

RCRA. Resource Conservation and Recovery Act, PL 94-580. Found at 40 CFR 240-271. EPA has jurisdiction. Enacted November 21, 1976, and amended since. RCRA's major emphasis is the control of hazardous waste disposal. It controls all solid-waste disposal and encourages recycling and alternative energy sources. In 1984 the USA generated 265 million tons of hazardous waste.

RCRA Hazardous Waste. A material designated by RCRA as hazardous waste and assigned a number to be used in record keeping and reporting compliance (e.g., D003, F001, U169).

Reagent Blank. Usually an organic or aqueous solution that is as free of analyte as possible and contains all the reagents in the same volume as used in the processing of the samples. The reagent blank must be carried through the complete sample preparation procedure and contains the same reagent concentrations in the final solution as in the sample solution used for analysis. The reagent blank is used to correct for possible contamination resulting from the preparation or processing of the sample. One reagent blank should be prepared for every analytical batch or for every 20 samples, whichever is greater.

Reagent Grade. Analytical reagent (AR) grade, ACS reagent grade, and reagent grade are synonymous terms for reagents which conform to the current specifications of the Committee on Analytical Reagents of the American Chemical Society.

Reagent Water. Water in which an interferent is not observed at or above the minimum quantitation limit of the parameters of interest.

Reconstructed Ion Chromatogram (RIC). (Organic Analysis) A mass spectral graphical representation of the separation achieved by a gas chromatograph; a plot of total ion current versus retention time.

Relative Percent Difference (RPD). To compare two values, the relative percent difference is based on the mean of the two values, and is reported as an absolute value, i.e., always expressed as a positive number or zero.

Relative Response Factor (RRF). A measure of the relative mass spectral response of an analyte compared to its internal standard. Relative Response Factors are determined by analysis of standards and are used in the calculation of concentrations of analytes in samples.

Remedial Investigation/Feasibility Study (RI/FS). Technical studies undertaken by the EPA (or by PRPs if permitted by the EPA) to investigate the extent of contamination (the RI) and the remedial alternatives (the FS) for a particular Superfund site. The RI/FS must be consistent with the National Contingency Plan.

Replicate. Repeated operation occurring within an analytical procedure. Two or more analyses for the same constituent in an extract of a single sample constitutes replicate extract analyses.

Replicate Sample. A replicate sample is a sample prepared by dividing a sample into two or more separate aliquots. Duplicate samples are considered to be two replicates. In cases where aliquoting is impossible, as in the case of volatiles, duplicate samples must be taken for the replicate analysis.

Residual. Amount of a pollutant remaining in the environment after a natural or technological process has taken place, as in the sludge remaining after initial wastewater treatment.

Appendix E

Residue. The solid materials remaining after completion of chemical or physical incineration or another disposal process. These materials include ash, ceramics, metal, glass, and unburned organics.

Resolution. Also termed separation or percent resolution, the separation between peaks on a chromatogram, calculated by dividing the depth of the valley between the peaks by the peak height of the smaller peak being resolved, multiplied by 100.

Resource Conservation and Recovery Act. See RCRA.

Retention Time Window. Usually defined as three times the standard deviation of the absolute or relative RT of an analyte standard injected over the course of a 72 hour period.

RI/FS. See Remedial Investigation/Feasibility Study.

Rounding Rules. If the figure following those to be retained is less than 5, the figure is dropped, and the retained figures are kept unchanged. As an example, 11.443 is rounded off to 11.44.

If the figure following those to be retained is greater than 5, the figure is dropped, and the last retained figure is raised by 1. As an example, 11.446 is rounded off to 11.45.

If the figure following those to be retained is 5, and if there are no figures other than zeros beyond the five, the figure 5 is dropped, and the last-place figure retained is increased by one if it is an odd number or it is kept unchanged if an even number. As an example, 11.435 is rounded off to 11.44, while 11.425 is rounded off to 11.42.

If a series of multiple operations is to be performed (add, subtract, divide, multiply), all figures are carried through the calculations. Then the final answer is rounded to the proper number of significant figures.

RRT. Relative Retention Time.

RSD. Relative Standard Deviation.

RT. Retention Time.

RT Window. See Retention Time Window.

Run. A continuous analytical sequence consisting of prepared samples and all associated quality assurance measurements as required by the Contract Statement of Work.

Salinity. The degree of salt in water.

Sample. A portion of material to be analyzed that is contained in single or multiple containers and identified by a unique sample number.

Sample Delivery Group (SDG). Unit within a single case that is used to identify a group of samples for delivery. An SDG is a group of 20 or fewer field samples within a case, received over a period of up to 14 calendar days (7 calendar days for 14-day data turnaround contracts). Data from all samples in an SDG are due concurrently.

Sample Number (EPA Sample Number). A unique identification number designated by EPA for each sample. The EPA sample number appears on the sample Traffic Report which documents information about that sample.

SARA. Superfund Amendments and Reauthorization Act. Signed into law October 17, 1986. Title III of SARA is known as the Emergency Planning and Community Right-to-Know Act of 1986. A revision and extension of CERCLA, SARA is intended to encourage and support local and state emergency planning efforts. It provides citizens and local governments with information about potential chemical hazards in their communities. SARA calls for facilities that store hazardous materials to provide officials and citizens with data on the types (flammables, corrosives, etc.); amounts on hand (daily, yearly); and their specific locations. Facilities are to prepare and submit inventory lists, MSDSs, and tier 1 and 2 inventory forms. The disaster in Bhopal, India, in 1987 added impetus to the passage of this law.

SD. Standard Deviation.

Semivolatile Compounds. (Organic Analysis) Compounds amenable to analysis by extraction of the sample with an organic solvent. Used synonymously with Base/Neutral/Acid (BNA) compounds.

Sensitivity. The slope of the analytical curve, i.e., functional relationship between emission intensity and concentration.

Serial Dilution. The dilution of a sample by a factor of five. When corrected by the dilution factor, the diluted sample must agree with the original undiluted sample within specified limits.

Serial dilution may reflect the influence of interferents.

SIM. Selected Ion Monitoring.

Sludge. A semisolid residue from any of a number of air or water treatment processes. Sludge can be a hazardous waste.

Solid Waste. (1) Nonliquid, nonsoluble materials ranging from municipal garbage to industrial wastes that contain complex, and sometimes hazardous substances. Solid wastes also include sewage sludge, agricultural refuse, demolition wastes, and mining residues. Technically, solid waste also refers to liquids and gases in containers. (2) Any garbage, refuse, or sludge from a waste treatment plant and other discarded material resulting from industrial, commercial, mining, and agricultural activities but not including discharges from point sources as described under the Federal Water Polution Control Act.

SOP. Standard Operation Procedure.

SOW. Statement of Work.

SPCC. System Performance Check Compound.

Specific Gravity. The ratio of the mass of a substance to the same volume of a reference substance, at a specified temperature. Specific gravity is a dimensionless number. Water (density 1 kg/l at 4°C) is the reference for solids and liquids, while air (density 1.29 g/l at 0°C and 760 mm Hg pressure) is the reference for gases. If a volume of a material weighs 8 g, and an equal volume of water weighs 10 g, the material has a specific gravity of 0.8 (8 ÷ 10 = .8). Insoluble materials with specific gravity greater than 1.0 sink to the bottom in water. Specific gravity is an important fire supression and spill cleanup consideration since most (but not all) flammable liquids have a specific gravity less than 1.0 and, if insoluble, float on water.

Spectral Interference Check Solution. (Inorganic Analysis) A solution containing both interfering and analyte elements of known concentration that can be used to verify background and interelement correction factors.

Standard Analysis. An analytical determination made with known quantities of target compounds; used to determine response factors.

Standard Curve. A standard curve is a curve which plots concentrations of known analyte

standards versus the instrument response to the analyte. Calibration standards are prepared by diluting the stock analyte solution in graduated amounts which cover the expected range of the samples being analyzed. Standards should be prepared at the frequency specified in the appropriate section. The calibration standards must be prepared using the same type of acid or solvent and at the same concentration as the samples following the sample preparation procedure. This is applicable to organic and inorganic chemical analyses. See Calibration.

Stock Solution. Standard solution which can be diluted to derive other standards.

Superfund Amendments and Reauthorization Act. See SARA, CERCLA.

Surrogate. Surrogates are organic compounds which are similar to analytes of interest in chemical composition, extraction, and chromatography, but which are not normally found in environmental samples. These compounds are spiked into all blanks, calibration and check standards, samples (including duplicates and QC reference samples) and spiked samples prior to analysis. Percent recoveries are calculated for each surrogate.

Surrogate Standard. A pure compound added to a sample in the laboratory just before processing so that the overall efficiency of a method can be determined.

Suspended. (Inorganic Analysis) Those elements which are retained by a 0.45 μm membrane filter.

Suspended Solids. Small particles of solid pollutants that float on the surface of, or are suspended in, sewage or other liquid. They resist removal by conventional means. See also Total Suspended Solids.

SV. Semivolatile.

SVOA. Semivolatile Organic Analysis.

System Monitoring Compounds. (Organic Analysis) Compounds added to every blank, sample, matrix spike, matrix spike duplicate, and standard for volatile analysis, and used to evaluate the performance of the entire purge and trap-gas chromatograph-mass spectrometer system. These compounds are brominated or deuterated compounds not expected to be detected in environmental media.

Appendix E

Target Compound List (TCL). A list of compounds designated by the Statement of Work for analysis.

TCL. Target Compound List

Tentatively Identified Compounds (TIC). Compounds detected in samples that are not target compounds, internal standards, system monitoring compounds, or surrogates. Up to 30 peaks (those greater than 10% of peak areas or heights of nearest internal standards) are subjected to mass spectral library searches for tentative identification.

TIC. Tentatively Identified Compound

Time. When required to record time on any deliverable item, time shall be expressed as Military Time, i.e., a 24-hour clock.

Total Metals. (Inorganic Analysis) Analyte elements which have been digested prior to analysis.

Total Suspended Solids (TSSs). A measure of the suspended solids in wastewater, effluent, or water bodies. See also Suspended Solids.

Toxic Substances Control Act. See TSCA.

Toxicity. The degree of danger posed by a substance to animal or plant life.

Toxicity Characteristic Leaching Procedure. Analytical method used to determine the mobility of organic and inorganic contaminants present in liquid, solid, and multiphase wastes. If an extract from a representative sample is shown to contain any contaminant in an amount exceeding the levels allowed by regulations, the waste is banned for land disposal unless properly treated.

Traffic Report (TR). An EPA sample identification form filled out by the sampler, which accompanies the sample during shipment to the laboratory and which is used for documenting sample condition and receipt by the laboratory.

Trip Blank. Usually an organic or aqueous solution that is as free of analyte as possible and is transported to the sampling site and returned to the laboratory without being opened. This serves as a check on sample contamination originating from sample transport, shipping, and from the site conditions. One trip blank should be analyzed with each analytical batch or every 20 samples, whichever is greater.

TSCA. Toxic Substances Control Act. Public Law PL 94-469. Found in 40 CFR 700-799. EPA has jurisdiction. Effective Jan. 1, 1977. Controls the exposure to and use of raw industrial chemicals not subject to other laws. Chemicals are to be evaluated prior to use and can be controlled based on risk. The act provides for a listing of all chemicals that are to be evaluated prior to manufacture or use in the US. (Call the EPA, Industry Assistance Office, [202] 554-1404.)

Tuning Solution. A solution which is used to determine acceptable instrument performance prior to calibration and sample analyses.

Turbidity. (1) Haziness in air caused by the presence of particles and pollutants (2) a similar cloudy condition in water caused by suspended silt or organic matter.

Twelve-Hour Time Period. (Organic Analysis) The twelve (12) hour time period for GC/MS system instrument performance check, standards calibration (initial or continuing calibration), and method blank analysis begins at the moment of injection of the DFTPP or BFB analysis that the laboratory submits as documentation of instrument performance. The time period ends after 12 hours have elapsed according to the system clock. For pesticide/Aroclor analysis performed by GC/EC, the twelve hour time period in the analytical sequence begins at the moment of injection of the instrument blank that precedes sample analyses, and ends after twelve hours have elapsed according to the system clock.

Type I Error. Also called alpha error, is the probability of deciding a constituent is present when it actually is absent.

Type II Error. Also called beta error, is the probability of not detecting a constituent when it actually is present.

Validated Time of Sample Receipt (VTSR). The date on which a sample is received at the contractor's facility, as recorded on the shipper's delivery receipt and Sample Traffic Report.

VOA. Volatile Organics Analysis.

Volatile. Description of any substance that evaporates easily.

Volatile Compounds. (Organic Analysis) Compounds amenable to analysis by the purge and trap technique. Used synonymously with purgeable compounds.

Water. Reagent, analyte-free, or laboratory pure water means distilled or deionized water or Type II reagent water which is free of contaminants that may interfere with the analytical test in question.

Wet Weight. The weight of a sample aliquot including moisture (undried).

Wide Bore Capillary Column. A gas chromatographic column with an internal diameter (ID) that is greater than 0.32 mm. Columns with lesser diameters are classified as narrow bore capillaries.

Appendix E

APPENDIX F

Should you need assistance interpreting a regulation or understanding the details of a method, you can get help from the Federal and state agencies listed in this appendix.

NATIONAL AND REGIONAL CONTACTS

ENVIRONMENTAL HOTLINES AND INFORMATION CENTERS

Acid Rain Hotline
The Acid Rain Hotline records questions and document requests covering all areas of the Acid Rain Program.
Phone... (202) 233-9620

Asbestos Ombudsman Hotline
All except VA .... (800) 368-5888
TDD machine .... (703) 305-6824
Provides information to the public sector, including individual citizens and community services on the handling and abatement of asbestos in schools, the workplace and the home.

Centers for Disease Control (CDC)
600 Clifton Rd., NE, Atlanta, GA 30333
Phone... (404) 639-3311
http://www.cdc.gov/

Center for Hazardous Materials (CHMR/CTC)
Phone .... (412) 826-5320
.... (814) 269-6888
Regulatory, toxic waste minimization, pollution prevention, publications and referrals.

Center for International Environmental Law
1367 Connecticut Avenue, NW Suite #300
Washington, DC 20036
Phone... (202) 785-8700
Fax.. (202) 785-8701
CIEL provides a full range of environmental legal services in both international and comparative national law, including: policy research and publication, advice and advocacy, education and training, and institution building.
http://www.econet.apc.org/ciel/

Chemical Manufacturers Association Hotlines
Main Line .... (202) 887-1100
Library .... (202) 887-1216
Non-emergency Information Line..... (800) 262-8200
CHEMTREC (for chemical emergencies only)
Emergency Line (*Do not call seeking info, use Non-emergency information line listed above.*)
.... (800) 424-9300

Control Technology Center (CTC)
National .... (919) 541-0800
Technical support and guidance on air pollution emissions and control technology. Air emissions and air pollution control technology for all air pollutants including air toxins emitted by stationary sources.

Consumer Product Safety Commission Hotline
National.. (800) 638-2772

Emergency Planning and Community Right-to-Know (Title III SARA) Hotline
National .... (800) 535-0202
Washington D.C. Area .... (703) 412-9877
Open Monday through Friday, 9:00 a.m. to 6:00 p.m. Closed on weekends and Federal Holidays. Provides regulatory, policy and technical assistance to federal agencies, local and state governments, the public and regulated community in response to questions related to the Emergency Planning and Community Right-to-Know Act (Title III of SARA). Information on reporting of hazardous substances for community planning purposes.

Environmental Business Council of the United States
National.. (617) 449-5600

Environmental Technologies Export Council (ETECH)
National.. (202) 466-6933

EPA Hotline
National.. (800) 346-5009

EPA Action Line
National.. (913) 551-7122
Referral service to appropriate program office.

EPA AIR QUALITY INFORMATION HOTLINE
National.. (800) 438-4318

EPA Air Risk Information Support Center
Provides technical assistance and information in areas of health, risk, and exposure assessment for toxic and criteria air pollution.
Phone.. (800) 541-0888

EPA Assistant Administrator for Enforcement and Compliance Monitoring
National..(202) 260-5145
Provides direction for the review and enforcement of compliance activities.

EPA Assistant Administrator for Research and Development
National..(202) 260-7676
Provides technical information for the EPA administrator on scientific and technical issues.

EPA General Information
New England Region ....(617) 565-3420
Referral service to appropriate program office.

EPA General Info - Environmental Issues
National...(303) 293-1603
Emergency Reporting ....(800) 424-8802
General information - environmental issues.

EPA Method Information Communication Exchange (MICE)
National...(703) 821-4690
Provides information using a voice-mail answering system about technical questions regarding testing methods contained in Test Methods for Evaluating Solid Waste Physical/Chemical Methods (SW-846).

EPA Public Information Center
National...(202) 260-7751
Provides non-technical documents and guidance about general environmental information for the public.

EPA Office of the Inspector General - Whistle Blower Hotline
National .... (800) 424-4000
....(202) 382-4977
Fraud, waste or mismanagement in EPA-funded activities.

EPA Stratosphere Ozone Information Hotline
Phone... (800) 296-1996
Fax.. (202) 775-6681
This hotline provides in-depth information on ozone protection regulations and requirements under the Clean Air Act.

EPA Superfund (Region II Hotline)
Restricted to following area codes: 809, 201, 609, 908, 906, 212, 315, 516, 518, 607, 716, 718, 914
.... (800) 245-2738
Enables the Superfund Civil Investigators to receive information relevant to specific Superfund Site Enforcement Investigations. Do not call with questions, this is an investigative unit.

EPA Technical Resources and Federal Superfund Hotline
National...(800) 535-0202
Provides information on technical questions

EPA Toxic Substance Control Act (TSCA) Hotline
National...(202) 554-1404
Online service ....(202) 554-5603
Information on regulations, precautions and health effects of toxic substances.

EPA Wetlands Information Hotline
National, VI and Guam...................(800) 832-7828
Responsive to public interest, questions and requests for information about the values and functions of wetlands and options for their protection. Provides referrals to callers when necessary. They also provide free literature and publications, call to get on their mailing list.

ERT Edison
USEPA Environmental Response Branch
2890 Woodbridge Avenue
Building 18
Edison, NJ 08837
.... (908) 321-6740

Food and Drug Administration (FDA)
General Information .... (301) 443-3170

Government Printing Office
Phone .... (202) 512-1800

Hazardous Materials Standards Information Line
National...(202) 366-4488
Information of DOT 49CFR regulations.

HAZMAT
National .... (800) 423-1363
IL only ....(800) 367-9592
Information Exchange, operated by the Dept. of Transportation and the Federal Emergency Mgt. Agency for questions about the transportation of hazardous materials,

Hazardous Waste Ombudsman
National .... (800) 262-7937
Washington, DC............................. (202) 475-9361
The hazardous waste management program established under RCRA is a highly complex regulatory program developed by EPA. It assists the public and regulated community in resolving problems concerning any program or requirement under the Hazardous Waste Program. The ombudsman handles complaints from citizens and the regulated community, obtains facts, sorts information, and substantiates policy.

Appendix F

Mercury Hotline
National..(800) 833-3505
Provides answers to questions. No emergency Service

Mobile Sources
New England states only ....(202) 260-7645
Complaints regarding auto emission tampering, emission, auto warranty, recall notices, fuel issues, CFC recycling, auto air conditioning.

National Institute for Occupational Safety and Health (NIOSH) Information System
National..(800) 356-4674

National Institute of Standards and Technology (NIST)
National..(301) 975-2000
....(303) 497-3000
Previously the National Bureau of Standards (NBS).

National Lead Information Center
General information ....(800) 532-3394
Technical assistance ....(800) 424-5323

National Technical Information Service (NTIS)
National..(703) 487-4600
Source of information services (technical documents and databases) and documents from federal agencies, industries, and universities.

Nuclear Regulatory Commission
National..(301) 415-8200
NRC provides information about technical questions and documents regarding hazardous nuclear materials and wastes.

National Pesticides Telecommunications Network (Pesticides and Herbicides Hotline)
National, incl. AK, PR, VI ....(800) 858-7378
....(806) 743-3091
Provides the medical, veterinary, professional communities and general public with information on pesticides and herbicides product information, recognition and management of pesticide poisonings, toxicology and symptomatic reviews, safety information, health and environmental effects, clean-up and disposal procedures.

National Radon Hotline
National..(800) 767-7236
Radon testing information. A message records names and addresses of callers and a brochure on radon is sent via 1st class mail.

National Response Center - U.S. Coast Guard Oil and Hazardous Material Spills
National except DC ....(800) 424-8802
DC and outside U.S. ....(202) 267-2675

For reporting of oil and hazardous material spills. NOTE: Please have ready as much relevant data as possible when calling.

Occupational Safety and Health Administration Hotline
National..(800) 321-6742
24-hour access line to report unsafe and hazardous work practices.

Public Health Service Information Office
National..(301) 443-2403
Agency for Publications ....(800) 358-9295

RCRA/Superfund/EPCRA/Community Right-to-Know (Title III) Hotline
International ....(800) 424-9346
International ....(703) 412-9810
TDD machine...................................(800) 553-7672
Answers factual questions from the regulated community, other interested parties and the public about EPA's RCRA regulations and policies; referrals for obtaining related documents. RCRA, Underground Storage Tanks (USTs), Superfund/ CERCLA and Pollution Prevention/Waste Minimization.

Safe Drinking Water Hotline
National..(800) 426-4791
Provides assistance and regulatory knowledge to the regulated community (public water systems) and the public on the regulations and programs developed in response to the Safe Drinking Water Act Amendments of 1986.

Substance Identification Hotline
National..(800) 848-6538
Identifies chemical by CAS number or Name.

US Army Corps of Engineers
National..(202) 761-0660

Water and Waste Water Information (600-series methods)
National..(202) 260-7120
Provides information on testing methods for water and waste water (600-series methods) contained in 40CFR part 136.

ENVIRONMENTAL DATABASE AND COMPUTER CONTACTS

ATSDR
Agency for Toxic Substance and Disease Registry
ATSDR
Office of the Assistant Administrator
1600 Clifton Rd., E28
Atlanta, GA 30333

Appendix F

Phone.. (404) 639-0700
Fax... (404) 639-0744
http://atsdr1.atsdr.cdc.gov:8080/

CEDAR
Central European Environmental Data Request Facility
Central European Environmental Data Request
Facility (CEDAR)
Marxergasse 3/20, A-1030 Vienna, Austria
Tel. +43/1/715 28 28-0, Fax. +43/1/715 28 28-19
http://pan.cedar.univie.ac.at/

CERCLIS - Helpline
National .... (703) 908-2066
Answering machine for all off-hour callers.
Technical Support and referrals to the users of
CERCLIS database, Waste LAN and Clean LAN.
http://www.epa.gov/enviro/htm/cerclis/cerclis_s ubj.htm

Environment Links on the Web
Environment Links on the Web. Includes energy,
conservation, wildlife, oceanography, Sierra Club,
EPA, newsgroup and more.
http://www.onlinenews.net/environment.htm

Environmental Organization WebDirectory
Environmental search engine and bulletin board
http://www.webdirectory.com/

Environmental Reference Data Base
An encyclopedic guide to environmental information
on the internet – over 500 public and
proprietary data bases reviewed and categorized.
http://www.csa.com/routenet/

EPA Office of Research and Development Electronic Bulletin Board (ORD BBS)
The ATTIC Bulletin Board System is currently
available through a direct modem connection or by
telnet. .... (513) 569-7610
You can access ATTIC using a PC, equipped with a
modem (1200, 2400, 9600, 14400, 28800, 33600
baud) and appropriate communications software
(e.g., CrossTalk, PROCOMM).
http://www.epa.gov/attic/accessattic.htm

To access ATTIC through the Internet, you can
telnet to **cinbbs.cin.epa.gov** or **204.47.188.79.**
For further information on connecting to the ATTIC
BBS call (513) 569-7272.

EPA Superfund Records of Decision System (RODS)
National Technical Information Service
Technology Administration
U.S. Department of Commerce
Springfield, VA 22161
Phone... (703) 605-6000
Fax.. (703) 321-8547

http://www.ntis.gov/envirn/rods.htm
EXTOXNET
Extension Toxicology Network - pesticide toxicology
and environmental chemistry information for the
general public - fully searchable and selectively
retrievable.
http://ace.orst.edu/info/extoxnet/

G7 ENRM
Environment and Natural Resources Management -
prototype global virtual distributed library of ENRM
data and resources.
http://ceo.gelos.org/

ORCA
Internet Information Service - Office of Ocean
Resources Conservation and Assessment (ORCA),
National Ocean Service (NOS), NOAA.
http://seaserver.nos.noaa.gov/

Right-to-Know Network (RTK NET)
RTK NET
1742 Connecticut Ave. NW
Washington, DC 20009
Voice .... (202) 234-8494
Fax .... (202) 234-8584
Access to numerous databases, text files, and
conferences on the environment, environmental
"toxics," housing, and sustainable development.
http://www.rtk.net/

STORET
EPA's oldest and largest computerized environmental data system. STORET is a repository for
water quality and biological monitoring data.
http://www.epa.gov/owow/STORET/

STORET User Assistance
Mail Stop 4503-F
U. S. Environmental Protection Agency
401 M St., S.W.
Washington, DC, 20460
Phone.. (800) 424-9067

Solid Waste Information Clearinghouse Hotline (SWICH)
Public Line .... (800) 677-9424
or .... (301) 585-2898, ext. 239
Members Only................................. (800) 67-SWICH
Online service, modem
SWICH computer system
All aspects of solid waste management, including:
source reduction, recycling, composting, planning
education and training, public participation, legislation and regulation, waste combustion, collection,
transfer, disposal, landfill gas and special waste.

Superfund & Resource Conservation and Recovery Act
National .... (800) 424-9346

http://www.epa.gov/epaoswer/hotline/index.htm

ToxFAQs (tm)
Quick Guide to Hazardous Substances Summaries about hazardous substances being developed by the ATSDR, Division of Toxicology.
http://atsdr1.atsdr.cdc.gov:8080/toxfaq.htm

Facility Index System (FINDS)
National .... (800) 908-2493
Technical user support; FINDS users only.

Data Processing Support Services National Computer Center
National...(919) 541-2385
User support; I.O. Control; mainframe IBM ES-9000-720 and VAX users.

Customer Technical Support National Computer Center (NCC)
National .... (800) 334-2405
NC only .... (919) 541-7862
Provides NCC customers with technical assistance, problem diagnosis, solution and tracking for ERA on IBM mainframes.

US EPA CONTRACT LABORATORY PROGRAM (CLP)

USEPA Analytical Operations Branch (OS-230)
401 M Street, SW/5204G
Washington, DC 20460
Phone... (703) 603-8870

USEPA Contracts Mgmt. Div. (MD-33)
Alexander Drive
Research Triangle Park, NC 27111

USEPA Environmental Monitoring Systems Laboratory (EMSL/LV)
944 East Harmon Avenue
Las Vegas, NV 89108

Mailing Address:
P.O. Box 93478
Las Vegas, NV 89193-3478

Data To:
EMSL/LV Executive Center
944 East Harmon Ave.
Las Vegas, NV 89119
Attn: Data Audit Staff

USEPA National Enforcement Investigations Center (NEIC)
Denver Federal Center 53, E-2
P.O. Box 25227
Denver, CO 80225
Phone... (303) 236-5111

USEPA Environmental Monitoring Systems Laboratory (EMSL/Cincinnati)
26 W. Martin Luther King Dr.
Cincinnati, OH 45268
Sample Management Office
Alexander Drive
Research Triangle Park, NC 27111

Mailing Address:
USEPA Contract Laboratory Program
Sample Management Office
P.O. Box 818
Alexandria, VA 22313
Phone.. (703) 557-2490

Street Address:
DynCorp Viar
300 N. Lee Street
Alexandria, VA 22314
Phone.. (703) 519-1000

US ENVIRONMENTAL PROTECTION AGENCY OFFICES

Headquarters
401 M. Street, S.W.
Washington, DC 20460
Main Number .... (202) 260-2080
Fax ....(202) 260-0279
National Public Information Center .. (202) 260-2080
National Response Center.............. (800) 424-8802
e-mail................Public-Access@epamail.epa.gov

US ENVIRONMENTAL PROTECTION AGENCY OFFICES BY REGION

Region 1
Connecticut, Massachusetts, Maine,
New Hampshire, Rhode Island, Vermont

US Environmental Protection Agency
Region 1 New England
John F. Kennedy Federal Building
Boston, MA 02203
Main Number .... (617) 565-3423
Fax ....(617) 565-3415
Haz. Waste Ombudsman .... (617) 565-3357
Public Affairs .... (617) 565-3423
Public Information Center .... (617) 565-3420
Small Business Ombudsman .... (800) 368-5888
Unleaded Fuel Hotline .... (617) 565-3220
www access: http://www.epa.gov/region01/

USEPA Region 1, ESD LAB
60 Westview Street
Lexington, MA 02173
Main Number .... (781) 860-4300
Fax ....(781) 860-4397

Appendix F

Region 2
New Jersey, New York, Puerto Rico,
U.S. Virgin Islands

US Environmental Protection Agency
Region 2
290 Broadway, 26th Floor
New York, NY 10007-1866
Main Number .... (212) 637-5000
Fax.. (212) 637-3526
External Affairs/Pub. Affairs............. (212) 637-3660
Haz. Waste Ombudsman .... (800) 262-4000
Public Info. Office (Niagara Falls) .. (716) 285-8842
RCRA & Superfund (region) .... (800) 424-9346
Small Business Ombudsman........... (800) 368-5888
www access: http://www.epa.gov/region02/

USEPA Region 2, ESD
2890 Woodbridge Ave., M/S 100
Edison, NJ 08837-3679
Main Number .... (908) 321-6754

Region 3
Delaware, Maryland, Pennsylvania, Virginia,
Washington D.C., West Virginia

US Environmental Protection Agency
Region 3
841 Chestnut Building
Philadelphia, PA 19107
Main Number .... (215) 566-5000
TDD (Hearing Impaired)................... (215) 580-2024
External Affairs................................ (215) 597-6938
Haz. Waste Ombudsman................. (215) 597-8181
Public Env. Education Ctr. .... (800) 438-2474
Public Affairs.................................. (215) 597-9370
Small Business Ombudsman........... (215) 597-8989
Spill Line for Emergencies .... (215) 566-3255
Spills in West Virginia...................... (800) 642-3074
www access: http://www.epa.gov/region03/

USEPA Region 3, CRL
839 Bestgate Road
Annapolis, MD 21401
Main Number .... (301) 266-9180

Region 4
Alabama, Florida, Georgia, Kentucky, Mississippi,
North Carolina, South Carolina, Tennessee

US Environmental Protection Agency
Region 4
Atlanta Federal Center
61 Forsyth Street, SW,
Atlanta, GA 30303-3104
Main Number .... (404) 562-9900
Environmental Accountability Div. ... (404) 562-9655
General Information Hotline............. (800) 438-2474
Haz. Waste Ombudsman................. (800) 262-7937
Public Information Center .... (404) 562-8327
Small Business Ombudsman........... (404) 562-8280
www access: http://www.epa.gov/region04/

USEPA Region 4, Superfund Branch
345 Courtland Street, N.E.
Atlanta, GA 30365
Phone.. (404) 347-4727

USEPA Region 4, ESD (ASB)
Analytical Support Branch
College Station Road
Athens, GA 30613
Phone.. (404) 546-3136

Region 5
Illinois, Indiana, Michigan, Minnesota, Ohio,
Wisconsin

US Environmental Protection Agency
Region 5
77 West Jackson Blvd, Chicago, IL 60604
Main Number .... (312) 353-2000
Public Affairs (in IL) .... (312) 353-2072
Hotline (IN, MI, MN, OH, WI) .... (800) 621-8431
Haz. Waste Ombudsman................. (312) 353-9510
Small Business Ombudsman........... (312) 886-4571
www access: http://www.epa.gov/region05/

USEPA Region 6, WND
111 W. Adam
Chicago, IL 60604
Phone.. (312) 886-7579

Region 6
Arkansas, Louisiana, New Mexico, Oklahoma,
Texas

US Environmental Protection Agency
Region 6
First Interstate Tower at Fountain Place
1445 Ross Ave., Suite 1200
Dallas, TX 75202
Main Number.................................... (214) 665-6444
Env. Emerg. Hotline (w/in Region)... (214) 665-2222
External Affairs................................ (214) 665-2200
Haz. Waste Ombudsman................. (214) 665-6746
Small Business Ombudsman........... (214) 665-3161
www access: http://www.epa.gov/region06/

USEPA Region 6 Laboratory
10625 Fallstone Road
Houston, TX 77099
Phone.. (281) 983-2100
Fax.. (281) 983-2248

Region 7
Iowa, Kansas, Missouri, Nebraska

US Environmental Protection Agency
Region 7
726 Minnesota Avenue
Kansas City, KS 66101
Main Number .... (913) 551-7000
Emergency Response .... (913) 551-7050
Haz. Waste Ombudsman................ (913) 551-7050

Appendix F

Public Affairs .... (913) 551-7003
Regional Action Line (w/in Region) .. (800) 223-0425
Small Business Ombudsman .... (913) 551-7030
Spill Reports (24 hours) .... (913) 281-0991
www access: http://www.epa.gov/region07/

USEPA Region 7, Lab
25 Funston Road
Kansas City, KS 66115

Region 8
Colorado, Montana, North Dakota,
South Dakota, Utah, Wyoming

US Environmental Protection Agency
Region 8
999-18th St., Suite 500
Denver, Colorado 80202-2466

Main Number .... (303) 293-2466
Region 8 Toll Free .... (800) 227-8917
Emergency Response Hotline .... (800) 227-8914
Environmental Information Center (EISC)
.... (303) 312-6312
Hazardous Waste Ombudsman .... (800) 262-7937
Public Affairs Branch .... (303) 312-6780
Small Business Ombudsman .... (800) 368-5888
www access: http://www.epa.gov/region08/

Montana Operations Office
US Environmental Protection Agency
Federal Building
301 South Park Drawer 10096
Helena, MT 59626-0096
Main Number .... (406) 441-1123
Administration .... (406) 441-1120
Superfund .... (406) 441-1150
Water .... (406) 441-1140
Hazardous Waste .... (406) 441-1130
Air .... (406) 441-1130

USEPA Region 8, Lab
Denver Federal Center
Building 53
8TMS - LAB
W1 Entrance, 2nd Floor
Denver, CO 80225
Phone .... (303) 236-5073

Region 9
American Samoa, Arizona, California,
Guam, Hawaii, Nevada

US Environmental Protection Agency
Region 9
75 Hawthorne Street,
San Francisco, CA 94105

Main Number .... (415) 744-1305
24 Hour Hot Spill Number .... (415) 744-2000
External Affairs .... (415) 744-1015

Public Affairs .... (415) 744-1020
Haz. Waste Ombudsman .... (415) 744-2110
Small Business Ombudsman .... (415) 744-1635
Public Inquiry Response Line .... (415) 744-1500
www access: http://www.epa.gov/region09/

USEPA Region 9 Laboratory
944 East Harmon Avenue
Las Vegas, NV 89119

Region 10
Alaska, Idaho, Oregon, Washington

US Environmental Protection Agency
Region 10
1200, 6th Avenue
Seattle, WA 98101

Main Number .... (206) 553-1200
External Affairs .... (206) 553-1107
Haz. Waste Ombudsman .... (206) 553-6901
Public Information Center (PIC) .... (206) 553-4973
Regional Toll Free .... (800) 424-4372
Small Business Ombudsman .... (206) 553-2634
Telecommunications Device for the Deaf (TDD)
.... (206) 553-1698
www access: http://www.epa.gov/region10/

USEPA Region 10
P.O. Box 549
Manchester, WA 98353
Phone .... (360) 871-0748

SATELLITE REGIONAL OFFICES

EDISON LABORATORIES
2890 Woodbridge Ave., MS100
Edison, NJ 08837-3679
Phone .... (732) 321-6754
Fax .... (732) 321-4381

CARIBBEAN FIELD OFFICE
Centro Europa Building
1492 Ponce Deleon Avenue, Suite 417
Santurce, PR 00907-4127
Phone .... (787) 729-6951
Fax .... (787) 729-7747

NIAGARA FALLS PUBLIC INFORMATION
CENTER
345 Third Street, Suite 530
Niagara Falls, NY 14304
Phone .... (716) 285-8842
Fax .... (716) 285-8788

VIRGIN ISLANDS COORDINATOR
Federal Building and Courthouse
550 Veterans Drive
Room 142
St. Thomas, US Virgin Islands 00802
Phone .... (340) 714-2333

Appendix F

STATE CONTACTS

Alabama

Alabama Dept. of Environmental Management
Chief, Land Division
1751 W. L. Dickinson Drive
Montgomery, AL 36109
Phone: (334) 271-7730; Fax: (334) 271-7950

Alabama Dept. of Conservation / Game and Fish
Division
64 N. Union Street
Montgomery, Alabama 36130-1456
(334) 242-3465
http://www.dcnr.state.al.us/agfd

Alaska

Alaska Dept. of Environmental Conservation
Solid and Hazardous Waste Mgt.
Pouch 0
Juneau, AK 99811
Phone: (907) 465-5150; Fax: (907) 456-5362

Alaska Dept. of Environmental Conservation
Chief, Spill Planning and Prevention
Pouch 0
Juneau, AK 99811
Phone: (907) 465-5250

Alaska Dept. of Environmental Conservation
Contaminated Sites Section
Pouch 0
Juneau, AK 99811
Phone: (907) 465-2630

Alaska Department of Fish and Game
P.O. Box 25526
Juneau, Alaska 99802-5526
(907) 465-4100
**http://www.state.ak.us/local/akpages/FISH.GAME/adfg
home.htm**

Arizona

Arizona Dept. of Environmental Quality
Assistant Director, Office of Waste Programs
3033 N. Central Avenue, 7th Floor
Phoenix, AZ 85002
Phone: (602) 207-2300; Fax: (602) 257-6874

Arizona Dept. of Environmental Quality
Hazardous Waste Section
3033 N. Central, Room 403C
Phoenix, AZ 85002
Phone: (602) 257-6995; Fax: (602) 257-6948

Arizona Game & Fish Department
2221 W. Greenway Rd., Phoenix, AZ
85023-4399
Phone: (602) 942-3000
http://www.gf.state.az.us/welcome.htm

Arkansas

Arkansas Dept. of Pollution Control and Ecology
Chief, Hazardous Waste Division
P.O. Box 8913
Little Rock, AR 72219-8913
Phone: (501) 562-0831 or (501) 682-0580
Fax: (501) 682-0880 or (501) 682-0707

Arkansas Department of Pollution Control and
Ecology
8001 National Drive, Little Rock, AR 72209
Phone: (501) 682-0744
http://www.state.ar.us/

California

California Dept. of Toxic Substances Control
Haz. Waste Management Program
P.O. Box 806
Sacramento, CA 95812-0806
Phone: (916) 323-6042; Fax: (916) 372-4495

California State Water Resources Control Board
Chief, Div. of Clean Water Programs
2014 T Street, Suite 130
Sacramento, CA 95814
Phone: (916) 227-4400; Fax: (916) 227-4349

State of California
Department of Fish & Game
1416 9th Street
Sacramento, CA 95814
Phone: (916) 653-7664
CalTIP: (888) DFG-CALTIP
http://www.dfg.ca.gov/

Colorado

Colorado Dept. of Health
Hazardous Materials and Waste Mgt. Div.
4300 Cherry Creek Dr. South
Denver, CO 80222-1530
Phone: (303) 692-3300; Fax: (303) 759-4355

Public Utilities Commission
Hazardous Materials Transportation Permits
1580 Logan Street, Off. Level 1
Denver, CO 80203
Phone: (303) 894-2000; Fax: (303) 894-2065

Connecticut

Connecticut Waste Management Bureau
Bureau Chief
79 Elm Street
Hartford, CT 06106
Phone: (860) 424-3023; Fax: (860) 424-4059
http://dep.state.ct.us/

Connecticut Resource Recovery Authority
President
179 Allyn St., Suite 603
Hartford, CT 06103
Phone: (860) 549-6390; Fax: (860) 522-2390

Delaware

Delaware Dept. of Natural Resources and
Environmental Control
Mgr., Hazardous Waste Management Branch
P.O. Box 1401/89 Kings Highway
Dover, DE 19903
Phone: (302) 739-3689; Fax: (302) 739-5060

Delaware Emergency Management Agency
Chemical Hazard Section
P.O. Box 527
Delaware City, DE 19706
Phone: (302) 326-6000
http://www.state.de.us/

District of Columbia

Dept. of Consumer and Regulatory Affairs (DCRA)
Pesticides, Haz. Waste, and Underground
Storage Tank Division
2100 Martin Luther King, Jr. Ave., SE Suite #203
Washington, DC 20020
Phone: (202) 645-6080

Florida

Florida Dept. of Environmental Regulation
Administrator, Solid and Haz. Waste
Twin Towers Office Bldg
2600 Blair Stone Road
Tallahassee, FL 32399-2400
Phone: (904) 488-0300; Fax: (904) 921-8061

Underground Storage Tanks (Same address as
above) Phone: (904) 488-3935

Florida Center for Solid and Hazardous Waste
Management
2207 NW 13th Street, Suite D
Gainesville, FL 32609
Phone: (352)392-6264
Fax: (352)846-0183
http://www.eng.ufl.edu/home/fcshwm/

Georgia

Land Protection Branch
Dept of Natural Resources
EPA Division, Land Protection
4244 International Park Way, Suite 114
Atlanta, GA 30354
Phone: (404) 362-2671
Fax (Underground Storage Tanks): (404) 362-2654
Fax (Solid Waste Div.): (404) 362-2693

Hazardous Waste Management Branch
Floyd Towers East
205 Butler Street, SE
Atlanta, GA 30334
Phone: (404) 656-2833
http://www.dnr.state.ga.us/

Hawaii

Hawaii Dept. of Health
Mgr, Solid and Hazardous Waste Branch
919 Ala Moana Blvd.
Room 212
Honolulu, HI 96814
Phone: (808) 586-4226; Fax: (808) 586-7509

Hawaii Dept. of Health
Mgr., Hazard Evaluation and Emerg. Response
919 Ala Moana Blvd.
Room 206
Honolulu, HI 96814
Phone: (808) 586-4249; Fax: (808) 586-7537

Division of Conservation and Resource
Enforcement
1151 Punchbowl Street, Room 311
Honolulu, Hawaii 96813
Phone: 808-587-0077
FAX: 808-587-00
http://www.hawaii.gov/dlnr/Welcome.htm

Idaho

Division of Environmental Quality
RCRA Programs
1410 North Hilton Street
Boise, ID 83706
Phone: (208) 373-0502; Fax: (208) 373-0417

Illinois

Illinois Environmental Protection Agency, Dir.
2200 Churchill Road
Springfield, IL 62706
Phone: (217) 782-3397; Fax: (217) 782-9039

Illinois Environmental Protection Agency
Public Information Officer
Division of Land Pollution Control
2200 Churchill Road
Springfield, IL 62706
Phone: (217) 782-3397; Fax: (217) 785-7725

Hazardous Waste Research and Information Ctr.
Illinois Energy and Natural Resources
1 E. Hazelwood Drive
Champaign, IL 61820
Phone: (217) 333-8941; Fax: (217) 333-8944
http://www.hazard.uiuc.edu/wmrc/news.htm

Indiana

Indiana Dept. of Environmental Management
Branch Chief, Office of Hazardous Waste Mgt.
Indiana Government Center North
100 N. Senate Ave./P.O. Box 6015
Indianapolis, IN 46206-6015
Phone: (317) 232-3292; Fax: (317) 232-3403

Appendix F

Kansas

Bureau of Air and Radiation
Kansas Dept. of Health and Environment
Director
Forbes Field, Building 283
Topeka, KS 66620
Phone: (785) 296-1593; Fax: (785) 296-1545

Bureau of Waste Management
Kansas Dept. of Health and Environment
Director
Forbes Field, Building 740
Topeka, KS 66620
Phone: (785) 296-1600; Fax: (785) 296-1592

Kansas Department of Health and Environment
Division of Environment
Forbes Field, Building 740
Topeka, Ks. 66620-0001
Phone: (785) 296-1535; Fax (785) 296-8464
http://www.state.ks.us/public/kdhe/environ.htm

Kentucky

Kentucky Dept. of Environmental Protection
Frankfort Office Park
14 Reilly Road
Frankfort, Kentucky 40601
Phone: (502) 564-2150; Fax: (502) 564-4245
http://www.state.ky.us/agencies/nrepc/dep/dep2.htm

Louisiana

Louisiana Dept. of Environmental Quality
Solid Waste Division
P.O. Box 82178
Baton Rouge, LA 70884-2178
Phone: (504) 765-0249; Fax: (504) 765-0299
http://www.deq.state.la.us/oshw/sw/sw.htm

Louisiana Dept. of Environmental Quality
Administrator, Ground Water Protection Division
P.O. Box 82215
Baton Rouge, LA 70884-2215
Phone: (504) 765-0585; Fax: (504) 765-0602
http://www.deq.state.la.us/oshw/gw/gw.htm

Maine

Department of Environmental Protection
Bureau of Haz. Mat'ls and Solid Waste Control
Director - Maine Dept of Environmental Protection
State House Station #17
Augusta, ME 04333
Phone: (207) 287-2651; Fax: (207) 287-7826
http://www.state.me.us/dep/mdephome.htm

Maryland

Hazardous and Solid Waste Management Admin.
Director - Maryland Dept. of the Environment
2500 Broening Highway
Baltimore, MD 21224
Phone: (301) 631-3304; Fax: (301) 631-3321

Massachusetts

Massachusetts Dept. of Environmental Affairs
Director, Executive Office
100 Cambridge Street, 20th Floor
Boston, MA 02202
Phone: (617) 727-9800; Fax: (617) 727-2754

Massachusetts Department of Environmental
Protection, Western Regional Office
436 Dwight Street, 5th Floor
Springfield, MA 01103
Phone: (413) 784-1100; Fax: (413) 784-1149
http://www.state.ma.us/dep/dephome.htm

Michigan

Michigan Department of Environmental Quality
P.O. Box 30457
Lansing, MI 48909-7957
Executive Division: 517-373-7917
Office of Administrative Hearings: (517) 335-4226
Special Assistant S.E. Michigan: (313) 953-0241
Environmental Assistance Center: (800) 662-9278
Pollution Emergencies: (800) 292-4706
http://www.deq.state.mi.us/

Minnesota

Minnesota Pollution Control Agency
Director, Hazardous Waste Div.
520 Lafayette Rd. North
St. Paul, MN 55155
Phone: (612) 297-8502; Fax: (612) 297-8676

Hazardous Waste Division
Minnesota Pollution Control Agency
Chief, Program Development
520 Lafayette Rd. North
St. Paul, MN 55155
Phone: (612) 297-8355; Fax: (612) 297-8676

Minnesota Technical Assistance Program (MnTAP)
Director
1313 5th Street, SE - Suite 207
Minneapolis, MN 55414
Phone: (612) 627-4646 or (800) 247-0015;
Fax: (612) 627-4769

Minnesota Office of Environmental Assistance
520 Lafayette Road North
St. Paul, MN 55155-4100
612-296-3417 or 800-657-3843
http://www.moea.state.mn.us/

Mississippi

Mississippi Dept. of Environmental Quality
Chief, Hazardous Waste Division
P.O. Box 20305
Jackson, MS 39289-0385
Phone: (601) 961-5171
http://www.deq.state.ms.us/domino/deqweb.nsf

Missouri

Missouri Dept. of Natural Resources
Dir., Hazardous Waste Program
205 Jefferson St./P.O. Box 176
Jefferson City, MO 65102
Phone: (573) 751-3176; Fax: (573) 751-7869
http://www.state.mo.us/dnr/homednr.htm
Division of Environmental Quality
P. O. Box 176
Jefferson City, MO 65102
Phone: (573) 751-6892; Fax: (573) 751-9277
http://www.state.mo.us/dnr/deq/homedeq.htm

Montana

Montana Dept. of Health and Env. Quality
Solid and Hazardous Waste Div.
2209 Phoenix Ave.
P.O. Box 200901
Helena, MT 59620-0901
Phone: (406) 444-2821; Fax: (406) 444-1499
http://www.deq.state.mt.us/

Nebraska

Nebraska Dept. of Environmental Quality
Hazardous Waste Sec., CERCLA Unit Supervisor
P.O. Box 98922
Lincoln, NE 68509
Phone: (402) 471-2186; Fax: (402) 471-2909

Nebraska Dept. of Environmental Quality
Hazardous Waste Sec., RCRA Unit Supervisor
P.O. Box 98922
Lincoln, NE 68509
Phone: (402) 471-2186; Fax: (402) 471-2909

Nevada

Division of Environmental Protection
Nevada Dept. of Conserv. and Natural Resources
Chief, Waste Mgt. Bureau
Capitol Complex, 123 W. Nye Lane
Carson City, NV 89710
Phone: (702) 687-5872; Fax: (702) 885-0868
http://www.state.nv.us/indexe.htm

New Hampshire

New Hampshire Dept. of Environmental Services
Director, Waste Mgt. Division
6 Haven Drive
Concord, NH 03301-6509
Phone: (603) 271-2906; Fax: (603) 271-2456
http://www.state.nh.us/des/descover.htm

New Jersey

New Jersey Dept. of Environmental Protection
Asst. Commissioner, Site Remediation Program
401 E. State Street
CN 028-6th Floor East
Trenton, NJ 08625
Phone: (609) 292-1250; Fax: (609) 633-2360

New Jersey Dept. of Env. Protection and Energy
Dir., Div. of Responsible Party Site Remediation
401 E. State Street
CN 028-5th Floor East
Trenton, NJ 08625
Phone: (609) 633-1408; Fax: (609) 633-1454

New Mexico

New Mexico Environment Dept.
Chief, Hazardous and Radioactive Mat'ls Bureau
Physical Location:
2044A Galisteo St.
Santa Fe, NM 87505
Phone: (505) 827-1564

Mailing Location:
P.O. Box 26110
Santa Fe, NM 87502
Phone: (505) 827-1557; Fax: (505) 827-1544

New York

New York State Dept. of Environmental Conserv.
Div. of Solid & Hazardous Materials
50 Wolf Rd., Room 488
Albany, NY 12233-7250
Phone: (518) 457-6934; Fax: (518) 457-0629
http://www.dec.state.ny.us/

North Carolina

N. Carolina Dept. of Environmental, Health,
and Natural Resources
Dir., Div. of Solid Waste Mgt.
P.O. Box 27687
Raleigh, NC 27611-7687
Phone: (919) 733-4996; Fax: (919) 733-4810

Division of Pollution Prevention and Environmental
Assistance (DPPEA)
P.O. Box 29569
Raleigh, NC 27626-9569
Phone: (919) 715-6500 Fax: (919) 715-6794
**http://www.ehnr.state.nc.us/EHNR/files/division.
htm**

North Dakota

N. Dakota Dept. of Health and Waste Mgt.
P.O. Box 5520
Bismarck, ND 58506-5520
Phone: (701) 328-5166; Fax: (701) 328-5200

Ohio

Ohio Environmental Protection Agency
Div. of Hazardous Waste Mgt.
1800 Watermark Drive
P.O. Box 1049
Columbus, OH 43266-0149
Phone: (614) 644-2917; Fax: (614) 644-2329

Appendix F

Ohio Environmental Protection Agency
Div. of Solid & Infectious Waste Mgt.
3205 Westbrooke Drive, Bldg. C
Columbus, OH 43228-9644
Phone: (614) 644-2621; Fax: (614) 728-5315
http://www.dnr.state.oh.us/odnr/divs.htm

Oklahoma

Dept. of Environmental Quality
Waste Mgt. Div. (Hazardous Waste)
1000 Northeast Tenth Street
Oklahoma City, OK 73117-1299
Phone: (405) 271-5338; Fax: (405) 271-8425
http://www.deq.state.ok.us/capnew.htm

Oregon

Oregon Dept. of Environmental Quality
Waste Management/Environmental Cleanup Div.
811 SW 6th Ave.
Portland, OR 97204
Phone: (503) 229-5913; Fax: (503) 229-6977
http://www.deq.state.or.us/

Pennsylvania

Pennsylvania Dept. of Environmental Resources
Dir., Bureau of Land Recycling & Waste Mgt.
P.O. Box 2063
Harrisburg, PA 17105-2063
Phone: (717) 783-2388; Fax: (717) 787-1904

Pennsylvania Dept. of Environmental Protection
Municipal & Residual Waste
P.O. Box 8472
Harrisburg, PA 17105-8472
Phone: (717) 787-7381; Fax: (717) 787-1904

Department of Environmental Protection
Rachel Carson State Office Building
400 Market Street
Harrisburg, Pennsylvania 17101
Phone: (717) 787-2814
http://www.dep.state.pa.us/

Rhode Island

Rhode Island Dept. of Environmental Waste
Dept. of Env. - Waste Management Division
235 Promenade Street
Providence, RI 02908-5767
Phone: (401) 222-2797
http://www.state.ri.us/dem/

South Carolina

S. Carolina Dept. of Health and Env. Control
Chief, Bureau of Solid and Hazardous Waste Mgt.
2600 Bull St.
Columbia, SC 29401
Phone: (803) 734-5000
http://www.state.sc.us/dhec/

South Dakota

S. Dakota Dept. of Env. and Natural Resources
Office of Waste Mgt., Joe Foss Building
523 E. Capital Avenue
Pierre, SD 57501-3181
Phone: (605) 773-3153; Fax: (605) 773-6035
http://www.state.sd.us/state/executive/denr/denr.htm

S. Dakota Highway Patrol, Commerce and
Regulation
320 N. Nicollet
Pierre, SD 57501
Phone: (605) 773-3105; Fax: (605) 773-6046

Tennessee

Tennessee Dept. of Environment and Conservation
Dir., Div. of Solid Waste Mgt.
401 Church St.
L&C Tower, 5th Floor
Nashville, TN 37243-1535
Phone: (615) 532-0780
http://www.state.tn.us/environment/

Tennessee Dept. of Environment and Conservation
Dir., Div. of Superfund
401 Church St.
L&C Tower, 4th Floor
Nashville, TN 37243-1535
Phone: (615) 532-0900

Texas

Texas Natural Resources & Conservation Comm.
Dir., Hazardous and Solid Waste Division
P.O. Box 13087, Capitol Station
Austin, TX 78711-3087
Phone: (512) 239-1000; Fax: (512) 463-8408
http://www.state.tx.us/agency/agencies.htm

Utah

Utah Dept. of Environmental Quality
Dir., Div. of Solid and Hazardous Waste
288 North 1460 West St.
Salt Lake City, UT 84114-4880
Phone: (801) 538-6170; Fax: (801) 538-6715
http://www.eq.state.ut.us/

Vermont

Vermont Agency of Natural Resources
Dir., Hazardous Mat'ls Mgt. Div.
103 South Main St.
Waterbury, VT 05676
Phone: (802) 241-3888; Fax: (802) 241-3296
http://www.state.vt.us/

Vermont Dept. of Health
Dir., Occupations and Radiological Health Div.
108 Cherry Street
Burlington, VT 05402
Phone: (802) 865-7730; Fax: (802) 865-7745

Appendix F

Virginia

Virginia Dept. of Waste Management
Div. of Regulation
629 East Main Street
Richmond, VA 23219
Phone: (804) 225-2667; Fax: (804) 762-4500
Washington

Washington Dept. of Ecology
Mgr., Solid and Hazardous Waste Program
P.O. Box 47600
Olympia, WA 98504
Phone: (360) 407-6000; Fax: (360) 407-6102
http://www.wa.gov/ecology/

West Virginia

Bureau of Environment
Division of Environmental Protection
Office of Waste Management
1356 Hansford St.
Charleston, WV 25301
Phone: (304) 558-5929; Fax: (304) 558-0256

Air Pollution Control Commission
1558 Washington Street, East
Charleston, WV 25311
Phone: (304) 558-4022
Public Information Line: (304) 558-3381

West Virginia Division of Highways
Secretary/Commissioner of Highways
Building 5, Room A-109
Charleston, WV 25305
Phone: (304) 558-3505

Wisconsin

Wisconsin Dept. of Natural Resources
Dir., Bureau of Solid and Hazardous Waste Mgt.
P.O. Box 7921
Madison, WI 53707
Phone: (608) 266-1327; Fax: (608) 267-2768
http://www.dnr.state.wi.us/

Wyoming

Wyoming Dept. of Environmental Quality
Solid and Hazardous Waste Div.
122 West 25th St.
Herschler Building
Cheyenne, WY 82002
Phone: (307) 777-7752; Fax: (307) 777-5973
http://deq.state.wy.us/

Appendix F

APPENDIX G

Source information used to compile this manual is the latest promulgated versions of methods commonly available to analytical testing laboratories and environmental engineering companies. In some cases newer proposed versions of methods exist. These are not included in this reference because their use is not universally accepted. The latest versions of the inorganic and organic Statements of Work for the Contract Laboratory Program were used, though some laboratories may still have contracts using an older Statement of Work version. SW-846 third edition, methods, including promulgated updates to December 1996, are included in this reference. Appendix D is a detailed explanation of SW-846 third edition and its updates.

REFERENCES

U.S. Environmental Protection Agency

National Primary Drinking Water Regulations – 40 CFR Part 141; July 1, 1988.

Contract Laboratory Program – User's Guide to the Contract Laboratory Program. EPA/540/P-91/002, January 1991, Office of Emergency and Remedial Response. Washington, DC

Contract Laboratory Program – Statement of Work for Organic Analysis, Multi-Media, Multi-Concentration. Document OLM3.2, 1996

Contract Laboratory Program - Statement of Work for Inorganic Analysis – Multi-Media Multi-Concentration. Document ILMO4.0, 1995

EPA100-400 Series – Methods for Chemical Analysis of Water and Wastes, EPA-600/4-79-020, Revised March 1983.

EPA100-400 Series – Methods for the Determination of Inorganic Substances in Environmental Samples, EPA/600/R-93-100, August 1993.

EPA 200 Series – Methods for the Determination of Metals in Environmental Samples, EPA/600/4-91-010, June 1991.

EPA 200 Series – Methods for the Determination of Metals in Environmental Samples. Supplement I, EPA/600/R-94-111, May 1994.

EPA 500 Series – Methods for the Determination of Organic Compounds in Drinking Water, EPA/600/4-88/039, December 1988.

EPA 500 Series – Methods for the Determination of Organic Compounds in Drinking Water, Supplement I, EPA/600/4-90/020, 1990.

EPA 500 Series – Methods for the Determination of Organic Compounds in Drinking Water, Supplement II, EPA/600/R-92/129, August 1992.

EPA 500 Series – Methods for the Determination of Organic Compounds in Drinking Water, Supplement III, EPA/600/R-95/131, August 1995.

EPA 600 Series – 40 CFR, Part 136, Revised as of July 1, 1995. Appendix A to Part 136 - Methods for Organic Chemical Analysis of Municipal and Industrial Wastewater.

EPA 600 Series – Methods for the Determination of Nonconventional Pesticides in Municipal and Industrial Wastewater - Volume I - EPA-821-R-93-010-A August 1993, Revision 1.

Variability in Protocols – Guy F. Simes, Risk Reduction Engineering Laboratory, Cincinnati, OH, September 1991

Test Methods for Evaluating Solid Waste Physical/Chemical Methods (SW-846) – Third Edition, September 1986; Final Update I, July 1992; Final Update IIA, August 1993; Final Update II, September 1994; Final Update IIB, January 1995; Final Update III, December 1996.

Standard Methods for the Examination of Water and Waste Water – 19th Edition 1995. American Public Health Association, American Water Works Association, Water Pollution Control Federation.

Appendix G